Laboratory
Animal
Management

Marine Invertebrates

COMMITTEE ON MARINE INVERTEBRATES

Institute of Laboratory Animal Resources
Assembly of Life Sciences
National Research Council

NATIONAL ACADEMY PRESS
Washington, D.C. 1981

NOTICE The project that is the subject of this report was approved by the Governing Board of the National Research Council, whose members are drawn from the Councils of the National Academy of Sciences, the National Academy of Engineering, and the Institute of Medicine. The members of the committee responsible for the report were chosen for their special competences and with regard for appropriate balance.

This report has been reviewed by a group other than the authors according to procedures approved by a Report Review Committee consisting of Members of the National Academy of Sciences, the National Academy of Engineering, and the Institute of Medicine.

The National Research Council was established by the National Academy of Sciences in 1916 to associate the broad community of science and technology with the Academy's purposes of furthering knowledge and of advising the federal government. The Council operates in accordance with general policies determined by the Academy under the authority of its congressional charter of 1863, which establishes the Academy as a private, nonprofit, self-governing membership corporation. The Council has become the principal operating agency of both the National Academy of Sciences and National Academy of Engineering in the conduct of their services to the government, the public, and the scientific and engineering communities. It is administered jointly by both Academies and the Institute of Medicine. The National Academy of Engineering and the Institute of Medicine were established in 1964 and 1970, respectively, under the charter of the National Academy of Sciences.

The primary support for development of these guidelines was provided by Contract N01–CM–53850 with the Division of Cancer Treatment, National Cancer Institute. Additional support was provided by Grant RC–1W from the American Cancer Society, Inc.; Contract APHIS 53–3294–9–2 with the U.S. Department of Agriculture; Contract DNA001–79–C–0087–P00001 with the Defense Nuclear Agency; Contract DE–AT02–76CH93011 (Task Agreement DE–AT02–76CH93011) with the Department of Energy; Contract N01–RR–9–2109 with the Division of Research Resources, Animal Resources Branch, National Institutes of Health; Grant PCM–7921163 with the National Science Foundation; Contract N00014–80–C–0162 with the Office of Naval Research; Contract N01–CP–65805 with the Division of Cancer Cause and Prevention, National Cancer Institute, National Institutes of Health; and contributions from pharmaceutical companies and other industry.

Library of Congress Cataloging in Publication Data
Main entry under title:

Marine invertebrates.

Bibliography: p.
Includes index.
1. Marine invertebrates as laboratory animals.
I. National Research Council (U.S.). Committee on Marine Invertebrates.
SF407.M37M37 639′.4 81-4015
ISBN 0-309-03134-6 AACR2

Available from:
NATIONAL ACADEMY PRESS
2101 Constitution Avenue, N.W.
Washington, D.C. 20418

Printed in the United States of America

COMMITTEE ON MARINE INVERTEBRATES

RALPH T. HINEGARDNER, *Chairman*, University of California, Santa Cruz

JAMES W. ATZ, American Museum of Natural History and Sea Research Foundation
RIMMON C. FAY, Pacific Bio-Marine Laboratories
MILTON FINGERMAN, Tulane University
ROBERT K. JOSEPHSON, University of California, Irvine
NORMAN A. MEINKOTH, Swarthmore College
JOHN W. MILLER, Baldwin-Wallace College
MARY ESTHER RICE, Smithsonian Fort Pierce Bureau

NAS Staff

NANCY A. MUCKENHIRN
VERONICA I. PYE

Preface

This manual summarizes contemporary methods for maintaining and rearing marine invertebrates in laboratories without immediate access to the sea. The first part contains general information. The second part consists of chapters written by investigators who are skilled in handling specific animals or groups of animals. Detailed directions are given on the animals' care and handling, on food requirements, and on ways to meet some of the difficulties that may arise. At best, this manual can provide specific recommendations for but a small number of species from among those potentially available.

We recommend that the reader also consult other sources, such as:

Needham *et al.*, 1937. *Culture Methods for Invertebrate Animals*. This is one of the first books on invertebrate culture. It covers a wide range of animals from protozoans to ascidians, both aquatic and terrestrial, free living and parasitic. Although methods are dated, the book, which has been reprinted by Dover Publications Inc., still contains useful information.

de Graff, 1973. *Marine Aquarium Guide*. Although intended principally for amateurs handling tropical fish aquaria, this colorful book includes a number of practical hints and suggestions useful for aquaria operated in the temperature range of 20–27°C. Principles of aquarium set-up and maintenance are clearly stated and generally applicable.

Smith and Chanley, 1975. *Culture of Marine Invertebrate Animals* covers a large variety of animals and has information on general aquarium management and feeding.

Spotte, 1973. *Marine Aquarium Keeping* primarily concerns fish husbandry, but also contains much information on general aquarium management applicable to invertebrates.

Spotte, 1979a. *Fish and Invertebrate Culture* deals with many specific techniques that are relevant to invertebrates. It emphasizes very large systems such as those in public aquaria.

Kinne, 1976–1977. *Marine Ecology*. Volume III. Part 1 covers water management and the culture of marine plants. Part 2 deals with the culture of marine animals from protozoans to vertebrates. The treatment is generally from the ecological research point of view, but highly rewarding.

Costello and Henley, 1971. *Methods for Obtaining and Handling Marine Eggs and Embryos* is devoted primarily to the large number of species that can be found in and around Cape Cod. Many of the methods are applicable to other localities.

Wilt and Wessels, 1967. *Methods in Developmental Biology* covers both plants and animals, including a number of marine invertebrates.

Contents

Part I

Introduction

With the exception of the insects, the majority of invertebrates live in the ocean. Some phyla are exclusively marine, and many species can be collected along coastlines. Despite their diversity and abundance, marine invertebrates have never played a large role in biomedical research, with the conspicuous exception of such species as sea urchins (eggs and early larvae), squid (nerve axon), and a few others. In part, this stems from the widely held assumption that discoveries made using invertebrates will have little relevance to human biology, and in part because the animals usually have been difficult to keep alive and in good condition once they were removed from the sea.

The situation is gradually changing. As information on invertebrates accumulates, it is becoming clear that animals, whatever the phylum, have a great deal in common. Furthermore, because invertebrates are very diverse, they often provide just the animal or organ that an investigator needs to address a particular question. At times, a given species seems almost made to order; witness the squid axon, horseshoe crab eye, or sea urchin egg as examples.

It is gradually becoming more feasible to maintain invertebrates in laboratories without access to the sea. Though some techniques have not yet been perfected, the task of maintaining a large collection is, in some cases, not much more difficult than keeping an equal number of mice. Obviously, not all marine animals are equally easy to keep, and some difficulties will be encountered. The maintenance of marine systems, as of most land animals, is still an art, and the methods for even the most-used species are based primarily on experience. Nonetheless, maintaining a marine animal colony is today both possible and practicable.

Neither an aquarium with seawater flowing through it from the sea, nor a closed system, can be fully equivalent to the open ocean or a wave-washed tide pool. In closed systems waste products are often in higher concentration than the animal naturally experiences. If the animals are maintained some distance from the ocean, available foods tend to be substitutes for what is usually eaten. Conditions are generally more crowded than in the wild, but of course there are fewer predators and usually less exposure to parasites. In many ways an aquarium is a wet cage.

The closed or recirculating system offers several advantages for the culture of marine invertebrates that are to be used in research and teaching. Physical and chemical parameters such as salinity, temperature, and ion composition are more easily monitored and controlled. It is possible to establish a greater variety of experimental conditions. Undesirable bacteria and planktonic forms that often arrive uninvited through seawater pipes can be excluded or reduced in number. A closed system can be highly stable.

A major task, often, is providing acceptable food. Inland food sources include agricultural products, frozen or fresh meats, animal chows, and laboratory concoctions. The feeding operation gives rise to many difficulties; this is, therefore, an area in which new and original work is still very much needed. Although it is now possible to maintain animals for extended periods, it is as yet not easy to breed many species in captivity or to have them grow. More often than not, the lack of a suitable food is the most troublesome problem.

The cost for setting up a small system varies, depending on the species involved, its size, its ability to withstand crowding, and whether or not the system has to be cooled. If refrigeration is not needed, $150 to $300 will suffice for the components of a 100-liter aquarium and supporting apparatus. Costs reach several thousand dollars for an aquarium system with numbers of resident animals equivalent to a modest-sized rat colony. In any case, the best equipment available should be used, because the investment in the system is small compared with that necessary for the research itself. Survival of the animals depends upon the equipment operating dependably; the impact of equipment failure is rapid and often disastrous.

Because culturing any marine animal is part science and part art, it is well to start with a small system and determine the animals' requirements before investing heavily in equipment that may not be wholly appropriate.

1

Seawater as a Biological Medium

CHEMICAL AND PHYSICAL PROPERTIES

Seawater contains a mixture of inorganic ions; the more abundant are listed in Table 1–1. Although present in relatively minute amounts, many of the minor elements are significant in the biology of marine invertebrates (Nicol, 1967; Prosser, 1973).

Total salt content, or salinity, of seawater varies considerably depending on geographic location, but the relative amounts of most major elements remain constant throughout the oceans. The salinity of water in the open sea is about 35‰ (g/kg). Marine invertebrates are found in salinities as low as 5‰ (estuaries) and as high as 220‰ (salt lakes).

Dissolved oxygen in seawater ranges from zero to 8.5 ml/liter. It is most abundant at the surface and decreases with depth, increasing temperature, and increasing salinity (Figure 1–1). Carbon dioxide concentrations range from 34 to 56 ml/liter; some exists as free carbon dioxide but most as carbonate and bicarbonate (Nicol, 1967). Seawater is typically alkaline (approximately pH 8) and is buffered.

There is little dissolved or suspended organic material naturally derived from the metabolism of living organisms and decomposition of dead tissue. However, the increased release of pollutants to the oceans and dumping of solid wastes has significantly increased the organic content of near-shore waters.

Average surface temperature varies from 27°C at the equator to −1°C in the Arctic and Antarctic. The open ocean rarely shows a seasonal

TABLE 1–1 Average Composition of Seawater at Salinity 35‰ (Nicol, 1967)

Component	Percent in Sea Salt	Concentration, g/kg
Chloride, Cl^-	55.04	18.980
Sodium, Na^+	30.61	10.556
Sulfate, $SO_4^=$	7.68	2.649
Magnesium, Mg^{++}	3.69	1.272
Calcium, Ca^{++}	1.16	0.400
Potassium, K^+	1.10	0.380
Bicarbonate, HCO_3^-	0.41	0.140
Bromide, Br^-	0.19	0.065
Boric acid, H_3BO_3	0.07	0.026
Strontium, Sr^{++}	0.04	0.0085
Fluoride, F^-	0.004	0.001

Minor elements—boron, silicon, nitrogen, phosphorous, iron, manganese, copper, zinc, molybdenum, vanadium, chromium, cobalt

temperature range of more than 10°C, but intertidal and estuarine areas experience much greater fluctuations.

BIOLOGICAL CONSIDERATIONS

The open sea is a constant and relatively stable environment. The sea's great volume, high specific heat, continuity, and constant mixing by currents preclude the wide fluctuations of environmental conditions that characterize intertidal, freshwater, and terrestrial environments.

The chemical composition of seawater, with the important exception of pollutants, is very uniform and rarely limits the occurrence of invertebrates. The oxygen, carbonate, and pH in the sea are usually sufficient to support their growth. The availability of nitrogen as nitrate or ammonia is generally considered a principal limitation to phytoplankton in the ocean and thus indirectly limits invertebrates. In aquaria, ammonia is almost always present in excess and can reach toxic levels.

Two of the more important factors affecting the distribution and success of marine invertebrates are *temperature* and, in coastal regions, *salinity*. Invertebrates found in the open sea tend to have a limited capacity to adust to or tolerate environmental fluctuations of either. These species are also more difficult to maintain in the laboratory. Most

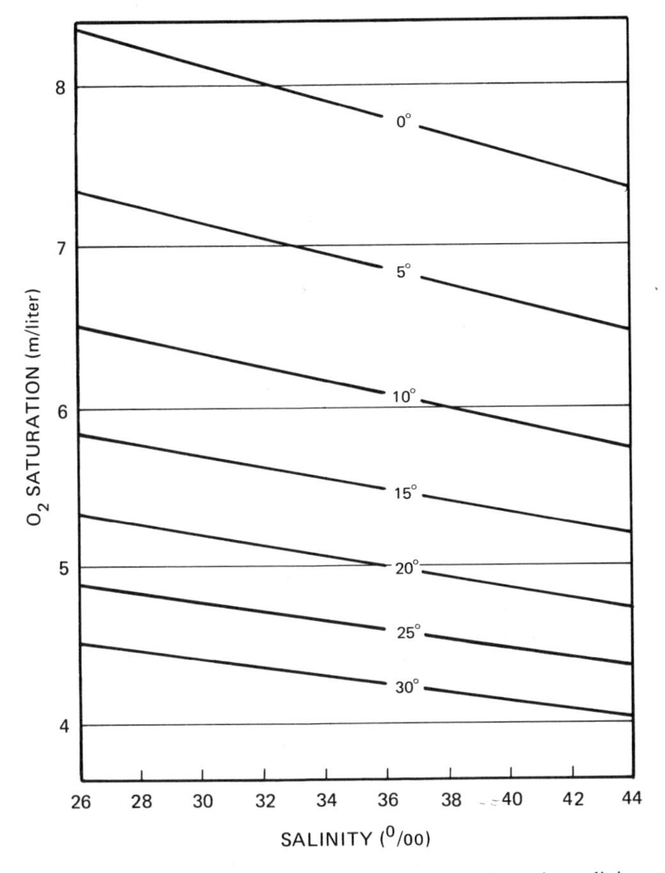

FIGURE 1–1 Oxygen saturation of seawater of varying salinity at different temperatures (°C).

types used in research and teaching are obtained from shallow, intertidal, and estuarine habitats, where they are regularly exposed to considerable environmental change. Animals from these more variable habitats tend to tolerate wider ranges of temperature and other factors typical of marine aquaria.

A plot of monthly average temperature against salinity, for a given locality, results in a polygon called a hydroclimograph. The hydroclimographs shown in Figure 1–2 demonstrate the pronounced seasonal variations characteristic of estuaries as compared with the stability of open-sea environments. Hydroclimograph data, along with data on the seasonal flourishing of certain marine invertebrates, are useful in predicting what laboratory culture conditions will prove successful.

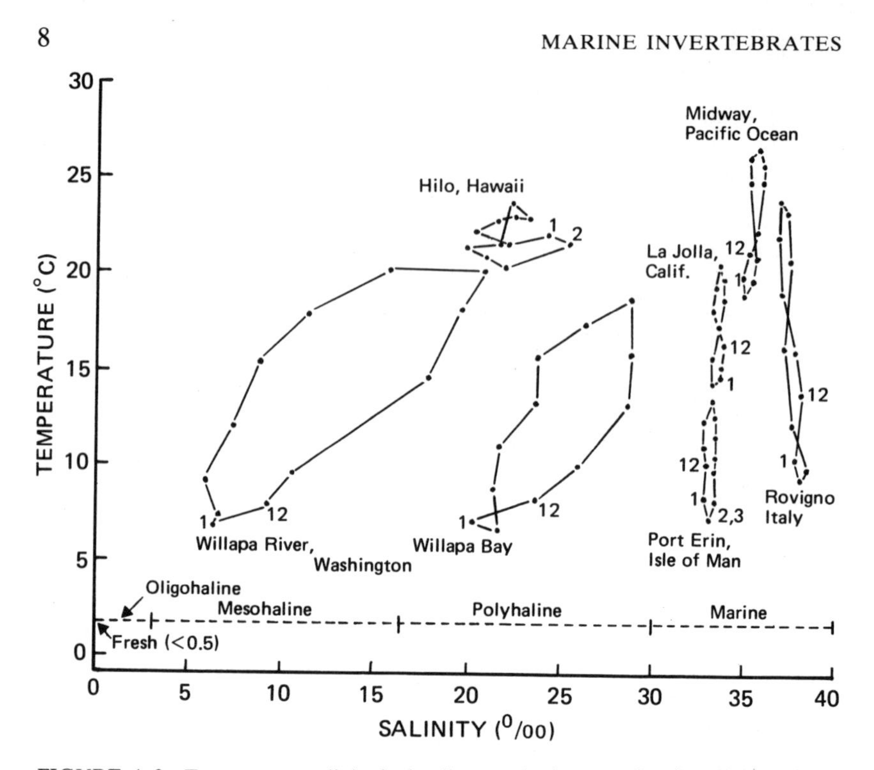

FIGURE 1–2 Temperature-salinity hydroclimographs for estuarine (brackish) and marine localities. Note that seasonal variation in both temperature and salinity is pronounced in estuarine habitats, whereas salinity is virtually constant in true marine habitats. (From Odum, 1971.)

TEMPERATURE

Temperature is critically important in determining the distribution, survival, and reproduction of marine invertebrates. The relationship is complex, varied, and in many cases not well understood. Prosser (1973) and Schmidt-Nielsen (1979) provide good discussions of temperature as a physiological factor, and Moore (1958) gives a useful account of temperature as a component of the marine environment.

Since marine invertebrates are poikilothermic (body temperature corresponding closely to the environment), the following generalizations concerning temperature and marine invertebrates may be useful in developing laboratory culture conditions:

- Increased temperature, within the tolerance range, speeds up metabolism and increases oxygen demand. As a general rule, metabolic rate doubles for each 10°C temperature rise.

• The solubility of oxygen in seawater decreases as temperature increases (Figure 1–1).

• Many marine invertebrates exist in the sea at temperatures closer to their upper temperature tolerance than to their lower. Maintenance at somewhat lower temperatures in the laboratory will minimize losses due to accidental increases in temperature and usually reduce the need for food.

• Developmental stages of marine species are usually less temperature tolerant than the adult form.

• Although many marine invertebrates respond to decreased temperatures with a reduction in metabolic rate and activity, in some forms metabolism sharply increases and the organisms remain active.

• Marine invertebrates from colder waters are generally more tolerant of temperatures close to the freezing point of seawater than are warm-water species. Such intertidal forms as mussels (*Mytilus*), barnacles (*Balanus*), and periwinkles (*Littorina*) from the northeastern coast of the United States can survive freezing at temperatures from −10°C to −20°C (Hill, 1976). However, some warm-water species may not survive at temperatures much below 10°C.

• The limits of temperature tolerance for many marine invertebrates are not fixed and vary significantly from season to season.

• Temperature change (usually a rise) is often a factor in the initiation of hormonal activity or reproductive behavior.

• A change in temperature can affect tolerance to other factors, e.g., when a reduction in temperature lessens tolerance to low salinity.

• When the temperature is abruptly raised or lowered, most poikilotherms show an initial shock reaction. The magnitude of this effect depends on the amount of temperature change and the type of organism.

SALINITY

The salinity of seawater influences the distribution and success of marine invertebrates through its effect on osmotic and ionic balance. Prosser (1973) and Schmidt-Nielsen (1979) present extensive discussions and data concerning osmotic and ionic control adaptations exhibited by various marine invertebrates.

Most marine invertebrates are *osmoconformers* in that their body fluids osmotically match the surrounding seawater. Such forms passively conform to osmotic changes in their environment, but actively regulate the composition of specific ions so that characteristic differentials are maintained between their body fluids and the environment. Echinoderms and cephalopods can tolerate only a narrow range of salinities,

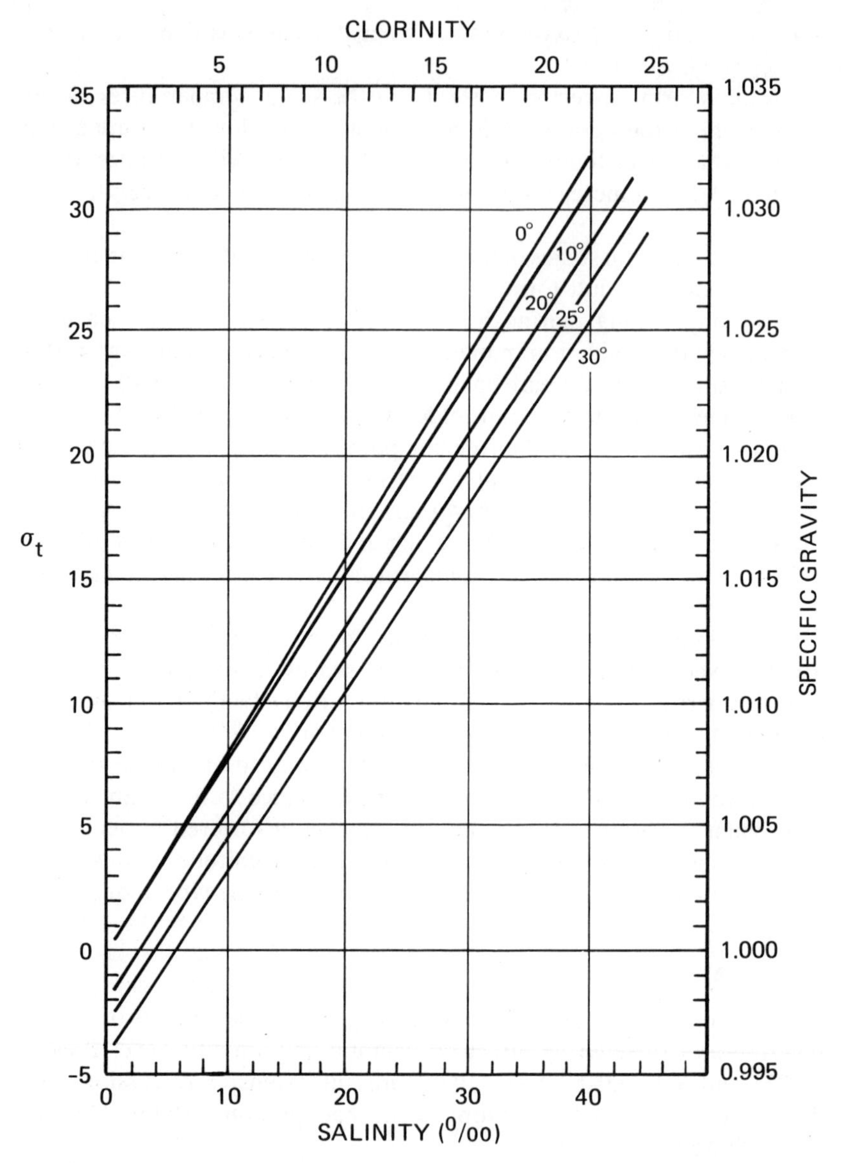

FIGURE 1–3 Graphs showing the relationship between the specific gravity of seawater at selected temperatures, and salinity and chlorinity. (From Nicol, 1967.)

usually about 30 to 35‰. Differences in salinity tolerance customarily correlate with habitat; some osmoconformers, particularly those found in estuaries, can tolerate wide ranges. The oyster (*Crassostrea*), for example, can be cultured in the low salinities (10‰) of the upper Chesapeake Bay as well as in coastal areas with salinities of 35‰. Even some osmoconforming coelenterates, such as the sea anemones *Aiptasia* and *Diadumene*, can tolerate significant salinity variations.

The alternative to osmoconformity is *osmoregulation*. Osmoregulators maintain their internal fluids at a relatively constant concentration, despite salinity changes in the environment, by active absorption and excretion, decreased permeability to water and salts, and behavior. Such crustaceans as the blue crab (*Callinectes*) and green crab (*Carcinus*) are noted for becoming hyperosmotic as they are exposed to decreased salinities. Other crustaceans, such as the brine shrimp (*Artemia*), show marked hyposmotic regulation and survive in salinities well above that of normal seawater. In general, osmoregulators can withstand a wider environmental range than can osmoconformers. Osmoregulation is often temperature-dependent and is less marked as the temperature departs from the optimum. Oxygen consumption is generally increased in dilute seawaters, where energy must be expended in active uptake of ions. Developmental forms are often more sensitive to salinity changes than are adults, and their ability to conform or regulate often varies with age.

It is important to maintain the salinity and temperature of the marine aquarium close to that of the natural habitat of the captive specimens. Salinity can be estimated by measuring the specific gravity with a hydrometer and applying that value in terms of Figure 1–3. However, temperature and salinity do not always directly limit an organism in the wild state. A species may, instead, be restricted to a particular habitat because of food availability or because certain physical or chemical factors discourage a predator or parasite. Many marine invertebrates are surprisingly tolerant of both temperature and salinity changes if they take place gradually.

PRESSURE

Inasmuch as deep-sea animals are rarely kept alive in captivity, pressure is an environmental factor that seldom needs to be considered. In recent years, however, high-pressure water systems have been constructed to maintain marine invertebrates (Avent, 1975; Quetin and Childress, 1980), and further development of this special technology is to be expected.

LIGHT

The penetration of light into seawater is indirectly important to marine invertebrates, as it relates to phytoplankton growth. Light drops off rapidly as depth increases, but absorption differs with wavelength. For example, in clear water at a depth of 4 m red light intensity is about 1 percent of that at the surface, whereas blue light is still at nearly 95 percent of surface intensity. At 70 m, only about 30 percent of the blue light will have been absorbed.

Most animals are light sensitive, responding in a variety of ways—diurnal vertical migration, phototropsim, photoperiodically controlled reproductive cycles, and protective coloration. A few are bioluminescent, producing their own light. Deeper-water species tend to have larger eyes than their shallower-water relatives, presumably an adaptation to the lower light intensities at greater depths.

Ordinarily, room light suffices for most captive animals, though the specific needs are not well established. Aquaria should be located where they will not receive direct sunlight, which may overheat the water or encourage excessive algal growth. For the same reasons, unusually intense artifical illumination should be avoided, except in the case of certain species of corals and sea anemones that are dependent on algal symbionts.

ACCLIMATION TO PHYSICAL AND CHEMICAL FACTORS

The effects of the various physical and chemical factors vary from one group of organisms to the next. Some are tolerant both to abrupt changes and to wide variations; others are sensitive to small changes and even these must be gradual. It is essential to avoid abrupt changes such as often occur when newly collected specimens are introduced into an aquarium or when organisms are transferred from a long-established system to one that has just been set up. The latter can often be surprisingly destructive. Subtle changes in water quality are particularly troublesome to deal with because they are so difficult to measure. The gradual acclimation of animals to a new environment should include a progressive matching of salinity and temperature and a mixing of the water from the old and new environments over a period of several hours. Experience will soon indicate what circumstances a given species will tolerate. For example, cephalopods are very sensitive and require careful acclimation, whereas sea anemones require very little.

2

The Laboratory Marine Aquarium

The marine systems that will be considered here are closed systems, without direct contact with the ocean. No aquarium of this kind can exactly duplicate the ocean habitat but, rather, represents an attempt, often successful, to provide conditions that marine animals readily tolerate. This goal can be achieved by providing the important environmental conditions discussed in Chapter 1—clean water and appropriate temperature, salinity, pH, and oxygen tension.

In contrast to a closed system, open-system marine aquaria have a fresh supply of seawater continuously passing through them. This arrangement generally provides clean, well-aerated, unpolluted seawater that removes biological wastes. However, even open systems may occasionally be forced to operate in a recirculating mode, either partially or completely, for indefinite periods of time. To that extent, information on closed systems is relevant to open systems as well.

Several closed-system marine aquaria are now commercially available (Appendix B). They are usually designed for varied use and, though expensive, most do their job well. Manufacturers generally provide detailed specifications for units they sell and offer advice as to which organisms can be maintained in them and under what conditions.

TANK CONSTRUCTION

Several important design criteria must be met. First, aquaria must be constructed of inert, nontoxic materials. Seawater is highly corrosive and can leach out toxins that might kill or debilitate the animals. Second, construction must be of strong, durable materials to withstand the weight of the water. Even a modestly sized aquarium containing 100 liters weighs about 103 kg (226 lb). Finally, the aquarium must remain watertight under moderate stress and should be designed to facilitate recirculation, aeration, and filtration.

GLASS–STAINLESS STEEL AQUARIA

Aquaria with inert surfaces in contact with seawater are strongly recommended. However, the low cost and ready availability of glass aquaria framed with stainless steel make them economically attractive. If such tanks are used, the framing must be of high-quality stainless steel. Despite this, the steel can be expected to rust, although it does no harm other than being unsightly and causing stains. If aquaria are refrigerated, rusting is accelerated because water condenses on the outside surfaces. Aquaria of this type have a comparatively shorter lifetime than is characteristic of those made of other materials.

ALL-GLASS AQUARIA

All-glass aquaria ranging from 20 to 200 liters (5 to 50 gal) are popular tropical fish aquaria and are readily available. They do not rust and offer an unobstructed view, but are vulnerable to breakage, either from stress or an accidental blow. If this type of aquarium is cooled, condensation can occur on all sides, and the dripping water will be a nuisance. Despite these hazards, an all-glass aquarium can give very long service at low cost, providing reasonable caution is exercised. No aquarium, all-glass or otherwise, should be moved when it is full of water.

CONSTRUCTION

Before constructing one of these aquaria, it is well to check prices for ready-made ones; often a well-made glass aquarium of 20–75 liters (5–20 gal) can be purchased at a price comparable to the cost of materials.

All-glass aquaria from 20 to 40 liters (5–10 gal) can be made rather easily with 3.18 mm (⅛ in.) double-strength glass. Tanks in the 40- to 75-liter (10–20 gal) range require 4.76 mm (3/16 in.) glass, while those in the 75- to 150-liter (20–40 gal) range require 6.35 mm (¼ in.) glass. Tanks exceeding the 150-liter (40-gal) size require 9.5 mm (⅜ in.) to 12.7 mm (½ in.) glass. The glass must be carefully measured and cut squarely, and the fit of all the parts checked by taping them together before any permanent sealing. Once the pieces fit closely, actual construction can commence by disassembling the aquarium, one piece at a time, and applying *clear* silicone rubber sealant to each of the abutting edges. Each piece should then be carefully repositioned and held firmly in place with tape. When the aquarium is completely reassembled, an additional bead of sealant should be applied to each of the inner joints to ensure a good seal and maximum strength. White-pigmented sealants should not be used because they often contain a fungicide. The sealant is most economically purchased in the 325-ml tube, is applied with a caulking gun, and will bond the glass in 24 h. At this time the tape should be removed and the excess sealant cut away with a sharp blade; an additional 24 h should elapse before the aquarium is used.

It is essential that an all-glass aquarium be positioned on an even surface to avoid stress and consequent breakage. This may be accomplished by placing the aquarium on a base of 2.5-cm (1-in.) styrofoam, which can be purchased as sheet insulation and cut to size. A more esthetically pleasing solution is to purchase the appropriate base molding at an aquarium or building supply store.

The top edges of the aquarium should be finished with plastic molding if possible. The best type has an inner, recessed flange that will support a plastic or glass lid. The lid is essential in that it greatly reduces water loss from evaporation, minimizes salinity change, and cuts down on damage to nearby objects from salt spray. Plastic lids are light, resist breakage, and can be readily cut or drilled to accommodate air lines and filter tubes. If plastic is used, it is advisable to attach knobs to both sides so that the lid can be turned over periodically and thereby counteract the warp that occurs as the consequence of differential water absorption. Aquaria containing such animals as an octopus should have heavy glass lids that fit tightly; otherwise, the animal will escape.

Figure 2–1 shows an all-glass aquarium with an easily constructed undergravel filter and airlift system. Undergravel filters can be purchased, but very effective filters can be constructed less expensively with readily available materials. The filter plate shown in Figure 2–1 is cut from a rigid styrene louver designed for a fluorescent lamp. These louvers, with a 1.25-cm^2 grid, can be purchased in sheets at building or electrical supply stores. The filter plate is supported by 7.6 cm × 2.5 cm × 0.6 cm (3 × 1 × ¼ in.) L-shaped pieces of Lucite plastic. The strips can be shaped by placing them in a vise, heating them with a propane torch, and bending them to the desired angle with pliers. The filter plate is covered with plastic screen to prevent the calcareous filtering medium from falling through. The screen can be tied to the filter plate with monofilament nylon fishing line. The air-lift is constructed by

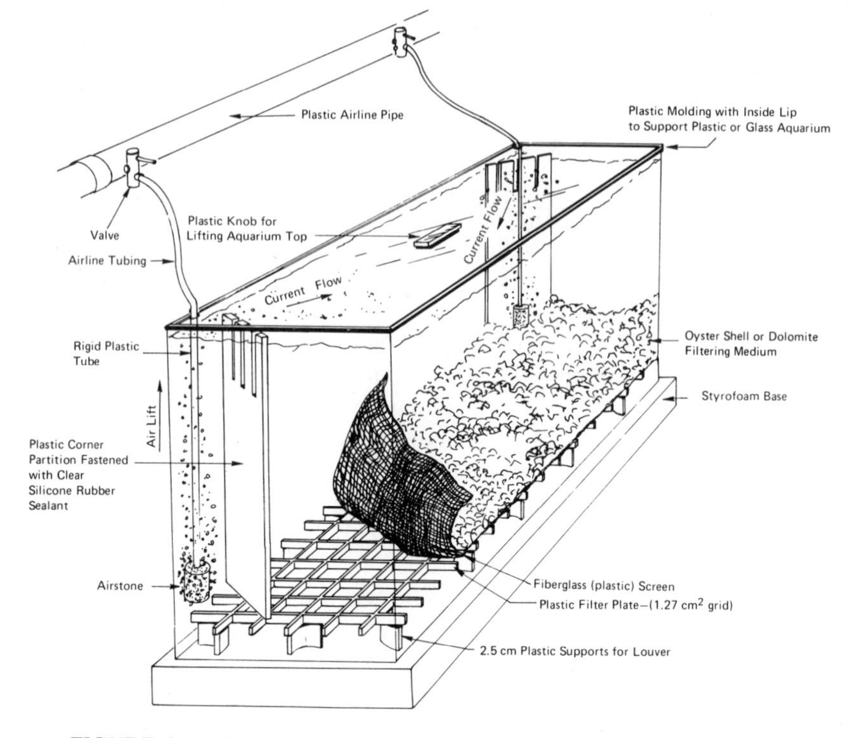

FIGURE 2–1 Glass aquarium with undergravel filter and airlift construction.

using silicone rubber sealant to attach 8-cm-wide plastic partitions in the rear corners of the tank. The partitions should be placed so that the bottoms are flush with the filter plate and so that the tops extend to a point just above the water level that will be maintained in the aquarium. This will prevent animals from crawling or swimming into the airlift corner and, in the case of sea urchins, chewing up the airstone. An airstone attached to a rigid plastic tube is placed in the sealed-off corner as shown in Figure 2–1. Three grooves measuring 2.5 cm × 0.6 cm are cut into the partitions to allow for the escape of the air–water mixture. If such a system is aerated vigorously, a strong current will be directed from each of the airlift corners and water will be constantly drawn through the filtering medium to replace that removed by the airlifts.

ALL-PLASTIC TANKS

Acrylic (Lucite), polycarbonate, polyethylene, polypropylene, polyester resin (fiberglass), or polyvinylchloride (PVC) are suitable for all-plastic aquaria. Acrylic and polycarbonate are transparent. As a practical matter, tanks made of these materials can be no larger than about 200 liters unless they are reinforced. Molded polyethylene and polypropylene may require considerable reinforcement, or thick walls, whenever capacities exceed about 100 liters. Fiberglass aquaria can be molded to any shape and can be pigmented. PVC may be glued, sealed, and worked in much the same manner as plywood to form excellent inert aquarium tanks. Again, the thickness of the material and the volume of the aquarium determine when reinforcement is necessary.

Plastic tanks are not as brittle as glass, but will scratch easily. Aquaria can be warmed to about 30°C or slightly above without losing shape and may be refrigerated without severe condensation. Plastic aquaria are generally easy to clean and maintain. They can be modified easily and drilled for inserting tubing or plumbing.

PLYWOOD TANKS

It was the development of plywood refrigerated marine aquaria, and suitable artificial sea–salt mixtures, that made possible the routine handling of marine organisms at low cost at inland laboratories.

These units may be assembled relatively inexpensively and provide reliable, reproducible conditions that most marine organisms will tolerate. Construction is of 127 to 190 mm (½ to ¾ in.) marine or exterior-grade plywood, which is screwed, glued, and then sealed with an inert covering or sealant. Stainless or galvanized steel screws should be used rather than brass, bronze, cadmium-plated, or iron screws. The glue should be waterproof, e.g., resorcinol, and the sealant polyester resin (fiberglass), epoxy, or polyurethane. Of the three, polyester resin appears the most satisfactory. Pigments may be included in the sealant; of these, titanium dioxide white has proven satisfactory. Tanks of about 200-liter capacity may be constructed from plywood with a minimum amount of reinforcing. Larger units may be fabricated from thicker plywood, with additional reinforcing. Windows may be added simply by sealing in a plate of glass of appropriate strength and thickness and allowing a sufficient margin of wood to assure the needed strength and rigidity in the tank wall. If refrigeration is provided, a double window (thermopane) is recommended to avoid the condensation that forms on a single layer.

Exterior-grade plywood will suffice for aquaria, since the layers of wood are joined with waterproof glue, although marine plywood is superior because it has a solid core, i.e., the holes and chinks are filled. Plywood 12 mm (½ in.) thick is adequate, even for larger marine tanks, provided it is braced or framed at intervals of about 50–60 cm (2 ft) to prevent bowing.

Specially treated plywood (e.g., Duraply) that has been impregnated with phenolic resins may also be used. This material is also referred to as Medium Density Overlay (MDO). Because phenolic resin is potentially toxic, care must be taken to cover the surface of the plywood completely with an inert substance such as epoxy paint or polyester (fiberglass) resin.

The surfaces of wooden aquaria in contact with seawater must be completely covered to be secure from such boring organisms as gribbles (a group of amphipods) and shipworms (*Teredo* and *Bankia*). Two other troublesome species in aquaria are sea urchins and chitons, both of which are capable of rasping through layers of plastic. They have not yet been observed to penetrate polyester (fiberglass).

Many tank designs are not escape-proof, although most animals will not climb out if there is a "lip" around the top of the tank. Of course, a "lip" will not deter octopuses; for them only an absolutely secure lid will suffice.

GENERAL PRECAUTIONS

All aquaria should be well constructed and must not be unduly stressed in use. Many fail because the base is not evenly supported, although aquaria can at times be placed on moderately irregular surfaces if they are first placed on a pad of styrofoam or polyethylene foam.

Some commercial aquaria are strong enough to be moved if partly filled, but under most circumstances at least 75 percent of the water should be drained off before attempting a move.

MANAGEMENT OF WATER QUALITY

Frequent, usually daily, careful inspection of the aquarium system and its organisms is essential. When disasters occur, they almost always do so rapidly. In general, the water must be kept clear and clean (filtered), well aerated, and at a suitable temperature (heated or refrigerated). In time the water may become lightly colored by pigments that resist oxidation and degradation; these pigments accumulate more readily if plants are provided as food. In moderate amounts, they cause no harm. If they do, they can be removed with activated carbon. Care must also be taken to maintain a suitable chemical environment (salinity, pH, etc.), to avoid overloading the aquarium with animals, and to select compatible species if more than one species is to be maintained in a single container.

Reserve seawater should always be on hand. If a salt mixture is being used, a supply sufficient to fill the largest unit of a multiaquarium system should be kept in reserve. An extra aquarium or several plastic barrels are appropriate for this purpose.

WATER SOURCE

Very clean seawater can be obtained from 5 to 10 mi offshore. It will usually be free of pollutants and have a salinity of 35‰. If unpolluted shore water is available, it can be readied for a closed aquarium system by first filtering it to remove sediment and large planktonic organisms and then allowing it to stabilize in a darkened container, with moderate or no aeration, for a period of 2 wk (Roberts, M. H., 1975; Spotte, 1979b). In some areas, water is of such high quality and low biomass that it can be used immediately in a system equipped with its own filtration. If natural seawater is used in small culture containers without built-in filtration or aeration, such as those commonly employed in short-term investigations of developmental phenomena, the water should be fine filtered using a 0.45-μm filter. Millipore® filtration (0.22 μm) and intense UV irradiation are commonly used to greatly reduce the biomass of organisms (Smith and Chanley, 1975). Millipore® filtration, followed by UV irradiation and/or treatment with ozone, is also effective in keeping the reservoir water clean and sterile, although these measures are usually not necessary. If seawater is to be collected at the shore, one should use plastic, hard rubber, or cast iron pumps and avoid brass pumps. All lines and vessels used in the transfer and transportation of seawater should be rinsed before filling the containers.

ARTIFICIAL SEAWATER

Artificial or synthetic seawater is increasingly used because it is convenient and because natural seawater is difficult for inland laboratories to obtain. Most of the published formulations seek to mimic water from the ocean; they succeed to varying degrees and, under some circumstances, offer a usable substitute. Ready availability, ease in mixing, freedom from living material, and standardized composition are significant advantages. The decision whether to use tap or deionized water in which to dissolve the artificial mixtures depends on locality and can only be answered by experience or extensive chemical analysis. Tap water is usually suitable for routine culture, but this point must be

TABLE 2–1 Formulations for Artificial Seawater

Constituent[a]	Formula I,[b] g/liter	Formula II,[c] g/liter	Formula III,[d] g/liter
Sodium chloride, NaCl	28.32	24.72	24.72
Potassium chloride, KCl	0.77	0.67	0.68
Hydrated magnesium chloride, $MgCl_2 \cdot 6H_2O$	5.41	4.66	4.78
Hydrated magnesium sulfate, $MgSO_4 \cdot 7H_2O$	7.13	6.29	6.30
Calcium chloride, $CaCl_2$	1.18	1.36	1.40
Sodium bicarbonate, $NaHCO_3$[e]	0.20	0.18	0.10

[a] The more abundant trace elements can be added if they seem to be necessary. Trace elements in grams per liter are:

Potassium bromide, KBr—0.89
Sodium fluoride, NaF—0.003
Hydrated strontium chloride, $SrCl_2 \cdot 6H_2O$—0.037
Boric acid, H_3BO_3—0.0024

[b] Developed by Mazia. Reported by Hinegardner (1967).

[c] Recommended by the Marine Biological Laboratory, Woods Hole (MBL Formula). This is isosmotic with Woods Hole seawater, which is 31‰ (Cavanaugh, 1975).

[d] Developed by Allen-Pantin (Cavanaugh, 1975).

[e] Add sodium bicarbonate after the other salts have dissolved.

clarified in relation to each particular source. Tap water that may have traces of copper should not be used.

Thus far none of the artificial mixtures fully duplicates real seawater, and animals kept for extended periods in artificial mixtures are more susceptible to a variety of maladies. No one really understands why salt mixtures do not work as well as natural seawater. The commercial mixtures are usually made from technical-grade chemicals that may contain elements the animals cannot easily tolerate. Some investigators have tried aging artificial seawater the way natural seawater is aged, but to no avail. However, Fridberger and Fridberger (1979) have found that if artificial seawater is kept in an illuminated (two 30-watt fluorescent lamps) and circulating, but otherwise unaerated, aquarium (75–100 liters) for 4 to 6 mo, it is much more satisfactory than a newly prepared solution. They recommend beginning with a salinity of about 25‰, to allow for evaporation. Possibly the algae and any other organisms that establish themselves in the illuminated aquarium detoxify the water or improve the composition. This biologically aged water can be mixed with freshly prepared artificial seawater in a ratio of two to one. Natural seawater can also be extended in this way. In any case, before a commitment to artificial seawater is made, it should be tried in comparison with natural seawater.

For short-term use, it is often unnecessary to duplicate seawater; simple salt mixtures often work equally well. A number of larval forms will develop for days in formulations containing only the more abundant elements. A few formulas of this kind are listed in Table 2–1. Other formulas for more complex salt mixtures can be found in de Graff (1973) and Kinne (1976). Special formulas have been developed for the laboratory culture of the hydroid *Bougainvillia* by Tusov and Davis (1971), the jellyfish *Aurelia* by Spangenberg (1965), the brine shrimp *Artemia* by Provasoli and D'Agostino (1969), the copepods *Euchaeta* by Lewis (1967) and *Acartia* by Gentile *et al.* (1974), the oyster *Crassostrea* by Zaroogian *et al.* (1969), the mussel *Mytilus* by Courtright *et al.* (1971), and various bryozoans by Jebram (1977b). Ready-made mixtures can be purchased from many aquarium stores or from suppliers (Appendix B).

The salt mixture obtained by evaporating seawater is no better than the synthetic mixtures and indeed is often inferior. Why this is so is unclear; possibly during evaporation compounds very different from the original are formed and new ionic interactions take place.

pH

Natural seawater and freshly mixed artificial formulations have a pH of 8.0–8.3. The metabolic activities of organisms in a closed system will cause a detrimental decrease in pH. It is essential that the pH of marine systems be stabilized by the use of calcareous material in the filter bed. Crushed oyster shell (large) and dolomite gravel with a grain size of 5 mm are recommended for their buffering qualities. The pH of water that has been in a closed system for a few months will normally drop to 7.5–7.9 without harming the animals. The pH of such older systems can be raised by partial replacement of the water. A sudden drop in pH or a level below 7.5 is an indication that something is wrong with the system. An overload of organisms, or death and decay, are usually responsible. The pH of marine systems should be monitored regularly.

SALINITY

Most marine invertebrates will do well in salinities ranging from 30 to 35‰. The constant circulation of water in a closed system causes considerable evaporation and a subsequent increase in salinity. A recessed lid on the system will minimize water loss, but weekly salinity measurements are recommended. Adjustments should be made with demineralized or copper-free, low-mineral tap water.

A NBS-approved hydrometer is the most common and least expensive device used to determine salinity. Specific gravity as measured by a hydrometer is temperature dependent and tables or graphs such as Figure 1–3 must be used to convert to salinity. A specific gravity of 1.025 at 20°C is equal to a salinity of 35‰. Hydrometers are fragile and somewhat limited by the relatively large volume of water required for flotation.

A salinity refactometer (Appendix B) is probably the most useful instrument for measuring salinity in culture systems. This hand-held optical instrument is temperature-compensated within a range of 15°C to 38°C and requires only a drop (0.02 ml) of water to read salinity directly. This is particularly important when working with test animals that may be isolated in small containers that will not accommodate a hydrometer.

Salinity can be most accurately determined by measuring freezing point depression or electrical conductivity, but most culture work does not require this degree of sensitivity.

OXYGENATION AND GAS EXCHANGE

Diffusion

Aeration may be provided in shallow tanks or trays merely by maintaining active circulation of the water (Figure 2–2). Provided water depths are on the order of 10 cm (4 in.) or less, oxygen diffusion from the air seems to be adequate.

Air Stones

Most aquaria are designed for viewing, and the air–water interface is small compared to the volume. Consequently, virtually all aquaria must have some form of forced aeration and circulation. Air stones accomplish this effectively and inexpensively. Some air stones are advertised as producing very fine and presumably efficacious air bubbles. Unfortunately, most of these special designs clog easily and require more attention to maintain than ordinary air stones and offer little advantage over them.

In some cases, air stones may have to be weighted and protected from crabs, octopuses, lobsters, sea urchins, and sea stars. They can be tied

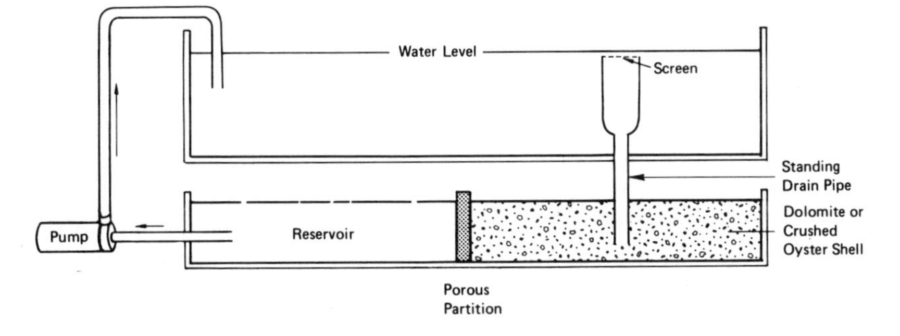

FIGURE 2–2 Shallow, recirculating aquarium.

to a rock or small shell, or fitted inside a length of plastic pipe. In the latter design, a piece of 37 mm (1½ in.) plastic pipe 7.5 cm (3 in.) long is suitable. A hole is drilled in the middle of the pipe to accommodate the stem of the air stone and its connecting air line.

Air-Lift Recirculating Devices

There are a number of commercially produced marine aquaria that incorporate air-lift recirculating devices similar to the one shown in Figure 2–1. Systems of this kind rely upon the production of bubbles that, in effect, lift water through a tubing or channel. An air-lift provides adequate aeration for a moderate-sized aquarium without need for additional air stones, provided enough air is used. As in Figure 2–1, these devices are connected to a bottom undergravel filter, which is usually a grid covered with gravel and elevated 1–3 cm above the bottom of the tank. Water is drawn through the gravel, thereby filtered, and then recirculated as it rises to the surface. Not all commercially produced air-lifts operate efficiently. Castro *et al.* (1975) and Spotte (1979a,b) discuss the principles of good air-lift design.

Air Pumps and Air Supplies

A variety of air pumps can be used. Reliability (service life) is of prime importance here and in all other parts of the system. Other considerations include capacity, noise, and cost. Two different types of pumps are listed in Appendix B.

As a general rule, a pump should have a capacity of approximately 2 liters of air per minute for each 100 liters of water. This rate is adequate for almost all systems. If the animals are relatively inactive, uncrowded, and held at low temperature, half that capacity will suffice. For most laboratory systems—100 to 200 liters up to a few thousand liters—there are a number of small, efficient air pumps available, ranging from large vibrating diaphragms, which are very quiet and provide a modest amount of air, to motor-driven pistons or diaphragms that are at times fairly noisy. For larger systems, carbon-vane rotary compressors have proved reliable, but they are very noisy and objectionable to most people and should be isolated in an uninhabited space. All pumps must be air-cooled and therefore should not be placed in small, closed containers.

At least one backup pump should be available. Whenever multiple systems are in operation, a minimum of one backup air pump for each four or five in service provides a reasonably safe margin. Oil-free pumps are recommended; if oil-lubricated compressors are used, an air filter is absolutely essential, although it is a wise precaution to include an air filter between the air pump and the aquarium in other cases as well. If the laboratory building is equipped with a built-in system for compressed air, the air should be filtered before being introduced into the aquarium. Usually, the air comes from oil-lubricated compressors located in a region of uncertain air quality. Although most of the oil can be filtered out, toxic compounds generally cannot. Therefore, if a remote air intake is utilized, great care must be taken to assure nothing will pollute the air delivered to the aquarium.

It is essential to position a local air pump above the aquarium, to protect it from the backflow of seawater, which occasionally happens when a pump fails. Backflow can be prevented in several ways. If the pump cannot be located above the water level of the aquarium, a one-way valve (obtainable at most aquarium stores) can be installed in the line or the air line may be looped several times at the point where it leaves the aquarium—the top of the loops should be above the water level, thus forming water traps that break only under substantial reduction in pressure. Obviously, backflow also occurs if an air pump is accidentally connected in reverse.

Compressed air is often hot and dry, which accelerates evaporation of the seawater. If evaporation is excessive, the air supply may be passed through fresh water before it is introduced to the aquarium. Care must be taken to assure that the salinity of the water in the aquarium remains at the desired level.

Multiple tanks may be provided with air by introducing a header or manifold, a length of plastic pipe into which the required number of holes are bored and tapped to receive aquarium needle air valves. An initial air supply may be boosted by inserting a second pump in series with the primary one. This arrangement increases the pressure, and thus the volume, of a central supply derived from low-pressure high-volume air compressors.

Air bubbling through an aquarium will disperse a fine salt mist in the area around the aquarium. Similarly, water splashed by recirculating pumps deposits salt on the edges of the aquarium and on nearby fixtures. Both of these troubles can be greatly reduced by covering all aquaria.

Although adequate oxygenation is absolutely essential, it should not be overdone. Massive aeration produces an underwater fog of fine bubbles that may become trapped inside digestive tracts, shells, and other partially closed spaces. This is usually harmful and sometimes fatal to the animals.

Related to this situation is the gas-bubble disease of bivalves and shrimp that is brought about by the supersaturation of seawater with oxygen and nitrogen (Lightner *et al.*, 1974; Goldberg, 1978). The introduction of water under pressure into the aquarium or a sudden increase in temperature are the usual causes of supersaturation.

FILTRATION

The best filter systems for marine aquaria accomplish three things simultaneously: (1) mechanical filtration of particulate matter, (2) chemical buffering to maintain a pH of 7.5–8.3, and, most importantly, (3) biological breakdown of metabolic wastes. Biological filtration refers to the action of various species of heterotrophic and autotrophic bacteria on waste compounds in the aquarium (Spotte, 1979a). These bacteria grow primarily on the surface of the filter material, and water circulation must be continuous to maintain aerobic conditions in the filter bed. Ammonia is the main excretory product of marine invertebrates and also results from the action of bacteria on organic nitrogenous compounds such as the protein in uneaten food. Ammonia is toxic and must not be allowed to accumlate in a closed system. The solution to this critical problem is the establishment of populations of nitrifying bacteria: *Nitrosomonas* sp., which convert ammonia to nitrite (NO_2^-) and *Nitrobacter* sp., which convert the nitrite to nitrate (NO_3^-). Denitrification of the nitrate to nitrous oxide or free nitrogen occurs at a very slow rate in closed systems, and nitrate concentrations will reach high levels in aquaria that have been in operation for a few months. Although most marine animals are quite tolerant of nitrate accumulation, the level should be reduced periodically by changing about 25 percent of the water. Spotte (1973) and Strickland and Parsons (1972) give useful descriptions of the nitrogen cycle and the chemical tests for ammonia, nitrite, and nitrate. These tests are tedious and time consuming and are not usually necessary if proper attention is given to aquarium ecology. The bacterial populations can be started in a new system by introducing a few pellets of fish food or a particularly tolerant organism. The addition of a small amount of filter material from an established system will also facilitate the development of bacterial populations. The establishment of this abbreviated nitrogen cycle is temperature dependent and takes approximately 2 wk at 20–25°C as compared to 4 wk at 15°C. The omission of this critical "run-in" period (King and Spotte, 1974) by individuals who are too eager to add their first organisms to the system, along with overfeeding and overcrowding, is a major cause of failure in the culture of marine invertebrates (Miller and Mazur, 1974).

Filtration systems can only accomplish these several functions and enhance gas exchange if they are powered by a reliable pumping system that is capable of continuous and fairly rapid recirculation of the water. These systems must be checked daily for malfunction and adequate backup equipment must be available.

Filter Design

Three designs will be described below. All operate on the same principles, but each has certain advantages.

- *Internal or bottom filters.* A simple filtration and recirculation system can be composed of an air-lift pump of the kind illustrated in Figure 2–1, with filter material supported on a grating above the bottom of the aquarium. The tank should be partially filled with water before placing bottom filter and filter bed. Various types are available at most aquarium stores and this is the system used in some commercially available marine aquaria. Though silicate sand is often used as the filter bed in freshwater tropical fish aquaria, it is unsuitable for most marine aquaria. Dolomite gravel or crushed oyster shell should be used instead.

Bottom filtration satisfies most requirements for small, simple systems used for maintaining a modest number of marine invertebrates, as well as certain algae and fishes. The major drawback is that the aquarium must be drained whenever it is necessary to wash and replace the gravel. This need not be frequent, however; depending upon the biomass and type of organisms held, these simple tanks can sometimes be used for a year or more without needing thorough cleaning.

- *Remote submerged filters.* Simple marine aquaria do not suffice for maintaining relatively large numbers of specimens. Such herbivores as sea urchins and sea hares are especially troublesome, because they

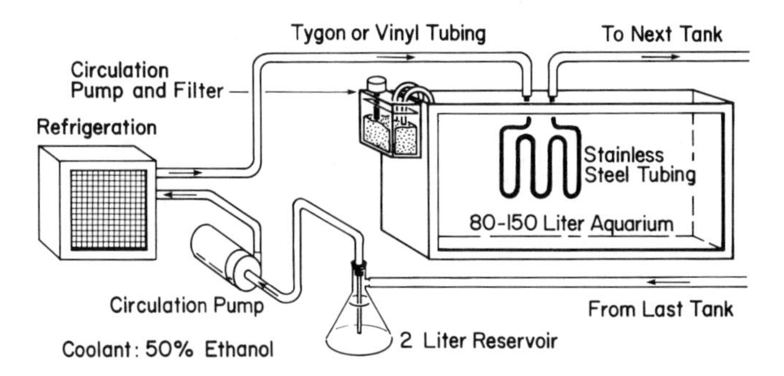

FIGURE 2–3 Aquarium with remote submerged filter and temperature control.

pass large quantities of partially digested fecal matter, which makes frequent filter cleaning necessary. This task is more easily accomplished in systems with remote or outside filters.

It is also convenient, and sometimes necessary, to have the bottom of the tank free of gravel. Not only is cleaning then much easier, but there are fewer places for rotting material to escape detection. Remote filters need to be no more distant than the outside of the tank (Figure 2–3), or they may be placed in filter boxes some distance from the tank. These larger, remote filters can be positioned above, below, or beside the tank, but they should be as close as practicable, to minimize plumbing. Water should be drawn from the tank through a screened outlet.

It is important to design the entire system so that unexpected flooding is made as unlikely as possible in the event that some element of the system fails or becomes clogged. Clogging commonly occurs when animals or plants lie over or get into improperly designed outlets.

Although freshwater aquaria often have filter beds made of fiberglass, wool, dacron, or similar material, these materials are less appropriate in marine aquaria, because fiber beds tend to clog quickly and are no better than gravel beds.

• *Remote trickle filters.* In filters of the kind illustrated in Figure 2–4, the water trickles through the filter material and around air spaces that establish a large air–water interface. The trickle compound filter must be overlain with a layer of material that will remove particles that would otherwise clog the filter. This layer also distributes the water evenly over the filter bed. Polyurethane foam (20 mm thick) works well for this purpose, but the prefilter must be checked daily and cleaned about once a week. A bypass or overflow drain is essential in the event

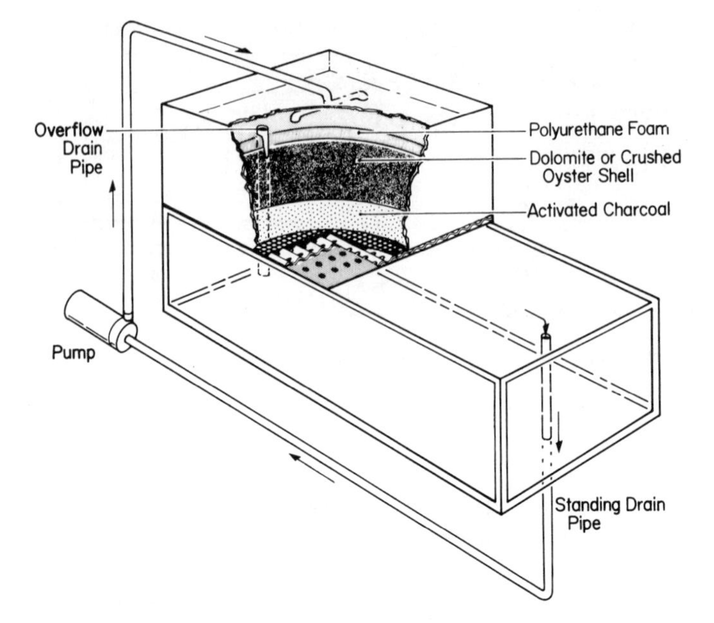

FIGURE 2–4 Diagram of a holding tank with superior remote tiered filter. Actual tanks can be larger than shown here (see section on filter capacity).

the filter becomes clogged. A second tier of activated carbon can be used in both the submerged and trickle filters. It may be useful to layer it on a screen of polyvinylchloride (plastic window screen) or on a polyester fiber mat.

Filter Beds and Cleaning

The best filter beds consist of crushed oyster shell (large) or dolomite gravel. Coral sand can sometimes be used. These substances help control pH, supply calcium, and provide surfaces on which microorganisms multiply. A good, well-populated filter can support far more animals than could otherwise be kept in the same aquarium. In time, any filter will become clogged and must be cleaned. In doing so, it is best to remove a portion of the gravel filter material, clean the remainder by repeated washing in fresh water, then add seawater and mix in the *unwashed* portion that had been set aside. This last step reinoculates the filter with the filter biota. It is unnecessary to reseed the activated carbon portion of a compound filter.

Whether or not activated carbon is used in addition to the gravel depends on the animals to be kept; usually it is not necessary, although it does remove the yellow-brown color from the water in aquaria containing feeding herbivores. Because carbon usually seems to cause no difficulties, it may be well to try it, but there are instances on record when fresh carbon caused sea urchins to spawn. Any dust present should be washed from new gravel and carbon before they are added to a filter.

Filter Capacity

Most submerged filters work well if their volume is between 2 and 5 percent of the water volume. A second rule of thumb suggests that submerged filters should be 30 times the bulk of the mass of the organisms in the aquarium (Saeki, 1964). Considerably less filter material is required for trickle filters because they provide much higher concentrations of oxygen. Consequently, a trickle filter at least twice as large as the biomass of the animals will suffice in most instances. In all cases an amount of water about equal to the volume of the aquarium should be circulated through the filter per hour. If carbon is used, it should about equal the biomass of the organisms. A grain size of 5 mm (3/16 in.) for the oyster shell or dolomite, and slightly smaller for the carbon, is satisfactory.

Water Pumps

Remote filters require water pumps. The simplest is an air-lift pump, but this is usually not adequate because of low flow rate and limited lift. A number of companies sell plastic filter–centrifugal pump combinations that can be hung on the side of the aquarium. The better-designed ones have the pump motor mounted above the box. There are also a number of nonmetallic centrifugal pumps that can be purchased as separate units. The best have the centrifugal pump magnetically coupled to the motor; other designs tend to leak. The impeller should be plastic, and in any case, pumps containing copper parts must not be used.

Sometimes experimental design requires that several aquaria receive the same water supply. Examples of such circulating water systems may be found in Breder (1964), Lickey *et al.* (1970), New *et al.* (1974), and Nair *et al.* (1978).

TEMPERATURE CONTROL

Under most uses, marine aquaria will require some degree of temperature control (Huguenin, 1976; Spotte, 1979b). Temperature control is sometimes critical, and failure of temperature control equipment is a common cause of serious trouble in marine systems. Most marine invertebrates from the coasts of the United States can tolerate at least short periods of low temperatures (5–10°C), such as occur if aquarium or building heaters fail in cold weather. However, even tropical species from Florida or Hawaii will be killed by high temperatures (30°C and above) that may occur when air conditioners or refrigeration units break down or when aquarium heaters fail to shut off properly. Even if the temperature is not high enough to kill the animals outright, the rate of chemical reactions will be accelerated and the level of dissolved oxygen will decrease to dangerous levels. If the increased temperature is not quickly detected and corrected, and some of the animals die, decay will aggravate the situation until the entire system fails. Because a research project may be completely lost if controls fail, careful attention must be given to the reliability of the control, and the system and its support equipment must be inspected daily.

A variety of solid-state temperature sensors are available and are recommended simply because they are more reliable than mechanical thermostats. If thermostats are used, the metallic parts should be protected from contact with seawater by a coating of plastic or rubber.

ELECTRICAL CONNECTIONS

Seawater is a very good conductor of electricity, and salt film will slowly accumulate on wires and sockets. If possible, therefore, all outlets into which electrical equipment is connected should be located above the aquarium. This minimizes splashing of water onto electrical connections and seawater flowing along wires and into connectors. Safety considerations also dictate that apparatus be plugged into the electrical supply with ground-fault interrupters.

HEATING

If the organisms require a temperature above ambient, a variety of aquarium heaters controlled by simple thermostats are available. Care must be taken that the devices are designed for use with marine aquaria and that none of the parts exposed to seawater will corrode and fail. Some form of overtemperature control should also be used, just in case the heater fails to turn off.

REFRIGERATION

If organisms are to be maintained below ambient temperature, refrigeration will be necessary, either by air conditioning or by cooling the water directly. The latter course is more usual.

In practice, refrigeration units of 1/6 hp rating have been found adequate under most ambient conditions for aquaria in the 100- to 200-liter (25–50-gal) range. Units of 1/4 to 1/2 hp capacity suffice for aquaria of 400 to 800 liters (100 to 200 gal) or for the simultaneous cooling of a number of smaller aquaria. These approximations apply to aquaria of plywood–fiberglass construction operating in "room temperature" environments. If the aquarium has a viewing window, the glass should be double-layer (Thermopane). The amount of refrigeration needed depends upon both the insulating properties of the materials used and the environment. It is wise to provide sufficient margin to keep the workload on the refrigerator well within its operating limits. In small rooms, or ones that do not have air exchange, it is advantageous to exhaust the heat from the refrigerator outside the room.

Transferring heat from aquaria is made more difficult because seawater is corrosive and because some otherwise convenient metals are very toxic. Copper tubing, for example, should be used only if it is well protected from any contact with seawater. In early models of refrigerated aquaria, copper coils were protected by placing them in a cooled freshwater reservoir that was in thermal contact with the seawater in the aquarium. In later designs copper tubing was sealed into the walls of fiberglass aquaria. Metal coils can also be coated, painted, or otherwise treated to prevent corrosion, or they can be encased in rubber or plastic tubing. The drawback in using covered tubing is that the covering acts as an insulator and, also, can be damaged by such animals as crabs, lobsters, and sea urchins. Stainless steel tubing can be used, provided the tubing is kept completely under water—stainless steel corrodes readily in air when it is covered with damp salt. Stainless steel tubing provides good heat exchange and is nontoxic. Plastic tubing must be used to bring the cooling fluid to the stainless tubing. Glass or nontoxic plastic tubing can also be used as heat exchangers; but some animals will eventually chew through the plastic, and glass is notoriously fragile.

Three to five m (10 to 15 ft) of 6-mm (1/4-in.) tubing should provide adequate heat exchange surface for most aquaria in the 100- to 800-liter range. If the tubing is exposed in the tank (rather than imbedded in the wall), it should be looped back and forth along one wall, thus occupying the least amount of space and facilitating inspection.

Coolant may be circulated through the aquarium in several ways. The simplest arrangement is to cool a container of water and then recycle that cold water through tubing in the aquarium. A somewhat more complex system utilizes a specially designed countercurrent cooling system that circulates Freon through a tube that is, in turn, inside a second tube containing the fluid being circulated through the aquarium. This system is shown in Figure 2–3. A 50 percent ethanol solution should be used as the secondary fluid, to avoid freezing and possible cracking of the tubing in the event the circulating pump fails. Finally, Freon can be circulated directly through the cooling coils, a scheme that works well for permanent setups but may be inconvenient if flexibility in design is desired. Since Freon seeps slowly through many kinds of plastic tubing, plastic refrigeration coils have been supplanted by models using copper tubing imbedded in the aquarium wall, or covered and isolated by protective screening.

A number of different refrigeration units are commercially available that can be used directly, or after modification. The only precaution needed is to select one that will operate effectively at temperatures above 0°C. Many of the cooling systems sold by laboratory supply houses are designed to maintain subfreezing baths and are inefficient in the 10°C to 15°C range most commonly maintained in laboratory aquaria.

If aquaria are close together, a single refrigeration unit can be used to cool several (Figure 2–3). Provided the room has a fairly even temperature distribution and the aquaria are the same size and construction, only one need be equipped with a sensor; this arrangement is almost as accurate as controlling each tank separately.

Close control over the temperature of an aquarium can be maintained by opposing constant refrigeration against a controlled heating unit. But this scheme has the added risk that if temperature control fails, the whole system fails. In fact, this level of precision is seldom necessary. It is advisable to back up all controls with a second unit that will prevent overheating or overcooling.

AQUARIUM ROOM

Care should be taken to place aquaria in rooms with a watertight floor and dependable floor drain. Aquaria must not be located above valuable equipment sensitive to damage from water even if these items are two or three floors below.

When constructing or remodeling a marine aquarium room, plans should specify that the floor and lower part of the walls be watertight. The floor drain must be at the lowest point on the floor, and electrical outlets at least 5 ft off the floor and well above the water level of the aquaria. Ground-fault connectors should be provided for all electrical apparatus in the room, and there should be a sink suitable for rinsing various aquarium accessories.

There should also be an alarm system (external to the room) to indicate when the temperature is abnormal or the system not working. Instructions should be posted for anyone who may observe an activated alarm; include telephone numbers of those to be contacted whenever there is a failure. The log of operation of each aquarium should record such factors as the number of animals per tank, water quality, temperature, filter changes and cleaning, and water changes.

PITFALLS AND SOURCES OF DIFFICULTY

Experience has shown that it is the unexpected that often leads to the downfall of the system, with consequent loss of valuable specimens and research time. Some common causes of trouble are:

- Use of toxins and solvents in or near aquaria, e.g., in spraying for insect control
- Use of plastics impregnated with insecticides or fungicides (e.g., Dow Bathtub Sealer) as sealing compound for aquaria
- Use of brass or copper valves or tubing that corrodes and sloughs off small amounts of toxic copper salts into the aquarium
- Overcrowding, which overtaxes the oxygen supply and the waste assimilation capacity of the aquarium
- Failure to isolate from the stock or holding aquarium those specimens that are to be manipulated surgically or used in pharmacological experiments
- Failure to remove decaying food or dead animals
- Failure to screen drains adequately
- Failure to provide bypass overflow pipes
- Failure to provide alarm warnings for the air supply, water circulation, or temperature change
- Failure to rinse dust from gravel, crushed shell, crushed dolomite, or activated carbon when recharging the filter
- Failure to age new carbon in the filter when working with ripe sea urchins
- Failure to place aquaria on sound foundations
- Failure to check carefully the specific gravity of newly-made solutions of synthetic seawater
- Failure to assign responsibility for aquaria to specific individuals

• Failure to separate certain species in the aquarium or when collecting in the field:

—Spiny animals such as sea urchins and sea stars should not be mixed with other animals.
—Noxious animals such as sponges should not be mixed with other animals.
—Toxic organisms such as acidic algae, e.g., *Desmarestia*, must be isolated.
—Organisms that produce prodigious amounts of slime must be isolated, e.g., cowries, some sponges, and some sea stars.
—Organisms that discharge ink must be kept in isolation, e.g., octopuses, squid, and sea hares.

Although most of these matters seem self-evident, they crop up again and again because appropriate precautions are not taken when the seawater system is set up or it is not properly operated.

DANGER SIGNALS

Daily attention to the marine aquarium and immediate attention to newly arrived specimens are critical to the successful operation of a marine system. A few warning signals that indicate potential trouble, but appear before the system collapses, are listed below.

• Turbidity
• An unusual amount of foam on the surface
• Excessive accumulation of yellow, organic residues
• Distress of the organisms, as indicated by unusual activity or inactivity, excess display of gills, drooping of spines, slow responses to prodding, labored breathing or lassitude (octopuses), or lowered threshold for inking (octopuses, squid, or sea hares), failure to right themselves (snails, sea stars), or poor display of tube feet (sea urchins)
• A dead animal, or continuing deaths of specimens of species known to do well in aquaria

The cause of the difficulty must be located promptly and measures taken to correct it. A series of questions should be asked about the equipment, commencing with the filtration system. Is it functioning? Is it overtaxed? Has the filter had time to acquire a stable population of microorganisms? In a compound filter, are all of the components still working? Is the rate of circulation of seawater adequate? Is the refrigeration working? Is the heater working? Is the aeration adequate? Which species of animals are not healthy? Is there adequate food? Has there been a significant change in the area around the aquarium?

TEMPORARY MEASURES AND MAINTENANCE

In dealing with aquaria of 100 or more liters, power filters charged with diatomaceous earth may be used as auxiliary filters whenever episodes of excessive turbidity occur, as, for example, when sea urchins spawn. An ultraviolet sterilizer can be used to reduce turbidity caused by bacteria; these devices are usually tubes containing a sealed ultraviolet light source and are available in most larger aquarium stores (sterilizers containing a small bulb are not very effective). The filter pump can be used temporarily to pass the water through the sterilizer before it is returned to the aquarium. Once the aquarium water is clear again, the auxiliary filter or sterilizer should be removed and reliance placed on the main aquarium filter. If the main filter proves inadequate, the aquarium loading should be reduced, the filter capacity increased, or the rate of recirculation and aeration increased, or the aquarium maintained at a lower temperature. It may be wise to invoke several of these adjustments simultaneously.

In most cases, close monitoring of the aquarium and its components is all that is necessary to avoid trouble. Chemical tests of the seawater for dissolved oxygen, ammonia, nitrites, or nitrate are usually not needed on a routine basis (Bower, 1980). Ammonia, for example, is generally less toxic to both larval and adult marine bivalves and crustaceans than it is to fishes (Wickins, 1976a; Armstrong *et al.*, 1978; Cardwell *et al.*, 1979). Specific gravity should be measured weekly or biweekly and fresh water added when needed.

Under even the best conditions and in the best laboratories, marine aquaria may become infested with pathogens. If this occurs, the aquarium and all its components should be emptied and dismantled, all parts thoroughly soaked with dilute sodium hypochlorite (bleach) and rinsed well (Blogoslawski *et al.*, 1978). The system can then be reestablished, with a break-in and leaching period.

At a minimum, maintenance should include daily inspection of the aquaria. Depending upon the intensity of use, periodic overhaul may be necessary on a schedule somewhat as follows:

- Weekly cleaning of the prefilter (urethane foam sponge), if one is used
- Quarterly cleaning of the filter and replacement of activated carbon

- Annual dismantling of the aquarium and complete cleaning, including any needed repairs

If there are pipes or tubing associated with the circulation system of the aquarium, they should be removed for cleaning annually. Various marine species (worms, mussels, barnacles, and tunicates) will tend to settle out in these tubes and eventually block them.

3

Aquarium Ecology and Biological Maintenance

An aquarium is a miniature ecosystem. There are complex and varying relationships between the animals and plants that may be present, and between these and the microfauna in the filter bed, along the walls, and in the water. Physical components such as volume of water per animal, temperature, light, and the concentration of various chemicals, particularly metabolic products, also affect the system, as does the condition of the animals and their tolerance of recirculating water. Finally, there are pathogens to take into account. Successful culture of marine organisms depends in part on an understanding of all of these.

VOLUME

Obviously, the larger the volume of seawater, the greater the biomass that can be supported. Larger numbers of animals may also be maintained when water quality, aeration, and temperature are carefully held at desirable levels. Just how large this biomass can be is best determined by experience. Some species require a great deal of space; others can be maintained at rather high densities. It is best to start with a few individuals, then increase the numbers until it becomes apparent that one is reaching a maximum population, as indicated by some of the danger signs noted in the preceding chapter.

Recommended biomass/volume ratios for some species are cited in Part II.

CONDITION OF ANIMALS UPON ARRIVAL

Transportation from the collecting site or supplier to the laboratory is traumatic for most animals. The temperature, oxygen tension, and container volume are seldom optimum. To avoid temperature shock when new collections arrive, opened containers should be brought gradually to temperature equilibrium with the aquaria before introducing the animals. New arrivals should be put in some form of quarantine until it is clear that they are neither moribund, diseased, nor physically damaged, and that they are reasonably certain to survive. While they are in quarantine, inspect them for parasites, or hidden predators (e.g., a collection of hydroids may introduce nudibranchs, their predators). Plastic dishpans or large glass jars are adequate as quarantine containers for many species; they must be aerated.

DENSITY OF BACTERIAL POPULATIONS

In general the bacterial population in a closed system will greatly exceed that of the natural habitat. Indeed, certain types are essential in a well-managed aquarium, particularly in the filter. But putrefactive species can multiply to the point where they reduce available oxygen to the detriment of the animals and beneficial bacteria. Their control depends partly on effective filtration, and even more on the prompt removal of all excess food and dead or moribund animals and plants. Organic material must not be allowed to remain and decay; if the water becomes cloudy, there is trouble somewhere.

PLANT GROWTH IN THE TANK

Under normal illumination, populations of various plants, particularly diatoms and certain filamentous algae, will appear on the walls of the tank and on other solid surfaces. These need not be removed, except perhaps from the window of an aquarium if visibility is impaired. Not only does such growth provide food for browsing species but these plants remove nitrates, nitrites, and ammonia, which they utilize in their own biosynthesis, and they release oxygen. Some algae (e.g., *Fucus*) do not survive well when continuously submerged, and soon begin to decay. Certain blue-green algae may appear and, if troublesome, should be controlled by scraping them off or reducing illumination. Very few organisms feed on blue-greens, and some species produce toxins that may be harmful. One genus that thrives particularly well in marine tanks and seems to be harmless is *Spirulina* (which is, paradoxically, red). It forms a maroon film on aquarium walls, rocks, shells, and on any sessile invertebrates that have a hard or tough exterior. At times, it appears as a surface film wherever air bubbling does not disturb it; this film can easily be skimmed off or removed with a suction device.

TOLERANCE OF SPECIES TO AQUARIA

The extent to which different species can tolerate life in recirculating seawater varies greatly. A few forms like hydrozoans can multiply so rapidly in aquaria that they become pests (Sandifer *et al.*, 1974a; Lawler and Shepard, 1978). Others are very fastidious and will not tolerate even relatively minor changes in seawater composition. Octopods and squid, for example, are sensitive to slight increases in the concentration of nitrogenous compounds, whereas confinement itself may damage soft-bodied forms (medusae, ctenophores) or excitable species (squid). Small zooplankters are difficult to retain and keep out of drain pipes. Squid, octopods, crabs (*Uca, Carcinus, Pachygrapsus*), and some snails (*Littorina, Tegula*) may well escape unless measures are taken to confine them. On the other hand, many species are very tolerant of the stresses of captivity and survive well. A list of the latter appears in Chapter 10.

DISEASES, SYMPTOMS, AND TREATMENTS

Most of the literature on diseases of marine invertebrates deals with species that are cultured for economic reasons. Apparently very little is known about the pathology of most invertebrates used in research. Investigators working with bivalves or decapod crustaceans should refer to Sindermann (1977) for the most convenient summary of shellfish diseases including their prophylaxis and treatment. Other useful reviews are those of Sindermann (1970), Pauley and Tripp (1975), Scarpelli and Rosenfield (1976), Johnson (1978), and Perkins (1979). The microbial diseases of lobsters have been reviewed by Fisher *et al.* (1978) and pathological conditions of fish and shellfish by Murchelano and Rosenfield (1978). Sparks (1972) has described the effects of physical and chemical trauma on invertebrates. Johnson (1968) prepared an extensive annotated bibliography on invertebrate pathology (except insects), beginning with the last century and ending with the mid-1960's. The first volume (Kinne, 1980) of a projected four-volume monograph on the diseases of marine animals covers the sponges, coelenterates, ctenophores, nonsegmented worms, and gastropods.

Clearly, considering the enormous diversity of marine invertebrates, much remains to be learned of their pathology. The *Journal of Invertebrate Pathology* is exclusively devoted to articles on invertebrates and may provide relevant information if disease emerges in the laboratory aquarium.

Because treating disease is problematical at best, it is wise to do everything possible to keep the animals healthy. Maintaining the health of marine invertebrates generally revolves around a few important requirements: adequate feeding, proper temperature, freedom from predators and obvious parasites, avoidance of overcrowding, and proper aquarium maintenance.

In general, disease is more likely to develop in those organisms held close to the upper tolerable thermal limit. The animals are often more stressed and more vulnerable to infection. For this reason alone, it is best to maintain any animals at temperatures somewhat above their minimal tolerable limits rather than near their upper limits. Additional precautions include: collecting animals from clean, unpolluted environments; taking special care in transit; and maintaining quarantine and close observation of newly arrived animals for 2 or 3 days to assure that they are healthy and adapting to captivity.

If control of disease proves troublesome, a few additional measures can be tried. The aquarium recirculation system can be fitted with an ultraviolet sterilizer, ordinarily downstream from the filter (Spotte, 1979a,b). This will reduce the number of bacteria in suspension without harming those in the filter bed. Ozone also may be used to reduce the number of microorganisms in aquaria (Sander and Rosenthal, 1975; Blogoslawski and Brown, 1979). This gas is produced by an electric ozone generator and is introduced into the water by means of an air supply (Spotte, 1979a,b). Ozone itself is highly toxic both to animals and bacteria, but in seawater it seems to decompose within about 5 minutes (Crecelius, 1979). The ozonation of seawater, however, produces long-lived toxic by-products such as bromate (Crecelius, 1979), and this residual toxicity is acute for developing oysters (MacLean *et al.* 1973; DeManche *et al.,* 1975) and presumably other tiny invertebrates. It can be eliminated by passing ozonated seawater through certain types of activated charcoal (DeManche *et al.,* 1975). Care must be taken to ventilate the aquarium room to avoid any accumulation of ozone in the atmosphere.

Occasionally, microbial diseases become rampant in an aquarium and will not yield to any treatment, conventional or otherwise. If this does occur, it will usually be necessary to discard the specimens being held and to sterilize the aquarium and start over, utilizing the sodium hypochlorite procedure described in Chapter 2.

SYMPTOMS

The following list shows, by phylum and class, some observable symptoms, and suggests methods of prevention or treatment when that is known:

Porifera—Sick sponges will decay and must be discarded.

Coelenterata—Loss of color and responses; at times an improved diet will aid recovery.

Sipuncula—Appearance of irregularities on the skin.

Annelida—Generally hardy, most annelid worms will survive well if provided with suitable habitat and food; antibiotics may be effective; discard if decay develops or if normal movements are not apparent.

Arthropoda—Failure to right themselves or exhibit normal movement; disease defenses of *Homarus,* the American lobster, fail at temperatures above 15°C, and it then becomes vulnerable to fatal bacterial infection.

Mollusca—

Class Bivalvia—Gaping shells indicate disease or stress; animals should be placed in appropriate sediments and fed with suitable phytoplankton, e.g., *Monochrysis, Isochrysis, Dunaliella,* etc.

Class Gastropoda—Symptoms include lethargy, slow retraction of foot, missing portions of epidermis; recovery should follow improvement in conditions. If rogue cannibalistic attack occurs, injured animals will usually bleed to death; the offender should be isolated.

Class Polyplacophora—Many species will crawl up and out of the water and die of dessication; chitons should be held in screened submerged containers.

Class Cephalopoda—Octopuses may attack each other fatally, especially if one is larger than the other. They will often escape and die out of water. Drooping legs, labored breathing, bleached coloration, and lethargy are symptoms of disease; no effective treatment is known.

Squid will often attack aggressively during mating; as a consequence, the females will deposit eggs and die. If it is possible to segregate squid by sex, this outcome may be avoided. Squid will also collide with the walls of the aquarium and abrade their delicate skin, leading to fatal infection. To the extent possible, squid should be held in large aquarium tanks and protected from rapid movements in the vicinity that may startle them.

Echinodermata—

Class Echinoidea—Generally require water saturated with dissolved oxygen. Care must be taken to avoid anaerobic deposits (rotting food) in the aquarium. Sea urchins may suffer from internal commensals that become parasitic when the former are fasted, though this hasn't been proven.

Class Asteroidea—Some starfish are especially vulnerable to bacterial infection, which can be treated with antibiotics, e.g., streptomycin–penicillin mixtures, or prevented by lowering the temperature of the seawater.

Class Holothuroidea—Species that burrow naturally and are continually covered by sediment occasionally suffer from microbial infection.

Class Ophiuroidea—The more hardy species appear relatively free of disease and adapt well to aquarium conditions.

Class Crinoidea—Only hardy species that adapt to the aquarium environment should be held.

Chordata—

Subphylum Tunicata—Many species of tunicates adapt readily to aquaria; many others refuse to accept confinement. Specimens should be inspected closely every day and any animals that appear weak or dead removed.

ANTIBIOTICS

There are risks in using antibiotics because one of the more important parts of the filter system is the filter biota, which are mostly bacteria. These are sensitive to antibiotics (Collins *et al.*, 1976; Levine and Meade, 1976). If an animal is to be treated internally with anything other than small amounts of antibiotic, or if antibiotics are to be added to the water, the individual should first be removed to a separate holding tank containing water from the home aquarium. Antibiotics should be used only in those instances when normally healthy animals become sick, not to compensate for slipshod techniques. Continuous use of antibiotics will produce resistant strains of pathogens, which can then become a major cause of biological breakdown in an aquarium (Christiansen, 1971; Martin, 1977). Certain kinds and concentrations of antibiotics are lethal to the larvae of marine invertebrates; for example, cephaloridine and rifamycin are toxic to *Paracentrotus* (de Angelis, 1977b) and kanamycin and neomycin to *Cancer* (Fisher and Nelson, 1978), and tetracycline and chloramphenicol to *Crassostrea* and *Panope* (Cardwell *et al.*, 1979). The adverse effect may not always be direct. Hofmann *et al.* (1978) found that a streptomycin/penicillin mixture prevented the development of polyps of cultured *Cassiopeia* by inhibiting an inducing factor produced by a species of *Vibrio*.

Many different antibiotics have been used on marine invertebrates, but the variables are so many and comparative studies so few that it is seldom possible to predict the best possible selection for a particular situation. The most widely used preparation is a mixture of streptomycin and penicillin (20–200 mg and 50–500 units per liter, respectively).* Stephens reported that a mixture of 40 μg/ml of tetracycline, 55 μg/ml of streptomycin, and 500 units/ml of penicillin was well tolerated by a wide variety of aquatic invertebrates (D'Agostino, 1975). Landau and D'Agostino (1977) have recommended dimethylchlortetracycline and de Angelis (1977b) ampicillin as more effective than these combinations. Fisher and Nelson (1978) found that chloramphenicol prevented microbial epibionts that fouled the appendages of the zoeae of *Cancer*.

Chanley (1975) has a fairly long discussion of bivalve diseases. Many of the bacterial infections, particularly those of larvae, can be controlled with streptomycin or aureomycin at a concentration of 50–200 mg/liter. The bacterial disease of lobsters called gaffkemia or red tail can be controlled with aureomycin (Hughes *et al.*, 1975). Antibiotics may also be injected into invertebrates to protect against infection following surgery. Postoperative disease of sea urchins was reported greatly reduced by penicillin in 0.05 ml of seawater at a concentration of 10,000 units/ml (Coffaro, 1979). de Graaf (1973) suggests several additional antibiotics; streptomycin (concentrations up to 100 mg/liter), aureomycin (3 mg/liter), or chloramphenicol (up to 50 mg/liter).

Reviews by D'Agostino (1975) and Le Pennec and Prieur (1977) provide general coverage of antibiotics in marine cultures.

OTHER DRUGS

Polyvinylpyrrolidine–iodine complex (PVP–I) at a concentration of 50 mg/liter for a maximum of 8 h has been used in treating clams. "Con-

*These combinations, some feel, are counterproductive, in that streptomycin and tetracycline are bacteriostatic, whereas penicillin is effective only against growing organisms.

centrations of 250 mg/liter for four hours were found to kill most bacteria, free-swimming as well as sedentary ciliates and even parasitic *Cyclops,* without harming young clams. The drug is worth trying on diseased invertebrates" (de Graaf, 1973).

Ectoparasites can sometimes be removed or controlled on adult specimens by bathing them in solutions of sodium hypochlorite (bleach) or formalin at a dilution of 0.25 ml/liter of seawater (1 ml/gal of seawater). Bathing animals in solutions of antibiotics or keeping them in antibiotic seawater for a day or so may be useful.

McCammon (1975) suggests commercial fungicide (not named) to treat fungal infections of tunicates.

The fact that fish are often bathed in solutions of copper sulfate, potassium permanganate, or vital stains to rid them of invertebrate ectoparasites suggests why these toxic solutions should not be used in marine invertebrate aquaria.

MONOSPECIFIC POPULATIONS

Monospecific populations imply maintenance of a single species of invertebrate, not an axenic (bacteria-free) culture. Although the latter is achievable for some species of invertebrates, it is achieved only by the most rigorous methods (Dougherty, 1959; Provasoli, 1977) and is not relevant to this manual. The general considerations dealt with above apply to both mixed and monospecific maintenance, but there are species that pose special problems over and above the physical environment and food. They are discussed below.

HERBIVORES

Assuming an appropriate food supply is available, herbivores are usually relatively easy to keep. There is little or no intraspecific conflict, and large numbers can be kept together. Some species (*Aplysia,* sea urchins) ingest large quantities of plant material, which they only partially digest, and produce correspondingly large amounts of feces that must be removed periodically to prevent excessive bacterial growth. Some of the more easily maintained herbivores include the opisthobranch gastropod, *Aplysia;* the prosobranch gastropods, *Acmaea, Haliotis, Diadora, Astraea, Tegula, Littorina,* and *Norrisia;* and the sea urchins, *Strongylocentrotus* and *Lytechinus* (which are also omnivorous). See Part II for specific directions on the culture of some of these animals.

CARNIVORES

Some carnivorous invertebrates are belligerent, and often turn cannibalistic in close quarters. Consequently, populations of such forms must be kept at low density. Predatory species whose prey is very specific (e.g., the sea star, *Asterias,* feeds on molluscs, whereas certain nudibranchs feed on particular cnidarians or sponges) may coexist peacefully, even in starvation, without resorting to cannibalism. If the food provided is nonliving, all uneaten portions must be removed after a suitable interval to prevent putrefaction.

Among the more easily kept carnivorous forms are the anthozoans, *Anthopleura, Metridium, Tealia, Corynactis,* and *Pachycerianthus;* the polychaetes, *Nereis* and *Glycera;* the gastropods, *Polinices, Lunatia, Busycon, Flabellinopsis, Hermissenda,* and *Hypselodoris;* the asteroids, *Asterias* and *Astropecten;* the ophiuroid, *Ophioderma panamensis;* the shrimps, *Pandalus, Crangon, Hippolysmata,* and *Sicyonia;* the lobster, *Homarus;* the true crabs, *Cancer, Carcinus, Callinectes, Ovalipes,* and *Hemigrapsus;* and various species of the hermit crab, *Pagurus.* Many of the crustaceans mentioned here will also feed on carrion and might properly be regarded as scavengers. The care and culture of a number of these species is discussed in Part II.

OMNIVORES

Much of what has been said above also applies to omnivores. Generally, such omnivorous species as the various urchins; the sea star, *Patiria;* and the mud snail, *Ilyanassa* (*Nassarius*) [Chapter 15], are easy to maintain because they have a broad spectrum of food choices available.

DETRITUS FEEDERS

Detritus-feeding species pose special problems. They require an adequate supply of detritus, but without having an excess that promotes undesirable bacterial growth. Frequent inspection is necessary. Adequate circulation of well-aerated water is especially important. Many detritus feeders are burrowing forms, in which case the tank should contain a suitable quantity of mud, muddy sand, or sand, as befits the species. Notable among these forms are many polychaetous annelids, the lugworm (*Arenicola*), bamboo worm (*Clymenella*), cone worm (*Pectinaria*), fiddler crabs (*Uca*), and holothuroids (*Parastichopus*). The annelids *Clymenella* and *Pectinaria* survive best if their tubes are intact.

SUSPENSION FEEDERS

Suspension feeders take their foods as fine particles, of either phytoplankton, bacteria, or detritus. These come directly from the water, which they process in large volumes. Many suspension feeders are sessile (various bivalves, such polychaetes as sabellids and *Chaetopterus,* barnacles, echiurans, phoronids, sponges); others are motile (e.g., the brine shrimp, *Artemia,* and other lower crustaceans, some sea cucumbers, and most larvae). Many of these can be kept in relatively large numbers. In feeding they crop off populations of single-celled algae, bacteria, and other unicellular forms as well as suspended detritus. In general, such species require less attention than do any of the other above-mentioned feeding types.

Burrowing bivalves should be provided with adequate sand or sandy mud into which they can burrow. Such forms as the razor clam (*Ensis*), jack-knife clam (*Tagelus*), and awning clam (*Solemya*) will soon gape and die if not provided with proper substrate. The first two species will survive much longer out of their preferred substrate if a rubber band that has a tension just adequate to keep their shells closed is placed around them. This device may be effective with some other species as well—but the shell of *Solemya* is too delicate. The obvious fact that pressure of the surrounding substrate in the natural habitat tends to keep the shells shut suggests that the extra work imposed on the adductor muscles, when the animal is out of its burrow for long periods, is debilitating.

PLANKTON

Large holoplanktonic zooplankton, those that spend their entire life suspended in the sea, are difficult to keep in monospecific culture (Paffenhöfer and Harris, 1979). However, see Chapter 10 for some species that can be maintained. Meroplankton are forms that spend only part of their life in suspension, typically at the larval stage. These species are covered in more detail in Chapter 11. In any case, it is important to have the outflow of the tank suitably shielded to retain and protect the animals. Such large, soft forms as jellyfish can easily be sucked up against a screen shield and be damaged, or cause the tank to overflow. Bubbling or other devices for aeration may damage soft-bodied forms; on the other hand, crustaceans (copepods) can stand considerable agitation.

Larvae are sometimes very difficult to maintain. Their small size makes it difficult to retain them in a recirculating system. Indeed, they are often best kept in noncirculating jars or aquaria in which the water is changed daily or that contains some form of gentle stirring device (Greve, 1968, 1975; Sorgeloos and Persoone, 1972; Kinne, 1976; Coleman, 1979). Larvae of many species are extremely fastidious, sensitive to subtle changes in environmental conditions, and demand careful control for successful growth. Aside from needing appropriate food, larvae are sensitive to bacterial levels, metabolites, oxygen tension, salinity, pH, environmental contaminants from the atmosphere, and sometimes even the detergents used to clean vessels. A number of aquaria have been developed specifically for culturing various larval forms, and a rapidly expanding literature is appearing (Costello and Henley, 1971; Salser and Mock, 1973; Sandifer *et al.*, 1974b; Ward, 1974; Kinne, 1976).

COMMUNITY TANKS

APPLICATIONS

The community tank houses a variety of species in a single container. Such an arrangement is common in classroom use as when a biology teacher wishes to study interspecific interactions, or when facilities are too limited to keep everything in monospecific culture.

It is entirely feasible to establish an artificial community of diversified and compatible forms that are tolerant of tank conditions. But it is difficult, if not impossible, to create a natural community—an aliquot of the ocean as it were—given the extreme variability in the tolerance of various invertebrates to captivity, and the almost indefinable biotic and abiotic conditions of the natural habitat. It is wiser to be pragmatic and to establish conditions that work, with species that survive readily. In this way one may maintain in the same tank, for example, species from such diverse sources as northern New England and California. In both areas temperatures of 13°C to 15°C are optimal.

ARRANGEMENT

A community aquarium can be arranged to simulate a given habitat, e.g., a rocky tide pool, sandy, or muddy bottom. For example, water systems that automatically simulate the tides have been described by Clark and Finley (1974) and Kinne (1976). Sessile forms, such as anemones, mussels or tunicates, should be introduced along with a piece of the substrate to which they are attached or placed so they can easily attach themselves. Scavengers and suspension feeders are desirable, to maintain water quality. If the tank is for display purposes, it may be well to arrange the sessile forms and rocks in a way that enhances visibility of all inhabitants of the tank.

COMPATIBILITIES

Compatibilities among species in the community tank can best be determined by experience, but a knowledge of the feeding habits and pugnacity of various species will suggest promising ways to start. Obviously, most herbivores, detritus feeders, and suspension feeders do not interfere much with each other, unless one of them releases a noxious substance. Carnivores and omnivores are more likely to show interspecific competition, and they may attempt to prey upon tankmates. Lobsters, larger crabs, and octopods will, predictably, kill many other forms. Sea stars, such as *Asterias* and *Pisaster,* will prey upon various molluscs, and *Pycnopodia,* the sunflower star, will swallow molluscs and sea urchins whole. Sea urchins eat sand dollars. Nudibranchs will consume hydroids, sponges, or ectoprocts and predacious shelled gastropods will consume other molluscs.

Forms that plow through the sandy bottom (the whelk, *Busycon;* the moon snail, *Polinices;* the horseshoe crab, *Limulus*) can upset sand dollars, sand stars, and other burrowing forms and destroy tubes imbedded in the sand.

Some anemones (*Anthopleura*) will not tolerate other anemones close by, and will attack and sting them, possibly fatally. Large *Anthopleura* may also scoop up and consume a passing crab, worm, or snail.

Crustaceans are soft and vulnerable immediately after ecdysis (molting); unless they are provided with a place to hide, they may be eaten by their own or other species.

4

Collecting, Transport, and Shipping

Those who live within a reasonable distance of a coastline will often find it convenient to collect whatever animals are needed. This chapter assumes these animals will be maintained in the laboratory over an indefinite period. The following comments include some general considerations that should be taken into account and some of the measures necessary to capture and transport marine invertebrates.

COLLECTING

Successful collecting begins at the receiving end and depends heavily upon having suitable aquarium facilities ready when the specimens are brought to the laboratory. In some cases it will be necessary to review the literature on the species in question, or related forms, and observe the animal in the field before it is captured. Observations of organisms in their native habitat provide clues to the requirements for successful culture. This is especially true with respect to environmental factors and such physical requirements as the subtrate, light, water, temperature, or water movement.

Once the proper aquarium conditions have been set up, equipment for capture and transport must be obtained, although in its simplest form, this may need to be no more than a bucket made of inert material, a pair of boots, a shovel, and a sieve. For collecting in deeper water, a swimsuit, face mask, snorkel, fins, and sack are appropriate. In its most elaborate form, collecting requires an expensive and well-equipped research vessel, which provides both a working platform and devices for the capture of the organisms. For collecting of any sort, it is essential to plan ahead and equip oneself properly. Because there is no substitute for experience, the advice or participation of an experienced collector is an important factor in the success of any expedition.

Aspects that must be taken into consideration include the following:

• Information as to where and when the organisms might be found, taking into account habitat, season, tide, time of day, weather, and depth. In most, but not all, situations, no dangerous or noxious species will be encountered. However, there are animals that can inflict painful bites, stings, or other wounds, particularly in tropical or subtropical habitats. Whenever one is working in an unfamiliar area, it is well to seek local advice on potentially dangerous marine organisms and to avoid these creatures.

• Appropriate equipment and good judgment are important in capturing specimens effectively. Species such as limpets, sea urchins, and anemones are often difficult to remove without a spatula or similar tool. Burrowing animals must be dug out and recovered in a sieve. A small net makes capture of crabs easier. A pair of heavy cloth gloves protects the hands from all but the most needlelike spines. If one is working from a boat, nets, diving gear, traps, and other essentials should be taken along.

If animals are collected by trap, net, or hook and line, the time during which they are restrained should be minimal. Trawling should ordinarily be limited to 15-min tows. Traps should be checked at least every day. Set lines and gill nets should be checked about every 2 h—first thing in the morning if set overnight.

• Factors of importance during holding include whether the species in question will accept confinement, crowding, the slime produced by other organisms, temperature change, and handling. It is often helpful to isolate specimens by size. Some animals have a tendency to eviscerate or to autotomize; others have a tendency to attack nearby organisms.

• If collecting is done offshore, it must be recognized that work at sea is at times dangerous.

• It is important to appreciate the special characteristics of the particular area from which material is to be collected. For example, estuarine mudflats are fragile environments that will not tolerate very much intrusion. Mudflats may also contain treacherous "quicksands" from which it is difficult to extricate oneself. Swimming in strange bodies of water is particularly risky; winds, currents, and tides may change suddenly. Work that must be done on a slack tide calls for careful advance planning. On occasion, rain or snow will quickly reduce the salinity of seawater below a tolerable limit for sensitive stenohaline species. If this is a possibility, it is well to bring along extra seawater in the boat or truck from which the collectors are working. Finally, it is essential to know the limits of the equipment being used—cable strengths, logistics of vessels and vehicles, capacity of holding tanks and buckets, and working time under water.

• In addition to all the above, account must be taken of such matters as permits and licenses (see Appendix A), permission from private landowners, and either liability insurance or waivers therefor.

TRANSPORTATION

Every effort should be made to get the animals back to the laboratory without undue damage. Some may die from injuries inflicted during collecting, but many more usually die from the stresses of transportation. Death is often not immediate, but occurs a few days after the specimens arrive in the laboratory.

The length of time that animals will tolerate confinement during transport depends upon temperature, the quality and quantity of water and oxygen present, and handling methods. In general, 4 to 5 h is the limit for many species carried simply in buckets of water. The water in the buckets can be kept cool by floating plastic bags or bottles of ice in the water (provided there are *no* leaks). If species that live in warm waters are being transported, they should probably not be cooled below about 10°C unless it is certain they can tolerate a lower temperature. Buckets should be checked frequently; if the animals are going to be in transit for a day or more it is preferable to pack them, generally segregated by species, in a liberal quantity of seawater in double plastic bags under an atmosphere of air or oxygen. The bags should be no more than half full, thus leaving space for an adequate volume of gas. Each should be sealed separately with a goose-neck closure, i.e., the bag twisted shut, bound with a rubber band, the twist folded over and bound with a second rubber band, or the same rubber band can be used for both steps. If properly closed, the bags should be airtight and watertight.

Sealed bags should be placed in watertight containers, e.g., styrofoam boxes, buckets, or plastic trash cans. Approximately 0.5 kg of ice per 50 liters (1 lb/10 gal) of seawater should suffice for 24 h in warm conditions. The watertight container may then be loaded onto the transport vehicle and protected from sunlight or freezing. Properly packed, hardy specimens can be expected to survive 24 to 48 h in transit under these conditions. If practicable, it is beneficial to add fresh seawater, replace ice, and aerate or replace oxygen after 24 h.

Fragile organisms must be protected from vibration and jolting during transport in boats or trucks. Large sea urchins, which superficially do not appear fragile, are subject to internal damage if they are not handled with care. Some species can be cushioned on or packed in layers of suitable seaweeds such as *Fucus*.

Species that excrete excessive slime must be transported separately, or the respiratory apparatus of other species will become clogged. Sponges and tunicates generally should be isolated during transport; some tunicates will leak acidic fluids into seawater if they are torn during collection. Some algae, such as the Desmaresterales, are very acidic, and should not be mixed with other organisms in collecting containers. Octopuses, squid, and sea hares may discharge ink and foul the water in the transport container.

Many species—snails, crabs, lobsters, and sea stars—have sharp, hard shells or surfaces; others—such as sea urchins—have spines. Obviously, these organisms should not be mixed with soft-bodied ones during collecting and transport. Crabs, also, will often grab and pinch or crush other organisms in the collecting bucket or aquarium.

Some species, such as most of the sea urchins, barnacles, and bivalves, are best transported "dry," i.e., cool and moist. If wrapped loosely with saltwater-moistened newspaper, they will remain viable for 4 to 5 h (sea urchins) or up to 24 h (barnacles and bivalves).

It must be kept in mind that most boats are plumbed with brass pipes and that some species are sensitive to even the minute amount of copper in holding buckets filled with water from the deck hose. Besides this, water taken from the collection site may differ qualitatively from that where the collecting vessel docks. Water from inshore areas may be of poor quality. Enough high-quality offshore water for transporting the specimens back to the laboratory must be secured.

EQUIPMENT

The following items of equipment represent a reasonably complete list for collecting operations and should be available to institutions where collections of a wide variety of animals are to be made. Any one collecting trip will generally require only selected items. To be successful, most trips will require either two or three experienced collectors, or one expert and much less-skilled labor.

COLLECTING EQUIPMENT:

Transportation: ¾- to 1-ton pickup truck; ¾- to 1-ton covered van; skiff (16 ft, built for rough water, e.g., Boston Whaler); work boat (24 to 30 ft or larger if heavy equipment and a 40-ft trawl net are to be used, or if long distances are to be covered).

Safety equipment, apparel, and information sources: radio receivers (for weather reports); telephone or radio communication; scuba equipment for three divers, wet suits, and auxiliary equipment; foul-weather gear; hip boots; life jackets; flashlights (preferably waterproof); waterproof watches; gloves; tide books; navigation charts; waterproof ink.

CATCHING GEAR: shovels; spatulas; sieves; wrecking bars; plankton nets (fine and coarse); geologist's hammers; dip nets (bait, aquarium, bait brailes, etc.); common sense seine; beach seine—30- to approximately 90-m (100- to 300-ft) long; semibaloon shrimp trawl—5.0-, 7.5-, or 12.0-m (16-, 25-, or 40-ft) wide; midwater trawl; rock dredge; bottom grab (recommended—Van Veen No. 2); traps (crab, fish, special purpose).

HOLDING AND TRANSPORTING EQUIPMENT: collecting sacks—recommended "Goodie Bags," canvas; plastic buckets—16–20-liter (4- or 5-gal) capacity, with bails and lids, 30 plus; plastic bags—2 to 4 mil: 8 × 12 in., 8 × 16 in., or larger, hundreds of each; insulated boxes—such as styrofoam ice chests or plywood-lined containers of 20- to 30-liter capacity, 20 to 100; rubber bands—No. 10, No. 64, hundreds; aluminized duct tape; anesthetics; fish transfer tank or similar (optional); large trash cans—100- to 120-liter capacity, plastic, with lids.

A collecting trip really ends only about 2 days later, when the adaptation of organisms to the aquarium has been ascertained.

A log of collecting activities or a list of the specimens collected, may be necessary in certain areas. In any event, a written record will be useful, provided it is maintained faithfully. It is extremely important to collect only what is needed and in quantities no larger than necessary.

The interval between capture and release into a suitable aquarium is often the most traumatic episode in the life of specimens brought in from the sea. This period of extraordinary stress must be taken into account and everything possible done to minimize its impact. Despite the best of care and equipment, some marine species will not tolerate capture and transport; obviously, these must be studied in their natural habitat.

SHIPPING

It will sometimes be necessary to ship animals from collecting sites to inland laboratories. This is best done by professional collectors who have experience both with the animals and with available transportation facilities. If, however, shipping is by the individual investigator, the following points should be considered.

Most invertebrates can be shipped in polyethylene bags in insulated boxes (styrofoam). They are generally held for a few days before shipping, which allows them to recover from the shock of collecting. Fasting prior to shipment is strongly recommended, because it reduces the amount of food in the digestive tracts and minimizes the amount of waste that is eliminated into the shipping containers. Marine invertebrates from tropical and subtropical environments should be kept at about the temperature of the water from which they are captured. Organisms from temperate environments usually do well if packed with a supply of ice, at a rate of approximately 0.5 kg of ice per 50 liters of seawater (1 lb/10 gal). Animals from boreal or subboreal environments may require even more cooling.

The success of a given shipment depends upon proper planning. Recipients must know when to expect the animals to arrive and be prepared to care for them immediately. With a little experience and proper packing, shipping times of 24 h or less are entirely feasible. In some cases, however, shipments in transit as long as 5 days have arrived with negligible loss of specimens. As a rule, therefore, shipments should not be refused even if delayed; after all, it will do no good to return it to the shipper and to do so will present considerable difficulty to the carrier who must then dispose of it.

5

Geographic Sources and Implications Thereof

In general, animals should be maintained in the laboratory under conditions that approximate those where the species naturally occur. Two attributes, temperature and salinity, must always be approximated if the species is to be maintained successfully. However, it is usually not necessary, and sometimes not even desirable, to duplicate the normal environment exactly. For example, to facilitate cleaning, reduce uncontrolled variables, and isolate individuals for identification, it may be preferable to keep individual animals in bare-walled containers rather than in something resembling an artificial tidepool. As noted earlier, a temperature somewhat lower than that of the natural environment may be advisable.

The inshore environments of the United States, of adjacent portions of Canada and Mexico, and of Puerto Rico are the source of almost all marine animals used in laboratories in the United States. The characteristics of these areas suggest the best laboratory conditions for the animals originating from them. For convenience, primarily, the coastal areas are here divided into the eight regions illustrated in Figure 5–1, which gives surface water temperatures as monthly averages, for some selected locations. The lower value is the average temperature for the coldest month, usually February; the higher value is the average temperature for the warmest month, usually August. The temperature data have been taken from extensive tables in two federal government publications: *Surface Water Temperature and Density, Atlantic Coast, North and South America* (NOS Publication 31–1), and *Surface Water Temperature and Density, Pacific Coast, North and South America and Pa-*

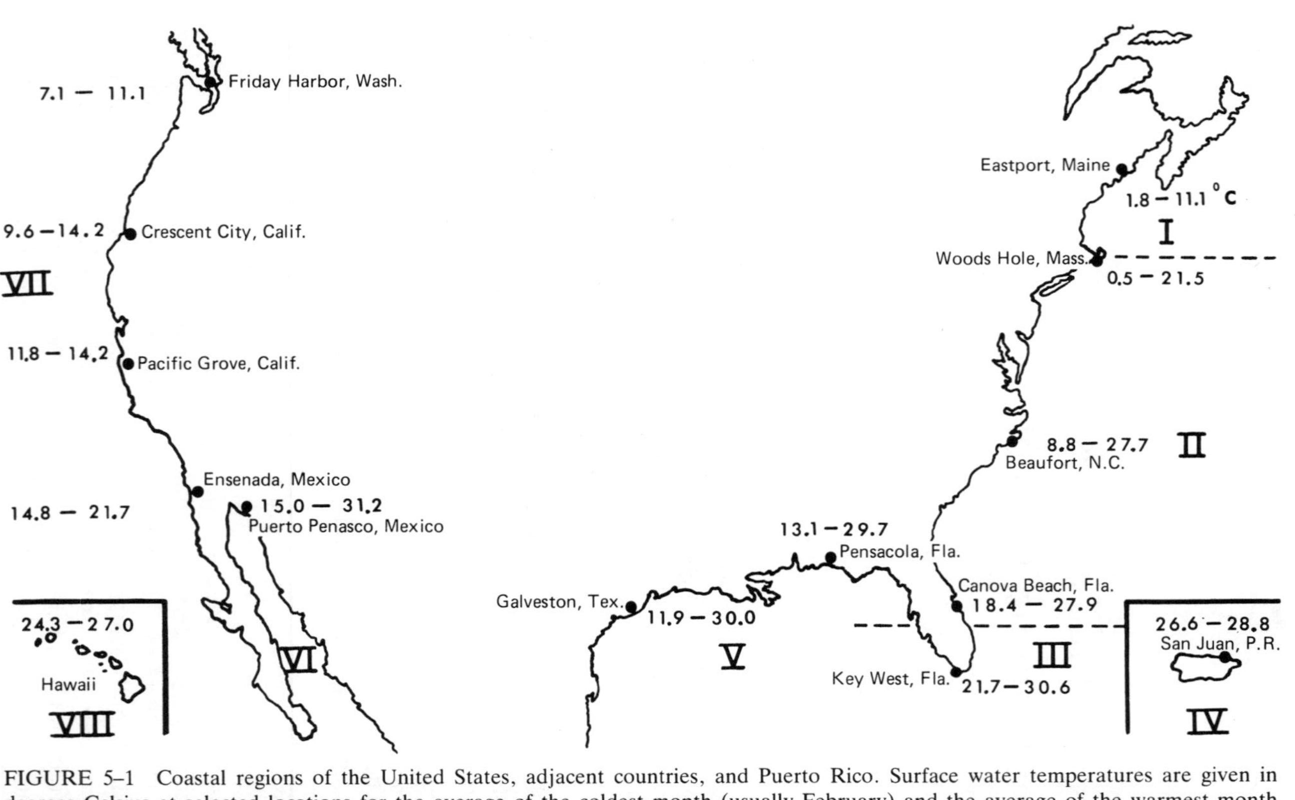

FIGURE 5–1 Coastal regions of the United States, adjacent countries, and Puerto Rico. Surface water temperatures are given in degrees Celsius at selected locations for the average of the coldest month (usually February) and the average of the warmest month (usually August).

cific Ocean Islands (NOS Publication 31–3), which are available for a small charge from the Superintendent of Documents, U.S. Government Printing Office, Washington, DC 20402.

REGION I: NOVA SCOTIA TO CAPE COD

This area is washed by waters of the Maine Current, a limb of the Labrador Current, which flows southward along the coast to Massachusetts Bay. It is a cold-water current, summer temperatures seldom rising above 17°C in Massachusetts or above 15°C in Maine. Winters are often severe, with temperatures near freezing in inshore waters and with the formation of intertidal ice. The shoreline is dominated by granite outcrops, interrupted by occasional sandy and rubbly beaches and estuaries, limited *Spartina* beds, and salt marshes typical of a "drowned" coastline. Tidal fluctuations are great, ranging from about 3 m at the southern limit to more than 10 m in the upper portion of the Bay of Fundy. There is a rich fauna of sessile forms on rocks, burrowing forms in the sand and mud flats of estuaries, and motile species in each habitat. Intertidal collecting is relatively easy. The winter and summer faunas may vary greatly, because ice abrasion takes a heavy winter toll on intertidal sessile forms, and many arthropods retreat to deeper waters. More northerly species may appear in winter, and some forms (for example, certain nudibranchs) reach sexual maturity at that time and disappear by spring.

Tanks for such fauna must be cold. A temperature of 12°C in the warmer months is satisfactory, although some species, especially the estuarine forms, can tolerate warmer conditions. In winter, the water temperature should be lowered several degrees. If the tank is supplied with suitable rocks, preferably those collected from the intertidal zone along with some attached flora, many invertebrates will establish themselves. A layer of sand or sandy mud to a depth of a few centimeters will provide a suitable habitat for many burrowing forms. Useful guides for identification of animals from this area, as well as the south side of Cape Cod, include Miner (1950), Smith (1964), and Gosner (1971, 1979).

REGION II: ATLANTIC COAST FROM CAPE COD TO CENTRAL FLORIDA

Summer water temperatures on the south side of Cape Cod usually exceed 20°C on the open shore; they are higher farther south. In winter, the inshore temperatures may fall below 10°C, with ice forming intertidally when winters are particularly severe. The nearby Gulf Stream ameliorates the temperature in all seasons.

From Cape Cod to New York Harbor, the shoreline rather resembles that of northern New England, dominated by rocky outcrops. But from Sandy Hook, New Jersey, to Cape Hatteras and beyond, one finds an uplifted coast, characterized by long sandy beaches, barrier islands, bays and lagoons, and extensive *Spartina* flats, with scarcely any rock outcrops. Sessile forms abound only on such solid substrata as stone jetties, bridge abutments, sea walls, and wharf piles. Tidal excursion here is moderate, on the order of 1–2 m, but may be much less in bays and well-protected estuaries. The continental shelf is wide in this region, and there are no major ocean currents (the Gulf Stream flows parallel to the coast, but is well offshore). As a consequence, surface water temperatures show marked seasonal variation, reflecting temperature changes in the adjacent continental land mass. In the northern part of this zone, winter water temperatures approach zero; indeed, intertidal animals may be exposed to air temperatures well below freezing, whereas summer water temperatures exceed 20°C. At the southern limit, the temperature fluctuations are not as extreme, but the low and high temperatures may still differ by 13–15°C in northern Florida. Some of the species in Region II are seasonal, thriving during part of the year and migrating or becoming dormant at other times.

THE WARM-TEMPERATURE ZONE

This subregion extends along the eastern coast of the Atlantic from Cape Hatteras, North Carolina, to the vicinity of Cape Canaveral, Florida, and encompasses the northern Gulf of Mexico. Although this portion of the Gulf is spatially separated from the remainder of the warm-temperature zone, it has many of the same faunal components. The shoreline of the warm-temperature zone typically consists of long stretches of sandy beaches interrupted by bays and lagoons. The southern Atlantic coast has much the same appearance as the more northerly region that adjoins it, its barrier islands on the outer coast separated from the mainland by a series of inland waterways. The more protected waters of these bays and lagoons are bordered by well-developed salt marshes, associated tidal flats, and meandering tidal creeks. As in the zone north of Cape Hatteras, there is little hard substratum other than structures such as pilings and jetties. The majority of benthic invertebrates are burrowing forms, dwelling in mud and sand; others may live on or attached to shells, blades of salt-marsh grass, or whatever debris or solid substratum is available. Surface temperatures of near-shore waters generally range from 10°C to 28°C through the year, although extremes may vary under local conditions.

Tank temperatures of 15–17°C are appropriate for most animals from the northern portion of Region II; higher values, up to room temperature (20–23°C), suit the more southern animals. There are exceptions. Species that flourish in the winter will do better in cooler water; summer species are more likely to thrive in warmer water.

As Figure 1–1 shows, oxygen solubility is inversely related to temperature; tanks maintained at room temperature should be well aerated to assure adequate supply of dissolved oxygen. Sudden temperature changes should be avoided. Unless experience indicates otherwise, specimens collected during the winter should be warmed slowly over days or weeks before being put into holding tanks with temperatures much higher than that of the water from which the animals came. Sand or sandy mud should be provided for burrowing forms. Sessile forms from hard substrata may be provided with rocks. The identification guides noted for Region I are also applicable to the northern portion of Region II, but coverage becomes spotty for the southern part, especially south of Cape Hatteras. The species in the area around Delaware Bay have been treated in detail by Watling and Mauer (1973). Dowds (1979) provides a comprehensive bibliographic guide to the literature on the identification of the littoral marine invertebrates from Chesapeake Bay to Florida.

NORTHERN GULF OF MEXICO

Although separated from the rest of Region II, the sandy beach communities of the northern Gulf of Mexico are similar to those along the southeastern Atlantic Coast. They have many of the same or closely related species, and there are large bottom communities of bivalves in the sandy areas of the northern gulf. For example, *Donax variabilis* occurs in large beds. But the characteristic inhabitant of the sand beach is the ghost crab, *Ocypode quadrata,* which during daylight hours inhabits burrows well above the high-tide line. Sedentary invertebrates such as the tubiculous polychaetes are more common in muddy regions. Shrimping and oystering are important commercial activities. The area is exceptionally poor in rocky shoreline and, except for man-made sites, there is no extensive hard-bottom community. Rock jetties and piers do provide a habitat for a limited fauna consisting principally of barnacles, snails, isopods (*Ligia exotica*), crabs (*Pachygrapsus transversus*), hydroids, bryozoans, and encrusting sponges.

During the winter, the surface water of the northern Gulf is cool, 12–14°C, but it becomes very warm in the summer. Farther south, the water is uniformly warm. Animals from this region do well in tanks at 18–25°C. Generally neither heating nor cooling of the aquarium water is needed. Seawater with a salinity of 30‰ and a pH of 8.3 is satisfactory.

The identification guides suggested for the northeastern Atlantic coast are of some use in this region, but a more useful guide to the animals of the northern Gulf coast is that of Fotheringham *et al.* (1980). Menzies and Frankenberg (1966) have provided a handbook on isopod crustacea of Georgia. Galtsoff *et al.* (1954) have prepared a descriptive volume on the Gulf.

REGION III: SOUTHERN FLORIDA

Southern Florida has a rich and diverse tropical fauna. It has long been a favorite source of animals for hobbyists who are attracted by the colorful fishes and invertebrates, as well as a source of material for research and teaching.

The tropical zone embraces the Atlantic coast of Florida south of Cape Canaveral, the southern portion of the Gulf of Mexico, Florida Bay, and the Florida Keys. The fauna is to be found in a great variety of habitats, ranging from sandy beaches, mangrove swamps, mud flats, grass beds, and coral reefs to rocky shores composed of limestone formations and beachrock. North of Miami, the outer coast is typically composed of barrier islands and sandy beaches. Along this shore, intertidal and subtidal outcroppings of coquina limestone provide a hard substratum that supports a variety of encrusting, sessile, and boring invertebrates, as well as many that seek refuge under ledges and in the interstices of rocks. Sabellariid worms constitute a significant community on these outcroppings, cementing sand grains together to form aggregates of sand tubes that build massive reefs. Numerous small invertebrates find habitats on and among these tubes. Bays and lagoons on the landward side of the barrier islands are fringed by mangrove swamps that replace the salt marshes of the warm-temperature zone and form important communities on both sides of the Florida Peninsula. Mangrove swamps occupy one-third to one-half of the area of the Florida Keys. Other habitats found in the more protected shallow waters of the tropical zone are the mud flats and the sea-grass beds, each with its characteristic fauna. Rocky shore environments, such as limestone bedrock, fossil coral reefs, and beachrock occur in the Keys and in certain areas of the Gulf of Mexico. They are inhabited by rock-boring organisms and provide a substratum for encrusting and sessile invertebrates.

A complex and diverse community is associated with the coral reefs, which stretch from just south of Miami along the windward side of the Keys. Separated from the Keys by a channel a few miles in width, these reefs lie inside the 10-fathom line and are mostly submerged, reaching the surface in only a few places. The tidal excursion is small—1–2 m in central Florida, 1 m or less in southern Florida and the Keys. The temperature range in the tropical waters is more restricted than in the regions farther north, ranging from 22–31°C at Key West.

Most invertebrates from both the warm-temperature and tropical zones of the southeastern United States survive well in tanks maintained at room temperature (22–25°C). Although these species may live well at room temperatures, conditions may not be optimal for reproduction or certain other physiological activities. When the aim is to rear marine invertebrates in captivity or to carry out critical experiments, the temperature and other physical and chemical parameters of the natural environment should be ascertained and reproduced in the tank. If there is doubt as to the most desirable temperature, it is better to err on the low side. Tanks maintained at room temperature should be well aerated. There are numerous monographs treating individual groups of animals from this area, and a general field guide by Voss (1976).

REGION IV: PUERTO RICO

Puerto Rico is located within the tropics between the latitudes of 17°50′ and 18°30′N and is a rectangular island with a coastline of about 100 km. On the north it is bounded by the Atlantic Ocean and on the south by the Caribbean Sea. The coast encompasses a great diversity of environments from sandy shores, rocky coasts, mangrove thickets, and turtle grass beds to coral reefs. These habitats support a rich and varied marine fauna.

The exposed northern coast is subject to considerable wind and wave action and has expansive sandy beaches interspersed with rocky shorelines, which may take the form of cliffs, plateaus, or loose boulders. The rock provides habitats for numerous crabs, gastropods, sea urchins, sea cucumbers, and polychaetes that dwell in cavities, on the surface, or burrow in the rock. Biota on the surf-swept sandy beaches is, by contrast, relatively sparse, being dominated by burrowing mole crabs of the genus *Emerita* and the ghost crab, *Ocypode quadrata*.

The southern coast is lined by mangroves that intrude inland along canals and lagoons and also into shallow waters offshore, where they form small islands. In the mangrove community some species, such as crabs and gastropods, crawl in and out of the water over the limbs and twigs. Other sessile organisms, including an abundance of sponges, anemones, ascidians, bivalve molluscs, and sabellid polychaetes, are found attached to the submerged roots.

Turtle grass beds are common in the sandy–muddy substrata of the shallow waters off the southern coast. Various echinoderms and gastropods live in association with the grass beds, as well as some sponges, certain corals, the bottom-dwelling medusa (*Cassiopeia*), and sand-burrowing worms.

Coral reefs, with their vast assemblage of associated fauna, are well developed offshore along the southern and northeastern coasts. In near-shore shallow waters and around small islands, small fringing reefs are formed by corals of the genus *Porites*.

Surface temperatures in Puerto Rican waters range between 28 and 30°C throughout the year. The tidal excursion is approximately 1 m.

REGION V: THE GULF OF MEXICO

The Gulf of Mexico is a partially land-locked body of water joined to the Atlantic Ocean by the Straits of Florida and to the Caribbean by the Yucatan Channel. The tidal range in the Gulf of Mexico is small in most places, averaging only 30–60 cm. By contrast with the Atlantic and Pacific coasts, those of the Gulf experience only one high and one low water most tidal days (24 h, 50 min).

The northern portion of the Gulf of Mexico (Region II) is in the warm-temperature zoogeographical zone; the southern portion is in the tropical zone. The coast consists for the most part of stretches of sand and mud. In addition, marine marshes and swamps are scattered along the coast, the former particularly conspicuous along the Louisiana coast. Coral reefs are also present, almost all of them in the warmer, more southerly waters of the Gulf. Scattered coral patches occur commonly as bottom growths, particularly off the west coast of Florida. Along much of the shoreline, there are barrier islands separated from the mainland by navigable waterways.

REGION VI: THE GULF OF CALIFORNIA

The Gulf of California is a narrow, moderately deep basin bounded on the west by the mountains of Baja California and on the east by the coastal plains of Sonora and Sinaloa. The shoreline of the northern Gulf is mostly sand and mud, sediments deposited by the Colorado River;

farther south there are extensive rocky outcrops. The tidal excursion varies considerably along the length of the Gulf; it is 1.5 m or less in the southern portions, but reaches 6–10 m near the mouth of the Colorado River. At the north tip of the Gulf, low tides uncover vast sand and mud flats cut by tidal channels. There is a fairly large seasonal variation in water temperature. Winter water temperatures average 15°C in the north and 20°C in the south; the average summer temperatures are 29–32°C throughout the region. The Gulf fauna is a mixture of Californian, Central American, and endemic species. Animals from this region should do well at room temperatures or somewhat above. Brusca (1980) has provided a general guide to the intertidal animals of the Gulf of California. Steinbeck and Ricketts, although published in 1941, is still a useful, well-written book on the animals of the Gulf.

REGION VII: THE PACIFIC COAST OF NORTH AMERICA FROM NORTHERN MEXICO TO ALASKA

The temperate Pacific Coast of North America is washed by cool, nutrient-rich water. The fauna is rich and varied, both in numbers of species and total biomass. The outer coast is lined by wave-swept, rocky outcrops separating exposed sand or pebble beaches. There are only a few bays and estuaries in the southern part of the region, which become more numerous to the north. From Puget Sound to Alaska, offshore islands and glacial valleys form an interconnected series of protected straits and bays, with little wave exposure but frequently strong tidal currents. The tidal excursion is moderate, generally 2–3 m in the southern part of the region and 3–5 m in the north. Geographical variation in temperature is considerably less than on the East Coast of North America. For example, Eastport, Maine, and Canova Beach, Florida, are as far apart north–south as are Ensenada, Mexico, and Friday Harbor, Washington.

The difference in surface water temperature between Eastport and Canova Beach in February is 16.9°C, whereas in that same month the water temperature difference between Friday Harbor and Ensenada is only 8.2°C. The corresponding temperature differences for August are 16.0 and 10.6°C. The small variation in temperature along the Pacific Coast is a consequence of the prevailing current pattern. As the Japanese Current approaches North America near the U.S.–Canadian border, it splits into two branches, the Alaskan current flowing northward and the California current flowing southward along the coast. Thus, the northern portion of Region VII is warmed by water from the south, and the southern coast is cooled by a northern current. Furthermore, as the California current moves south, the angular velocity of its water, due to the earth's rotation, becomes less than that of the adjacent land. As a consequence, the current tends to lag to the west and is replaced onshore by upwelling deeper water that enhances the cooling effect of the current. There is relatively little seasonal temperature change along the Pacific Coast. An extreme instance of temperature stability, even for the Pacific Coast, is to be found at Pacific Grove, California, where the average temperature for the coldest month differs from that of the warmest by only 2.4°C. Although the temperature gradient along the coast is not steep, there are distinct differences between the faunas of the north and south. Point Concepcion, California, is generally considered the dividing line between northern and southern faunal assemblages.

Animals from the Pacific Coast should be kept in cool water, 9–18°C, biasing the temperature toward the lower end of this range if it is practicable to do so. Many of the specimens obtained from the Pacific Coast are collected by diving or dredging and therefore come from environments somewhat cooler than the surface water above them.

There are several identification guides relevant to marine invertebrates of the Pacific Coast. That of Smith and Carlton (1975) emphasizes the central California coast. Brusca and Brusca (1978) describe marine animals and plants along the northern California coast and Kozloff (1973) deals with the Puget Sound area. Hinton (1969) is more suitable for the beginner than the specialist; Allen (1976) provides a more complete species list and simplified keys treating the invertebrate biota of the southern California coast. Ricketts and Calvin (1968) describe the many environments along the coast and the species likely to be found in each. A book edited by Morris *et al.* (1980) also covers the California coast.

REGION VIII: HAWAII

The Hawaiian Islands are volcanic in origin, lying across the Tropic of Cancer. Surface water temperatures are uniformly warm, almost always between 22 and 28°C. Many different marine habitats are represented, including rocky cliffs, sandy beaches, coral reefs and reef flats, mud flats, and mangrove swamps. Portions of the islands are famous for their wave exposure, which varies seasonally. The tidal excursion is small, seldom exceeding 1 m.

Hawaii lies on the northeastern edge of the Indo-Pacific faunal province. Taken as a whole, the Indo-Pacific fauna contains the richest, most diverse assemblage of marine species on the earth today, but the Hawaiian Islands are partially isolated and their fauna is poorer than that found in the western tropical Pacific. Nevertheless, the Hawaiian fauna is very rich compared with most temperate areas and has numerous representatives of all major phyla and classes, including many species of demonstrated, or potential, usefulness for biological research. Hawaiian species are tropical and should be maintained in aquaria at slightly above ordinary room temperature.

The standard field guide for Hawaiian marine invertebrates is Edmonson (1946), but it is both out of date and out of print. This book is being replaced by a projected series of six volumes. The ones on sponges, coelenterates, and ctenophores (Devaney and Eldredge, 1977) and on molluscs (Kay, 1979) have been published.

6

Foods and Feeding

The lack of a continuous supply of appropriate food sometimes places limits on the marine animals that may be successfully maintained in captivity. Since the best foods are those eaten in the natural habitat, the farther an installation is from the ocean, the greater the supply problem becomes. Under these latter circumstances, it is necessary to seek other more readily available foods that the animals will accept and thrive on.

NUTRITION

Little is known of the actual nutritional requirements of most marine invertebrates. Only rarely have investigators been able to establish the experimental conditions essential to such analysis—i.e., an edible, chemically defined ration and an axenic environment—and but one general reference is available (Rechcigl, 1977). These difficulties are well exemplified by the crustaceans, on which a considerable amount of effort has been expended, due in part to their importance in aquaculture. New (1976), Provasoli (1976, 1977), Kinne (1977), and Conklin *et al.* (1978) have reviewed present-day knowledge of crustacean nutrition. Despite the paucity of explicit information on dietary requirements, significant progress has been made in formulating diets that support growth, notably for the decapods, *Callinectes* (Sulkin, 1978), *Cancer* (Hartman and Letterman, 1978), *Carcinus* (Adelung and Ponat, 1977), *Homarus* (Gallagher *et al.*, 1976), *Libinia* (Bigford, 1978), and *Penaeus* (Venkataramiah *et al.*, 1976; Kittaka, 1976; Aquacop, 1978; Deshimaru and Kuroki, 1979; and Magarelli *et al.*, 1979). The closest approach to a defined marine crustacean diet, however, has been achieved with certain copepods and *Artemia,* commencing with the work of Provasoli and his collaborators (Provasoli *et al.*, 1959, 1970; Shiraishi and Provasoli, 1959; Provasoli, 1977). The only other marine multicellular invertebrates for which there are comparable data are nematodes (see references at the end of Chapter 10). Despite the widespread commercial culture of bivalves, there is little definite information on their nutrition, either for larvae (Ukeles, 1976a; Helm, 1977) or for juveniles and adults (Epifanio, 1976).

That many marine multicellular invertebrates can take up dissolved organic matter directly from seawater is now recognized (Jørgensen, 1976; Sepers, 1977; Stewart, 1979), but the significance of this process under natural conditions has not yet been determined. Recent studies and reviews have dealt with anemones (Schlichter, 1978), annelids (Jørgensen, 1979), bivalves (Ukeles, 1976a; Bunde and Fried, 1978; Stewart, 1978), and echinoderms (Kinne, 1977). The critical role that bacteria may play in this process has been pointed out by Castille and Lawrence (1979).

FOODS COLLECTED FROM THE SEASHORE

Those who live near the seashore will find it a relatively simple matter to collect food for captive animals. Various algae, such as *Macrocystis, Laminaria, Fucus, Ascophyllum, Egregia, Porphyra, Ulva,* and *Enteromorpha* can be easily collected (*Fucus* is eaten by only a few herbivores); typical forms are illustrated in Figures 6–1 and 6–2. Some can be collected at low tide from rocks or other solid objects; a few can even be used if they are collected shortly after they wash up on the shore. Though the algae listed are commonly used, other readily available species can be tried.

Before being used, the plants should be inspected, and sessile and clinging animals should be removed to avoid accidentally introducing them into the aquarium. It is advisable to rinse the plants in seawater before offering them to the animals. Some algae (*Ulva, Enteromorpha, Ascophyllum*), if left uneaten, will remain in good condition for a few days, but others (*Macrocystis, Fucus*) soon decay and foul the water. Species of algae that may be toxic to invertebrates, e.g., the acidic algae,

FIGURE 6–1 Examples of some common East Coast algae used as food for marine herbivores. Similar species can also be found on the West Coast. a. *Laminaria digitata;* b. *Laminaria agardhii;* c. *Fucus vesiculosus;* d. *Ascophyllum nodosum;* e. *Porphyria umbilicalis.* (From Taylor, 1962.)

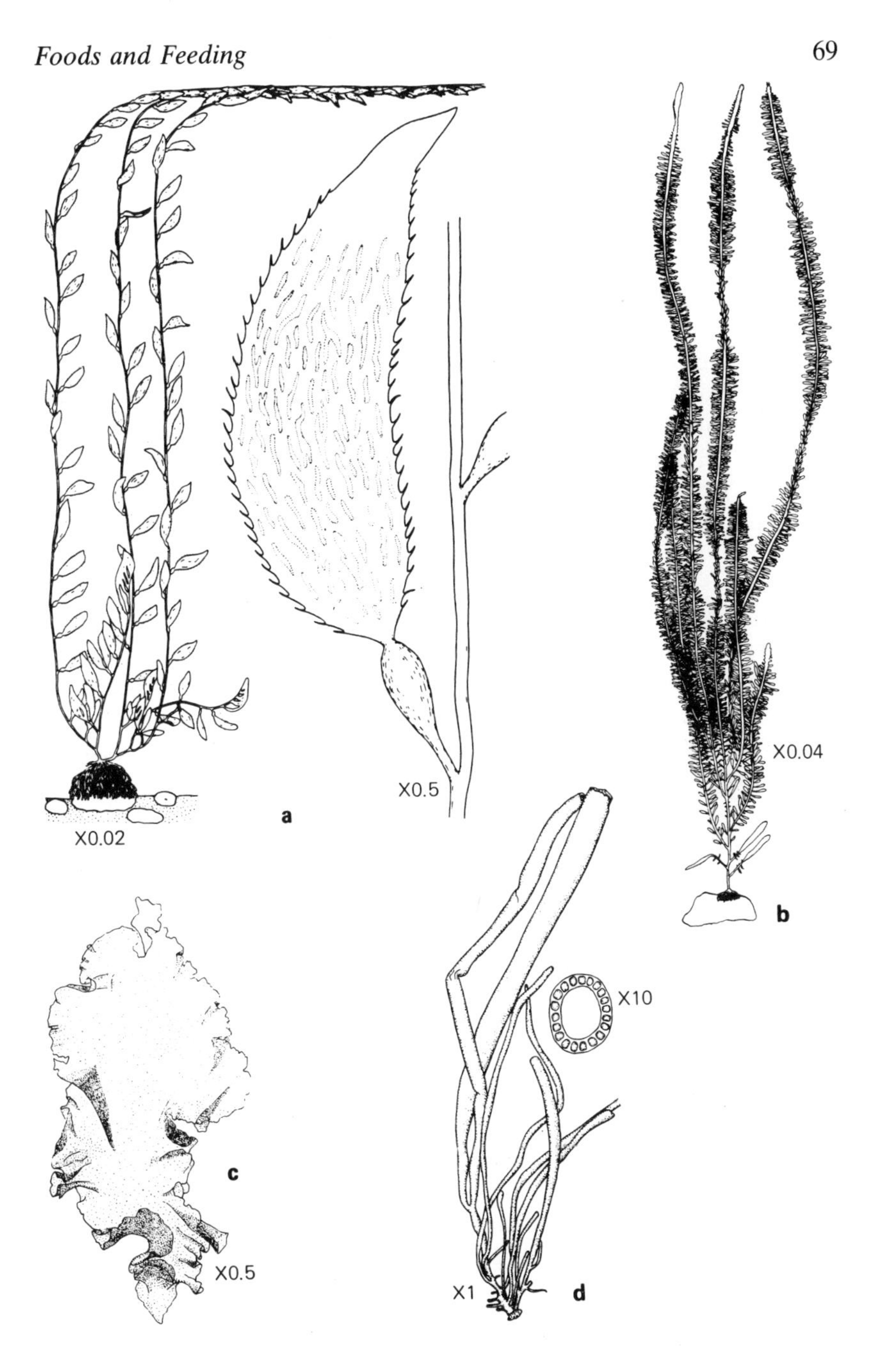

FIGURE 6–2 Some common West Coast algae that are used as food. There are species of *Ulva* and *Enteromorpha* on the East Coast also. a. *Macrocystis pyrifera;* b. *Egregia laevigata;* c. *Ulva lactuca;* d. *Enteromorpha* sp.; insert shows the hollow structure of a small branch. (From Dawson, 1966.)

Desmarestia, and most blue-green algae should be avoided. Even if blue-greens were not toxic, they often thrive so well under tank conditions as to overgrow everything else.

Some algae can be collected in quantity and kept either frozen or dried if frequent collecting trips to the seashore are not possible. *Ulva, Enteromorpha, Fucus,* and *Ascophyllum* keep well if drained, packed in plastic freezer boxes and frozen; others, however, deteriorate into mush after thawing. *Macrocystis* and *Laminaria* are two of several genera that can be dried and stored, and used after rehydration in seawater. Frozen or dried algae will rot more rapidly in an aquarium than freshly collected material.

Diatom films may be obtained by collecting small rocks and shells, especially from protected sites where wave action is limited. Besides serving directly as food sources for some grazers, diatoms also seed the aquarium and initiate the growth of diatom films on the tank walls and filter bed.

Natural foods for many carnivorous species can be collected in quantities large enough to last over extended periods if space to maintain them is available in the seawater system. Mussels (*Mytilus* and *Modiolus*) readily survive in circulating seawater and are acceptable to many carnivorous forms, including anemones, polychaetes, crustaceans, sea stars, and carnivorous gastropods. Furthermore, they are usually easy to collect, growing as they do in dense masses in the intertidal zone. Other bivalves, which may be locally abundant, are quahog clams (*Mercenaria*), long-necked clams (*Mya, Macoma, Tresus, Spisula*), and oysters (*Ostrea, Crassostrea*). Some clams (e.g., *Mercenaria*) can be kept alive for a month or more if drained and refrigerated at approximately 4°C. Many of these are readily available at fish markets.

WARNING: *Some bivalves as well as other invertebrates are legally protected animals; it may be necessary to obtain appropriate permits before collecting them.*

Small shrimp and small brachyuran and anomuran crabs, e.g., hermits and fiddlers, make excellent food for octopuses, squid, some anemones, and the larger crustaceans. Amphipods and isopods, which are frequently abundant in masses of seaweed, can be fed to various cnidarians (coelenterates). Polychaete annelids (*Nereis, Glycera, Clymenella*) may be found under rocks or by digging in sand or mud. Large nereids may be kept alive a week or more if refrigerated in damp, drained sand in shallow trays, or in damp seaweed, e.g., *Fucus.* Sponges, hydroids, anemones, ectoprocts, and compound tunicates may be collected as food for nudibranchs.

PREPARED FOODS

If collecting food at the seashore is not feasible, local markets may provide suitable alternatives. Fish, squid, scallops, shrimp, clams, or oysters are usually available in fish markets or grocery stores and are easily stored in a deep freeze. Their flesh may be chopped into pieces of appropriate size to feed cnidarians, errant polychaete worms, some gastropods (especially the carrion-feeders), squid, various crustaceans, many sea stars, and small brittle stars and sea urchins. Sometimes it is possible to find a free source of fish heads and skeletons left over after filleting at a local fish market. Bait stores are good sources of polychaete worms, live or dead shrimp, crabs, squid, or bait fish. Dried algae can be obtained at Oriental food stores. Flesh from freshwater fishes is usually accepted by marine carnivores. Sea urchins may be maintained on cakes of agar in which pieces of shrimp, dried algae, or bits of hard-boiled egg have been imbedded.

In addition to the items of marine origin, bits of meat (beef, chicken), heart, or liver are readily eaten by some carnivores, particularly carnivorous crustaceans. A few trials will establish which specimens will accept what food. Some herbivores will eat lettuce and fresh or frozen spinach leaves, although it is sometimes necessary to scald the leaves before they will sink.

Pet stores and aquarium supply companies carry such tropical fish foods as frozen or dried adult brine shrimp, fish meal, and other products that may be acceptable to marine invertebrates, especially to detritus and suspension or filter feeders such as annelid worms. Suspension feeders also often accept dried powdered algae, lettuce, grass or alfalfa, hard-boiled egg yolk, or yeast. Pet stores usually carry live dried "eggs" of the brine shrimp, *Artemia,* that can be hatched in salt water and the nauplii fed alive to such invertebrates as hydroids, anemones, carnivorous barnacles, and small crustaceans (see next section).

Beef heart extract is reported to promote the well-being and feeding activity of anemones, crinoids, fan worms, bivalves, brachiopods, and hermit crabs (McCammon, 1975). A 3–4-cm tube of fresh beef heart should be homogenized in distilled or seawater, strained through cheese cloth, and then double filtered; this preparation may be added to a 25-gallon aquarium twice a week.

Some investigators have devised artificial foods held together with a binder to form a solid sheet on which animals can browse. Chapter 19 describes a mixture of dried egg, dried yeast, soya flour, and corn meal bound in agar and dried in thin sheets as a food for the sea urchin,

Lytechinus pictus. de Graaf (1973) suggests algal meal mixed with an unspecified binding agent (agar?), and poured in thin sheets as a food for such herbivores as sea urchins and molluscs. Various ingredients can be prepared as above, and a few trials may well lead to satisfactory products.

According to Carefoot (1980), *Aplysia* can be maintained on an artificial diet of chemicals and *Ulva* extract set in agar.

The commercial culture of shrimp has stimulated the development of food pellets that will remain stable in seawater long enough to provide food for these slow, "messy" feeders. Although there are few, if any, really good pellet formulations currently available, accounts of the technology involved may enable individual investigators to develop their own artificial foods (Meyers *et al.*, 1972; Balazs *et al.*, 1973; Forster, 1973; Meyers and Zein-Eldin, 1973; Kinne, 1977; Meyers, 1980).

Feed companies also market food in pellet form for fish aquaculture. Some of these may contribute to the diet of invertebrates. The drawback to using some pellets is that they tend to float on the surface where they are unavailable to most species. It is often useful, therefore, to crush and sift the pellets into graded sizes.

A commercial preparation of vitamins and amino acids, Aminoplex®, is marketed for veterinary use. This same preparation has found a measure of application in aquarium husbandry and is now marketed under a variety of brand names. It may be used as a food supplement by adding it directly into the aquarium, but this is generally not recommended. It is preferable to soak the food, including brine shrimp, in the Aminoplex; the marinated food can then be fed to the aquarium animals. Feeding every other day—or once per week—with food supplemented with Aminoplex is usually adequate.

Because many suspension-feeding organisms are held in well-filtered aquaria, feeding may be difficult. One approach is to set up a nonfiltered feeding tank. The suspension feeders can then be placed in this tank, with suitable food, e.g., dinoflagellates, or brine shrimp (*Artemia*) nauplii, and allowed to feed for 30 min to an hour daily, or every other day. This technique has been used successfully with the hydroid *Coryne,* which accepts newly hatched brine shrimp.

Microencapsulation makes it possible to prepare artificial diets for suspension and/or filter feeders (Jones *et al.*, 1974, 1976, 1979; Gabbott *et al.*, 1976; Kinne, 1977), but no complete formulation for such invertebrates has yet been developed. The attempts of Conklin and Provasoli (1978) to formulate a chemically defined medium represent the most promising advance in studying the nutrition of these animals.

CULTURED FOODS

BRINE SHRIMP, *Artemia*

The brine shrimp, *Artemia salina,* is the most commonly used live food for marine invertebrates; newly hatched nauplian larvae are virtually indispensible in feeding many cnidarians and small crustaceans.

Shrimp are distributed worldwide and are found in salt lakes and man-made salterns where seawater is evaporated. A number of genetic varieties exist, and there is good evidence that the nutritional value of the larvae varies with the geographical source of the eggs and the food consumed by the shrimp (Claus *et al.,* 1977; Kinne, 1977; Simpson, 1979). Contamination with insecticides or other deleterious environmental chemicals quite possibly contributes to variability in nutritional value. Of the *Artemia* available commercially in the United States, those from the San Francisco Bay area are generally considered to be nutritionally superior. This variety is also smaller (approximately 350 μm long) at hatching, which may be advantageous if particle size is important.

The so-called "eggs" of *Artemia* are not really eggs, but arrested early embryos, that is, cysts containing a dormant gastrula (Olson, 1979). The term eggs is, however, firmly established and will be used here. Although they are becoming increasingly expensive, the eggs are usually readily available. Periods of scarcity and of poor hatching are not uncommon, and it is important that investigators locate a reliable commercial source and secure an adequate supply for the completion of the proposed research. Vacuum-packed eggs are sometimes available and will remain viable for a number of years if not subjected to temperature extremes. If the eggs are purchased in large quantities, they can be apportioned into smaller batches and placed in tightly sealed containers.

Resting eggs can be activated by placing them in saltwater. Noniodized table salt is adequate for this purpose, but natural or artificial seawater is preferable. Inland laboratories often use water that has been set aside during the regular aquarium water changes. A disadvantage of this approach, however, is the likelihood of contaminating the larvae. The salt used for commercial water softeners is inexpensive, available in 80–100-lb bags (ca. 40 kg), and satisfactory.

The egg shells should be separated from the nauplii before the larvae are harvested for feeding. The cysts can also be decapsulated before hatching (Sorgeloos *et al.,* 1977b). A number of elaborate devices for hatching and separation have been described by Kinne (1977) and Smith

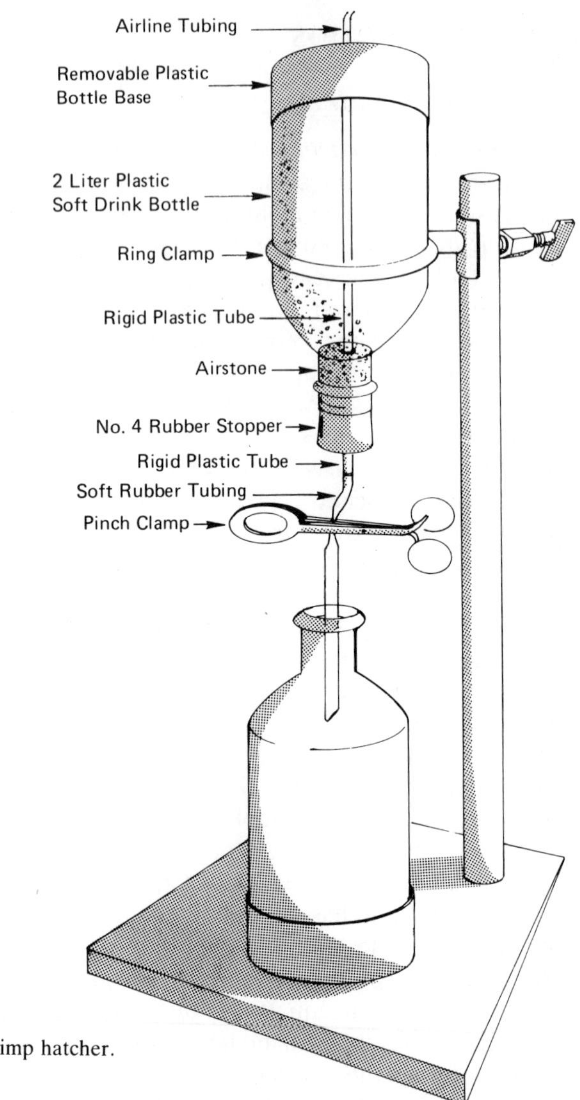

FIGURE 6–3 Brine shrimp hatcher.

et al. (1978), but these are unnecessary for most laboratories. Most of the separation techniques utilize either the positive phototaxis of the larvae or the tendency of the unhatched eggs and egg shells to float to the surface.

A small-scale, hatching-separation technique involves a 20- × 20-cm opaque plastic tray 6 cm deep. An opaque plastic partition divides the tray into two chambers—one 25 × 20 cm, the other 5 × 20 cm—and reaches only part way to the bottom, leaving a space of about 2 cm. A tray of this type suffices for hatching approximately 2 g (½ tsp) of eggs in 1 liter of seawater. The eggs are first floated in the larger chamber, which is covered to prevent the entrance of light, but the tray itself is placed under a continuous flourescent light. When the nauplii emerge they will be attracted by the light entering the smaller, uncovered chamber and swim under the partition. After they have congregated in the smaller chamber, they can be removed with a baster and collected in a fine mesh strainer. Although aeration increases the percentage of hatching, the tray method is adequate for smaller operations and an effective way to separate larvae from the egg shells.

A more efficient, but rather more elaborate, type of brine shrimp hatcher is illustrated in Figure 6–3. The principal pieces of apparatus are a 2-liter plastic soft-drink bottle and ring stand. The drawing shows a bottle that has a separate stabilizing base. This base is pulled off and the bottom of the bottle cut off. The base then serves as a cover. Bottles without these bases can be used if the bottom of the bottle is cut off in such a way that the bottom has a slightly larger diameter than the remainder of the bottle; thus with a little squeezing here and there, the cut-off bottom can be made to act as a lid. A hole is drilled or melted through this cover (base) to accommodate the rigid plastic tube. An airstone is attached to one end of this tube with a short piece of flexible tubing, and an air line is attached to the other end. The hatcher is then filled to within about 3 cm of the top (approximately 1,700 ml) with saltwater of salinity 35‰. Four grams (1 tsp) is about the maximum amount of eggs for a hatcher of this size. After the eggs have been added to the surface of the water, the hatcher is tightly covered with the modified bottle base, the airstone positioned as far down in the neck of the bottle as possible, and the egg–water suspension then aerated so that the eggs are in constant motion.

Shrimp are harvested by turning off the air and pulling up the airstone so that it no longer plugs the neck of the bottle. Swimming larvae will then settle to the bottom; egg shells and unhatched eggs will float to the surface. Good separation will be achieved in about 10 min, after which the concentrated shrimp can be drawn off by opening the pinch clamp. Shrimp can be collected in a fine-mesh strainer and washed; or collected in a second plastic soft-drink bottle, or flask. The first crop of shrimp can be concentrated and drawn off in about 500 ml of the water. The remaining 1,200 ml and unhatched eggs should then be again aerated and one or two subsequent harvests made during the next 24 h. This hatching technique is extremely efficient, and all but 100 ml of the original 1,700 ml can be drawn off without admixing larvae with shells.

Hatching time for *Artemia* eggs is variable, and larvae usually emerge continuously for a period of 24 to 48 h. To ensure maximum yield, more than one harvest should be scheduled during this period. Day-old nauplii are less nutritive than newly hatched ones. Hatching can be started every other day to ensure a continuous supply of fresh nauplii.

The larvae of *Artemia* can be reared to the adult stage by transferring them to larger containers and feeding them unicellular algae (*Dunaliella* works well), yeast suspensions, or whole wheat flour. The growth rate of such cultures is sometimes irregular, and consistent production of large quantities of adult shrimp is not always easily accomplished.

ROTIFERS, *Brachionus*

The euryhaline rotifer, *Brachionus plicatilis,* is easily cultured and represents a potentially useful food for a variety of marine invertebrates. It is most commonly used to feed decapod larvae, in which case its size (45–180 × 250 μm) makes it particularly suitable for the early larval stages that cannot capture or ingest the larger *Artemia* nauplii (300–400 μm). In most cases, however, *Brachionus* cannot support the complete larval development of decapods unless it is supplemented with *Artemia* (Sulkin, 1975). Although it has not been widely used for other invertebrates, this rotifer might be valuable for feeding small cnidarians or suspension feeders.

Brachionus can be reared in a concentrated culture of such unicellular algae as *Dunaliella* or marine-adapted *Chlorella.* These species will grow in the f/2 or ASP_2 media described in the following section. The cultures can be raised in cotton-plugged, 1- to 6-liter flasks under continuous fluorescent illumination. A concentration of $85–213 \times 10^4$/ml of *Chlorella* cells has been reported optimal for *Brachionus,* but determinations of population density this precise are not necessary for routine culture. An algal culture that is a rich green and so dense that one cannot see through it is appropriate for initiating a rotifer population. At 24°C and salinity of 30‰, the doubling time for *Brachionus* populations is 1 to 3 days. If 5 ml of a concentrated (100–200 rotifers/ml) *Brachionus* culture is added to 3 liters of algal culture, about 3 days will be needed to reach a population level suitable for harvesting. The rotifers should be harvested when the algal culture first shows signs of thinning. *Brachionus* is best harvested by pouring the culture through a 176-μm sieve to remove debris and then collecting the rotifers on a 44-μm sieve. Nylon mesh (Nytex) works well for this purpose. The rotifers should be rinsed while they are on the sieve and then resuspended in fresh seawater before being used. In a successful asexual *Brachionus* culture, 70–80 percent of the adults will bear eggs; a lack of food often causes a switch to sexual reproduction and eventual population decline.

PHYTOPLANKTON

Filter-feeding invertebrates (including most invertebrate larvae) and invertebrates being cultured as a food source require a steady supply of small particle-sized food. For many species, especially planktonic larval stages, the only appropriate diet at present is planktonic algae.

Among the common forms of phytoplankton usually cultured as food for marine invertebrates are the diatoms (Chrysophyta or Bacillariophycaeae), *Nitzschia, Phaeodactylum, Skeletonema, Thalassiosira,* and *Chaetoceros,* the golden-brown flagellates (Chrysophycaeae or Chrysophycophyta), *Monochrysis* and *Isochrysis,* and the green flagellates (Chlorophyceae or Chlorophyta), *Dunaliella* and *Platymonas.* Original inocula of these or other forms can be obtained from the collections, commercial sources, or can be isolated from nature. Samples obtained from established collections will be accompanied by detailed data on the history of the culture. Sources of marine unicellular algae are given in Appendix C.

The catalogs of commercial biological supply companies list marine phytoplankton species in culture. Carolina Biological Supply Company maintains most of the commonly used species.

Isolation of Unialgal Strains from Nature

Diatoms and autotrophic flagellates can be found in seawater from almost any source (although most successfully cultured species are neritic and littoral forms): plankton tows, scrapings from submerged shells, rocks, pilings, seaweeds and other solid objects, slides or other solids coated wtih agar or gelatin, and suspended for a time in a protected spot, e.g., a pier. The water and surfaces of aquaria are also well worth sampling. Cultures can be grown mixed, unialgal, or as clones. Single-celled algal cultures can be rendered bacteria-free by using antibiotics or other well-established microbiological techniques. In general it is not necessary to provide bacteria-free phytoplankton as food for marine invertebrates unless the research itself necessitates having bacteria-free experimental animals.

It may be necessary to concentrate the initial crude suspension by gentle centrifugation, collection on a millipore filter, or by taking advantage of the phototactic response of green flagellates. The tendency of pinnate diatoms to creep over the surface may be used to separate the desired species from others in the sample. Further refinement involves carefully pipetting a few cells (unialgal culture) or a single cell (clone culture) from a dilute sample with the aid of a dissecting microscope, and transferring the isolate into a selected growth medium.

Another useful isolation technique for many species (e.g., *Dunaliella, Platymonas, Phaeodactylum, Monochrysis, Isochrysis*) involves streaking the sample with a bacteriological loop on 1.5 to 2.0 percent agar in enriched seawater. The agar can be poured as slants in screw-cap test tubes or in Petri dishes, although the latter tend to dry out rather rapidly unless kept in a high humidity—they are, of course, easier to manipulate under the dissecting microscope. Colonies that appear to be unialgal can be removed by cutting out a block of agar with a sharp needle or loop, and then placing it in liquid growth medium.

Serial dilution of seawater into test tubes of culture medium is a good isolation technique. A modification of this method involves micropipette washing of the organisms, in which a capillary pipette with a bore several times the diameter of the cell is used to transfer the algae from a droplet of water under the dissecting microscope into a Petri dish containing sterile culture medium. This procedure is repeated several times, until finally a single washed cell is placed in a test tube of sterile culture medium. This technique has the advantage of greatly lowering the bacterial population (Ukeles, 1976b).

Media

Three general types of media can be used to culture algae: (1) filtered natural seawater, (2) enriched seawater, and (3) chemically defined synthetic mixtures. Seawater alone can support only a low-density of algae. The following are enriched and synthetic media that have been widely and successfully used.

- *Miquel's Seawater Medium* One of the earliest seawater enrichments, that of Miquel as modified by Allen and Nelson (Allen and Nelson, 1910; Allen, 1914; *in* Needham *et al.*, 1937; Cavanaugh, 1975), is still in use. It requires two separately prepared solutions, as shown below.

Solution A:

Potassium nitrate (KNO_3)	20.2 g
Distilled water	100.0 ml

Solution B:

Hydrated calcium chloride ($CaCl_2 \cdot 6H_2O$)	4.0 g
Sodium phosphate, dibasic ($Na_2HPO_4 \cdot 12H_2O$)	4.0 g
Hydrated ferric chloride ($FeCl_3 \cdot 6H_2O$) melted	2.0 ml
Hydrochloric acid, concentrated (HCl)	2.0 ml
Distilled water	80.0 ml

To prepare Solution B, dissolve the calcium chloride in 40 ml of water and add the hydrochloric acid. In a separate vessel, dissolve the sodium phosphate in 40 ml of water and add the melted ferric chloride. Then slowly mix the two solutions. To prepare the medium, add 2 ml of Solution A and 1 ml of Solution B to 1 liter of seawater. Sterilize by bringing it just to the boiling point. Cool and decant or filter to remove the slight precipitate. Aerate by agitation, pour appropriate amounts into sterile flasks, and cover. Inoculate with a few milliliters of an old culture and place under illumination.

• *Soil Extract Medium* Soil extracts, which are frequently used in culturing freshwater algae, can also be used to enrich natural or artificial seawater or in conjunction with Miquel's solution. The extract is prepared by placing 500 g of rich garden soil in 500 ml of distilled water in a flask and autoclaving it at 15 psi for 20 min. It is then cooled, filtered, and stored in a refrigerator. To prepare the medium, add 1 ml of the extract to each liter of seawater (Needham *et al.*, 1937).

• *Guillard's "f/2" Enrichment Medium* Guillard and Ryther (1962) and Guillard (1975) describe a successful and widely used enrichment medium for the culture of phytoplankton. The articles should be consulted for details of the culturing procedure. The enrichment itself is made up from the following stock solutions:

Major Element Stock Solution

	% (W/V)
Sodium nitrate ($NaNO_3$)[a]	7.5
Sodium phosphate, monobasic ($NaH_2PO_4 \cdot H_2O$)	0.5
Sodium silicate ($Na_2SiO_3 \cdot H_2O$)[b]	3.0

[a] 2.65 percent ammonium chloride (NH_4Cl) can be added if algae will not grow on nitrate.
[b] Omitted for all algae other than diatoms.

Trace Element Stock Solution

Dissolve 3.15 g of hydrated ferric chloride ($FeCl_3 \cdot 6H_2O$) and 4.35 g of sodium ethylene diamine tetra acetic acid (Na_2EDTA) in about 900 ml of distilled water. Add 1 ml from each of the following solutions and bring the volume to 1 liter. The final pH will be about 2.

	% (W/V)
Hydrated copper sulfate ($CuSO_4 \cdot 5H_2O$)[a]	0.98
Hydrated zinc sulfate ($ZnSO_4 \cdot 7H_2O$)[b]	2.2
Hydrated cobaltous chloride ($CoCl_2 \cdot 6H_2O$)	1.0
Hydrated manganous chloride ($MnCl_2 \cdot 4H_2O$)	18.0
Hydrated sodium molybdate ($Na_2MoO_4 \cdot 2H_2O$)	0.63

[a] Some species grow better if copper is not used.
[b] 1.5 percent zinc chloride ($ZnCl_2$) can be used instead.

Vitamin Stock Solution

Dissolve 20 mg of thiamine hydrochloride in 98 ml of distilled water, then add 1 ml of each of the following solutions:

Biotin	10 mg/100 ml
Vitamin B_{12}	5 mg/10 ml

This solution can be stored frozen in small lots and thawed as needed.

The medium is prepared by adding 1 ml of each of the major element stock solutions, 1 ml of the trace element stock solution, and 0.5 ml of the vitamin solution per liter of seawater. The result is a general purpose culture medium. If only one or a few species of algae are being cultured, several trials with varying the amounts of each of the stock solutions will often make it possible to tailor the culture medium to the species and achieve better or more rapid growth.

The algae should be centrifuged out of the culture media and resuspended in filtered seawater before they are added to the animal cultures. Some of the ingredients in the culture, and possibly algal metabolic products as well, can inhibit normal growth.

• *ASP_2 Medium* This is a completely synthetic medium that has proved successful in the culture of many species of phytoplankton. It is particularly useful for bacteria-free cultures (Provasoli *et al.*, 1957). It is composed as follows:

	Amount
Sodium chloride (NaCl)	1.8 g
Hydrated magnesium sulfate ($MgSO_4 \cdot 7H_2O$)	0.5 g
Potassium chloride (KCl)	0.06 g
Calcium chloride ($CaCl_2$)	10.0 mg
Sodium nitrate ($NaNO_3$)	5.0 mg
Potassium phosphate, dibasic (K_2HPO_4)	0.5 mg
Hydrated sodium silicate ($Na_2SiO_3 \cdot 9H_2O$)	15.0 mg
Tris hydroxymethyl aminomethan (Tris)	0.1 g
Vitamin B_{12}	0.2 μg
Vitamin Mix S3[a]	1.0 ml
Sodium ethylene diamine tetra acetic acid (Na_2EDTA)	3.0 mg
Hydrated ferric chloride ($FeCl_3 \cdot 6H_2O$)	0.08 mg
Zinc chloride ($ZnCl_2$)	15.0 μg
Hydrated manganese chloride ($MnCl_2 \cdot 6H_2O$)	0.12 mg
Hydrated cobaltous chloride ($CoCl_2 \cdot 6H_2O$)	0.03 μg
Cuprous chloride ($CuCl_2$)	0.12 mg
Boric acid (H_3BO_3)	0.6 mg
Water (H_2O)	100.0 ml

[a] Vitamin Mix S3 contains the following per ml: Thiamine hydrochloric acid 0.05 mg; nicotinic acid 0.01 mg; calcium pantothenate 0.01 mg; p-aminobenzoic acid 1.0 μg; inositol 0.5 mg; folic acid 0.2 μg; thymine 0.3 mg.

Ukeles (1976b) provides a detailed discussion of the properties and experimental results and the uses of various culture media.

Containers

The container selected for phytoplankton depends on the size and frequency of the harvest needed. Some situations can be met by using screw-cap test tubes or flasks of various sizes. A suitable surface–volume ratio must be maintained to ensure adequate diffusion of carbon dioxide from the atmosphere. For Ehrlenmeyer flasks, the volume of culture medium should equal about ¼ to ⅓ the total volume of the flask. Cultures may be aerated and kept in suspension by placing flasks on the reciprocal shaker or by bubbling filtered air through them. When larger containers are used it is necessary to bubble air, or an air–carbon dioxide mixture, through the medium.

Glass or plastic containers are also suitable. As Guillard (1975) pointed out, autoclaving media in glass containers often extracts a significant amount of silicon. Silicon is not known to be harmful to any algae, but if diatoms are accidentally introduced they will sometimes overgrow a nondiatom culture if silicon is present. This eventuality is greatly delayed if plastic containers such as polycarbonate are used instead of glass.

If a large volume of algae is required, carboy-sized containers will be needed. One successful continuous-supply method is that of Gentile and Sosnowski (1978) (Figure 6–4), which utilizes a 12-liter carboy with a drain arm at the base, to which is attached a tube with a pinch clamp for collecting the harvest. The carboy, containing a spin bar, is mounted over a magnetic stirrer. A triple-hole rubber stopper closing the carboy is secured by an aluminum clamp. The stopper is rigged with: (1) a cotton plug-filtered air supply with an airstone, (2) a vent tube with a filter to prevent contamination, and (3) a tube that delivers fresh sterile medium from a reservoir carboy mounted on a shelf overhead. The delivery tube can be separated from the tube attached to the drain arm of the reservoir carboy by means of a tube connector, on either side of which is a pinch clamp. The reservoir carboy has a volume of 40 liters, is closed by a single-hole rubber stopper equiped with a cotton-plugged air vent tube, and is held in place by an aluminum clamp.

In harvesting, a volume of culture is drawn off near the maximum log-phase cell density and replaced by an equal volume of fresh culture medium. Within about 24 h the culture will again reach the original cell density. This manner of harvesting can be continued indefinitely providing the culture does not become contaminated.

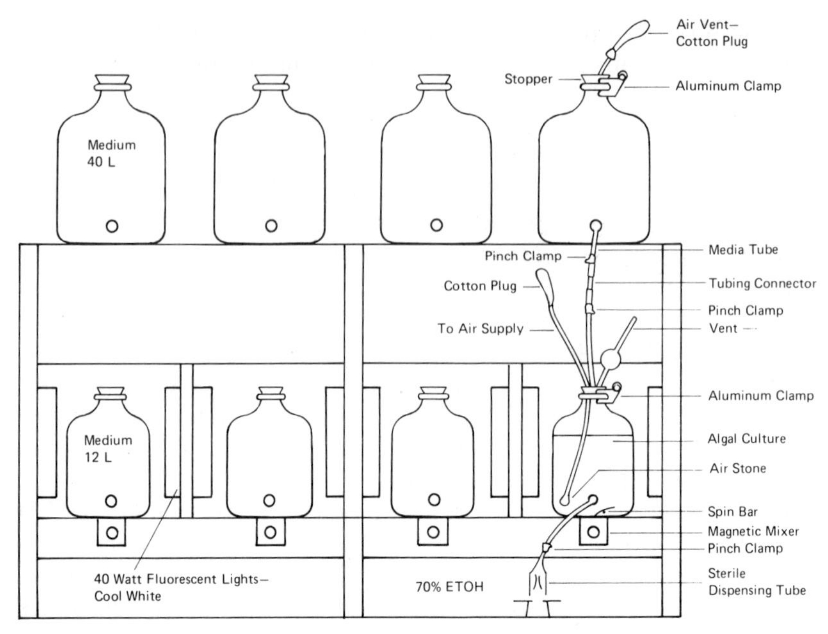

FIGURE 6–4 Carboy container for cultivation of phytoplankton. (From Gentile and Sosnowski, 1978.)

Culture containers should be made of good-quality borosilicate glass having low thermal expansion, or of autoclavable polycarbonate. Closures for the vessels must be such as to preclude contamination while allowing gas exchange. Nonabsorbent cotton, metal, or polypropylene caps, inverted shell vials (over rimless test tubes), or small beakers (over flasks) have been used successfully. Plastic screw-caps for test tubes should have the liners removed and should be screwed only partially shut after inoculation, to allow for gas exchange (Ukeles, 1976b). Another somewhat different type of culture apparatus that features rectangular tanks has been described by Sorgeloos *et al.* (1977a).

Cleaning

The extent to which cleaning is needed depends on the algae being grown. Often, ordinary laboratory washing is adequate. If it is necessary to go beyond this, then all vessels, closures, and tubing should be cleaned and sterilized before use. Such cleaning agents as potassium dichromate–sulfuric acid solution, concentrated sulfuric acid (H_2SO_4) saturated with sodium nitrate ($NaNO_3$), boiling solutions of soap–sodium hydroxide (NaOH), and commercial detergents are used to remove organic residues. These agents should be followed by 10 or more rinses in tap water or 3 or more rinses in distilled water. Glassware can then be oven-dried and stored dust-free under cover. Alternatively, the glassware can be autoclaved after the rinses, and then stored.

Illumination

Smaller cultures may be grown successfully using only daylight, but artificial illumination is more consistent and must be used for larger cultures. When daylight alone is used, the culture vessels should be placed near a window but not in direct sunlight, which may often be excessive. A north-facing window is best. Both fluorescent and incandescent lamps are suitable as artificial light sources. Cool-white fluorescent lights are generally satisfactory, as are warm-white and daylight fluorescent lamps, which are expressly designed for plants (e.g., Sylvania Gro-Lux, Westinghouse Plant Gro, Duro-Test Corp. Vita-Lite, etc.). The merits of the latter types for supporting algal growth have not yet been extensively tested. Incandescent lamps are usually less satisfactory because they generate heat, which must then be dissipated.

Temperature

Most phytoplankton thrive best at a temperature range of 15 to 22°C. They may be held at room temperature, providing their range is not exceeded. More reliable temperature management requires air-conditioned rooms or incubators; humidity is more easily controlled in the latter.

As a further resource, Ukeles (1976b) provides an excellent and comprehensive discussion on the cultivation of marine unicellular plants.

7

Recordkeeping, Marking, and Tagging

It may at times be necessary to identify and recover individual animals, either in the field or in the laboratory aquarium. One common method is to isolate the animals in marked containers, such as the plastic compartment boxes commonly sold to hold nuts, bolts, and similar hardware. The boxes should be constructed of plastic only and have a hinged lid and internal compartments. The sides and partitions must be drilled with numerous holes to ensure sufficient water circulation.

An alternative approach is to mark the animals themselves. Though there is an extensive literature on tagging vertebrates, relatively few methods have been devised for handling invertebrates. Hard-shelled species can easily be marked with an engraving tool, or a tag can be fastened to the shell. Of course the marking will be lost from crustaceans at the time of molting.

Biological stains that can be seen through the integument have been used to mark shrimp; these persist through at least one molting cycle (Dawson, 1957; Neal, 1969). It has been reported that Fast Green FCF, injected as a 0.5–1.0 percent solution in either 1:9 distilled water, seawater, or distilled water alone, remains at least 120 days, and through a molt. The dye—0.03–0.30 ml depending on the size of the shrimp—should be injected into the abdomen through a needle inserted through an articular membrane. In the case of shrimp, the dye becomes concentrated in the gills. Feder (1955) used solutions of either neutral red or nile blue sulfate to stain the rays of the starfish, *Pisaster,* where it was retained for a minimum of 5 months.

Sebens (1976) has devised a successful marking technique for sea anemones, wherein he prepared a thick paste of neutral red in fresh water. The anemones were blotted dry and the paste used to paint an identifying number on the column of the animal. The dye was allowed to dry for 15 min before the anemone was returned to the aquarium, where it remained legible for a minimum of 12 mo. This method is suggested for marking other soft-bodied animals.

Colored liquid latex, which solidifies but remains flexible, has been used successfully for marking crustaceans that have transparent exoskeletons. Phillips (1971) used latex (Diptex 319) obtained from Rubber Latex Limited, Harling Road, Withenshaw, Manchester 22, England, in red, blue, yellow, and green. This material makes possible color combinations and position coding by multiple injection. Liquid latex is also available from Carolina Biological Supply Co., Burlington, North Carolina 27215.

Plastic tags have been used for identification purposes (Meyer, 1974; Marullo *et al.*, 1976; Schwartz, 1977; Fannaly, 1978). An internal anchor tag that can be applied with a cartridge-fed tagging gun has been used successfully with both fishes and invertebrates. The tags and tagging gun are available from Floy Tag and Manufacturing, Inc., 4616 Union Bay Place Northeast, Seattle, Washington 98105.

Crustaceans can be "injected" with coded wire tags (West and Chew, 1968; Prentice and Rensel, 1977). Tag insertion can be verified by first magnetizing the tag and then using a detector to make certain they are sited properly. These ferromagnetic tags, which can be color-coded with epoxy inks, are commonly referred to as Bergman–Jefferts tags and were originally developed for use in studying Pacific salmon (Jefferts *et al.*, 1963). Inserted tags have also been used with sea stars (Birkeland, 1974; Dayton *et al.*, 1974). Dayton and his co-workers used spaghetti tags held by a nylon monofilament loop inserted through the base of one of the rays, whereas Birkeland used numbered Floy tags inserted into the interradius by means of a tagging gun.

Thus far no one has found a totally satisfactory method for permanently tagging sea urchins. Ebert (1965) suggested a procedure using a monofilament thread and colored beads, which he later modified (Ebert, personal communication, 1979) as follows: a hypodermic needle, somewhat larger in diameter than the nylon filament to be used, should be passed into the animal through a tubefoot hole on one of the ambulacral regions and then out through the shell, all the while twisting the needle such that it acts as a drill. The passage of the needle should be shallow to avoid harming internal organs. At this point, the nylon filament is passed through the needle, starting at the point, and the needle is then withdrawn. Colored beads can now be strung on the filament, which is then tied. A tag thus installed will last for a year or two, but apparently the urchin is able slowly to destroy the internal portion of the filament, with the result that the tag is eventually lost.

Olson and Newton (1979) described a tag for sea urchins that uses commercially available nylon clothing markers, as a modification of a fish-tagging procedure described by Thorson (1967). These tags have a "T" bar at one end and a flat portion at the other that can either be marked directly or used to affix a larger marking surface. Actual tagging is accomplished with a specially designed gun. After it has been inserted, the "T" bar is held snugly against the inside of the test by a small lead fishing sinker that is crimped over the shaft of the tag, just outside the test. In a closed seawater system, it may be necessary to substitute a less toxic material for the lead sinker. Under natural conditions mortality was fairly high—about 30 percent in 6 mo—but it may have been more the result of predation than the effect of the tags.

Several types of colored or marked tubes that slip over the spine have also been tried, but they remain for only a short time because the urchin either works the tube off the spine or drops the spine.

Oysters, *Crassostrea gigas,* have been tagged by drilling a hole through the umbo of their upward-directed left shell (Neudecker, 1977). A nylon thread with a plastic tag is then inserted; by this method the soft parts of the oyster are not injured. Ropes and Merrill (1970) successfully marked clams by notching the shell or attaching polyester tape to it with a sealant.

Freeze-branding has been applied to the mantle of the terrestrial slug, *Ariolimax columbianus,* by Richter (1973), a method that may have application for marine animals. In this case the branding head was cooled in liquid nitrogen and applied to the slug for 2–3 s, producing a mark that remained a minimum of 3 mo.

Snail shells have been marked by a grinding of a portion of the shell smooth and (1) writing a number directly on the shell, (2) applying an adhesive tape tag with a number written in India ink on it, or (3) gluing a numbered plastic marker (W. H. Brady Co., Milwaukee, Wisconsin) to the shell. The number, or tag, should then be covered with two coats of quick-drying plastic adhesive such as "Dekophane" (Rona Pearl Co., Bayonne, New Jersey) (Frank, 1965; Connell, 1970; Spight, 1974). Heald (1978) successfully tagged scallops by gluing plastic tags to the shell with alpha-cyanoacrylite, an adhesive that hardens in seawater in about 40 s. Many of the tags remained in place for more than a year. This method could probably be used on other animals that have smooth unincrusted shells or after a patch of shell has been cleaned.

Sonic transmitter "tags," which emit a high-frequency pulse periodically, have been used as means of tracking the movements of animals in the sea. Crustaceans, for example, have been tagged by wiring (Lund and Lockwood, 1970) or gluing (Herrnkind and McLean, 1971) the transmitter to the carapace.

A recent book, *Animal Marking,* edited by Stonehouse (1978), is almost completely devoted to tagging vertebrates. It may, however, suggest ideas that could be applied to invertebrates.

8

Anesthetics

Two terms—anesthetize and narcotize—are employed to describe the use of drugs to relax, immobilize, or desensitize marine organisms. They are frequently used interchangeably, but the recent practice is to identify anesthetics as drugs that temporarily reduce or eliminate an animal's response to stimuli, but from which the animal is expected to recover later, whereas narcotics relax the organism, often permanently, for subsequent fixation, and recovery is not anticipated. Obviously, there is a broad overlap. This chapter deals primarily with anesthetics.

Cori (1938) and Kaplan (1969) serve as general references on both anesthetics and narcotics as applied to invertebrates. More than 100 different method–species combinations for freshwater, marine, and terrestrial species have been tabulated. Another source, which includes a variety of other information as well, is a book on biological techniques by Knudsen (1966).

At times, cooling alone is adequate anesthetic. The specimen is cooled to about 2°C and worked with at that temperature. Not all species, particularly those from warm waters, survive prolonged cooling, whereas certain cold-water species that are not anesthetized can tolerate temperatures below 0°C.

The classic narcotizing/anesthetizing agents for marine invertebrates are magnesium chloride and magnesium sulfate. These have been used most often as relaxants prior to fixation; however, with proper care, some animals do recover. Although there are many variables, the general procedure is gradually to add isosmotic 7.5 percent weight per volume of hydrated magnesium chloride to the seawater containing the animal. The correct level has been reached when the animal no longer responds to prodding. An alternative procedure is to inject the animal with artificial seawater in which all the calcium has been replaced by magnesium. Because calcium is required for synaptic transmission and magnesium is a competitive inhibitor for calcium, the increased magnesium concentration dilutes the animal's calcium level and thus anesthetizes it.

A 5–10 percent solution of ethanol is at times effective as an anesthetic. Here again, different species respond differently and some are easily killed. In any case, the ethanol should be added gradually.

Chlorotone is a commonly used agent and often works well on crustaceans. As with magnesium chloride and ethanol, the concentration and period of time to which the animal should be exposed must be determined by experience. A final concentration of about 0.5 percent is suggested as a first approximation.

In recent years, the drug M.S. 222 (tricaine methanesulfonate or ethyl-*m*-aminobenzoate), which was first widely used in transporting fish, has been applied to marine invertebrates. Schwartz (1966) found that 0.5 g of M.S. 222 per liter of seawater was adequate to anesthetize the sand shrimp, *Crangon septemspinosa,* at 0–5°C and that most of the animals recovered after exposures of several hours. Ahmad (1969) used it to anesthetize the amphipod, *Gammarus pulex.* After trying a number of different concentrations, he found that those in the neighborhood of 0.5 to 1.0 g per liter worked satisfactorily. Sensitivity is affected by temperature as well; lower temperatures slow the process of relaxation but allow the animal to remain in the drug for longer periods.

Gamble (1969) used M.S. 222 to anesthetize the amphipods, *Corphium volutator, C. arenarium,* and *Marinogammarus obtusatus.* He found that, at 10°C, the optimum concentration both for relaxation and good recovery was 0.5 g per liter. M.S. 222 is not suitable for all crustaceans; for example, decapods respond poorly. Oswald (1977) investigated a number of other drugs for the crabs *Cancer pagurus* and *Carcinus maenas,* injecting the animals rather than depending on immersion. Injection was into the hemocoel, through the membrane at the base of a posterior leg. For short-term relaxation of about an hour, Propanidid (100 mg/kg of body weight) and Xylazine (70 mg/kg) worked well. Induction took 5 to 6 min and recovery was rapid. Pentobarbitone (250 mg/kg) induced anesthesia for about 1.5 h; Procain (25 mg/kg) induced anesthesia for several hours. Although the latter acts within 4 min, the recovery period is generally too long. These drugs permit close to normal water pumping through the gills, do not block heart action, and seem to do no permanent damage.

An excellent anesthetic for small crustaceans and crustacean larvae is 2-phenoxythanol. A fine drop from a full-strength solution is added to a depression slide containing seawater and larvae; the larvae regain normal swimming ability when restored to seawater without the narcotic (Bookhout, personal communication). Robinson *et al.* (1965) found that larvae of the brine shrimp, *Artemia salina,* could be anesthetized with the gas Halothane and that the animals recovered even after 10 h of exposure. The procedure is useful for the study of anesthetic action, but gases are difficult to use along with laboratory procedures.

There has been a limited amount of work on molluscs, primarily gastropods. Owen (1955) used propylene phenoxetol to anesthetize clams. Runham *et al.* (1965) investigated a number of narcotizing and anesthetizing agents, primarily on pulmonate snails and found that a mixture of 0.1 percent Nembutal and 0.3 percent M.S. 222 at 20°C worked well and permitted complete recovery. They also found that injection of 10 percent magnesium chloride in the vicinity of the cerebral ganglia induced a quick relaxation that lasted for 5 to 15 min; in this case recovery was variable. Runham *et al.* (1965) found that immersion in a 0.5–1.0 percent solution of propylene phenoxetol is suitable for anesthetizing certain marine snails.

Urethane has been used by O'Dor *et al.* (1977) to anesthetize the squid, *Illex illecebrasus*; a 3 percent solution in seawater was effective. They reported that after several minutes in the anesthetic the animals could be removed from the water and be marked or otherwise manipulated and that recovery took from 3 to 15 min, depending on the dosage.

The sea hare, *Aplysia californica,* as well as other gastropods, can be relaxed by injecting 0.5 mg/10 g live weight of succinyl choline, an analogue of acetyl choline, into the space near the cerebral ganglia (Beeman, 1969). The solution is made up fresh before use by dissolving succinyl chloride in seawater at a ratio of 5 mg/ml. The procedure is also suitable for nudibranchs, *Aeolidia papillosa* and *Hermissenda crassicornis,* the large prosobranch snail, *Strombus,* and the snail *Bulla gouldiana.* It has also been used on *Pleurobranchaea californica* and the limpet *Acmaea digitalis.* The latter relaxes almost immediately after the injection of one drop of succinyl choline solution into the foot. In certain individuals, however, succinyl choline sometimes gives inconsistent results, apparently due to poor circulation throughout the body. Circulation can be improved by lightly massaging the specimen after injection.

Most animals begin to respond to stimuli about an hour after injection, and recovery is good.

Although published reports are few, current trends in using anesthetics are toward injection rather than placing the animals in a solution. There seem to be no reports describing the action of stimulants to overcome or reverse the effects of anesthetics.

9

Use in Bioassays and Tissue Culture

INTACT ANIMALS

In bioassays, the organism is the indicator and the choice of a proper test animal is therefore critical. With increased interest in pheromones, toxins, potential pharmaceuticals, and other chemicals found naturally in the marine environment, as well as pollutants put there by human activity, has come a need for a greater variety of assay organisms. Fortunately, ever-better methods of rearing marine invertebrates in the laboratory give reasonable expectation that this need can be met.

Ideally, a given bioassay should be absolutely reliable (that is, indefinitely repeatable), simple, rapid, and inexpensive. To approach this ideal, the test organism must be readily available, uniform in reponse, and sensitive to the variable under study. These desiderata characterize many successfully cultured marine animals, particularly as regards uniformity and availability. If species that adapt well to life in the laboratory and regularly reproduce there are used, the likelihood of having stressed or diseased test animals is minimized and genetic variation can be controlled. To be sure, great sensitivity to the environment is, in some measure, antithetical to ease of maintenance in captivity, but careful control of culture conditions has made it possible to maintain some very delicate marine species.

An investigator wishing to become familiar with the technique and rationale of the bioassay should consult the Fourteenth Edition of *Standard Methods for the Examination of Water and Wastewater* (Rand *et al.*,

1976), the FAO's *Manual of Methods in Aquatic Environment Research* (Butler *et al.*, 1977), and the U.S. Environmental Protection Agency's *Bioassay Procedures for the Ocean Disposal Permit Program* (US EPA, 1978). Other useful sources are Perkins (1972), Portman (1972a,b), Stora (1972), Glass (1973), Cox *et al.* (1974), Swedmark *et al.* (1976), Kinne (1977), Peltier (1978), and Buikema and Cairns (1980).

EPA and the Army Corps of Engineers (ACE) have developed very complex procedures to assay the quality of sediments that are to be removed by dredging and disposed of elsewhere in marine environments. Guides to these may be found in *Ecological Evaluation of Proposed Discharge of Dredged or Fill Material into Navigable Waters* (U.S. Waterways Experiment Station, 1976), *Bioassay Procedures for the Ocean Disposal Permit Program* (US EPA, 1976, 1978), *Ecological Evaluation of Proposed Discharge of Dredged Material into Ocean Waters* (US EPA and US ACE, 1977), and *Methods for Measuring the Acute Toxicity of Effluents to Aquatic Organisms* (Peltier, 1978).

Most bioassays, especially those employing marine invertebrates, are black-box experiments because the animal's physiology, behavior and pathology are poorly understood. Even though a given technique meets the operational criterion of reliability, it must then be related to the real world. Although bioassays can detect and measure toxic substances, they can provide no more than first approximations toward determining acceptable medical or environmental risks (Stebbing, 1979).

Kinne (1977) has pointed out that little or no use has been made of turbellarians, rotifers, bryozoans, or nematodes, despite the fact that culture methods are well in hand for all of them (see Chapter 10). Bioassays based on specific marine invertebrates have been described for hydroids (Karbe, 1972; Stebbing, 1976), corals (Rand *et al.*, 1976), polychaete worms (Åkesson, 1970; Rand *et al.*, 1976; Reish *et al.*, 1976 and Chapter 13), a copepod (Rand *et al.*, 1976; Gentile and Sosnowski, 1978; Parrish and Wilson, 1978; Ward *et al.*, 1979), an isopod and amphipod (Lee, 1977), a mysid (Nimmo *et al.*, 1977), crabs, shrimp, and lobster (Rand *et al.*, 1976), and a shipworm (Karande and Pendsey, 1969).

Because they have acute sensitivities that reach into the parts-per-billion range, larvae of marine invertebrates are well suited for short-term bioassays. Undoubtedly the forms most widely used, especially for short-term bioassays, are the developing egg and larva of the sea urchin and sand dollar. The methodology involved has been described by several investigators (e.g., Bernhard, 1957; Okubo and Okubo, 1962; Karnofsky and Simmel, 1963; Kobayashi, 1971, 1974, 1977; Hagström and Lönning, 1973; Bidwell, 1976; Brown and Greenwood, 1978; Bougis *et al.*, 1979; Ikegami *et al.*, 1979). Generally, sand dollar larvae are more sensitive than those of sea urchins. Other larvae that have proved useful include oysters (Woelke, 1965, 1968, 1972; Rand *et al.*, 1976; Cardwell *et al.*, 1977, 1978, 1979), oysters and clams (Cardwell and Woelke, 1979), mussels (Courtright *et al.*, 1971; Lucas, 1976), coot clams and slipper limpets (Calabrese and Rhodes, 1974), barnacles (Tighe-Ford *et al.*, 1970; Karande and Thomas, 1971; Blundo, 1978), crabs and shrimps (Costlow and Bookout, 1965; Amiard, 1976; Mirkes *et al.*, 1978; Tyler-Schroeder, 1978b), brine shrimps (Michael *et al.*, 1956; Tarpley, 1958; Lüdemann and Neumann, 1961; Robinson *et al.*, 1965; Zillioux *et al.*, 1973; Kinne, 1977; Sorgeloos *et al.*, 1978), honeycomb worms (Wilson, 1968), and coral planulae (Rand *et al.*, 1976).

Not only are death or moribundity possible end-points of a bioassay, but growth or behavior often more accurately reflect environmental changes, especially in long-term studies. An extended discussion of behavioral measures of environmental stress may be found in Cox *et al.* (1974) and Eisler (1979). Scherer (1977) analyzed the principles and practices of aquatic behavioral bioassays. Maciorowski *et al.* (1977) and Ginn and O'Connor (1978) have described various avoidance-preference bioassays that use mobile aquatic invertebrates. Amiard (1976) found that the rate of movement toward light (photokinesis) of crustacean larvae was a sensitive indicator of metallic pollutants. Kinghorn *et al.* (1978) used an abnormal, spasmotic response of brine shrimp nauplii as a system to assay plant toxins. Davis (1978) and Lyes (1979) measured the disruption of reproductive behavior in amphipods to test pulpmill effluents and surfactants, respectively. On the other hand, Davenport (1977) showed that the behavioral response of the mussel to noxious chemical stimuli, i.e., quickly isolating itself by closing its valves, makes it an unsuitable indicator organism. Devices that automatically record activity are obviously of considerable help in conducting behavioral assays. Several have been described, e.g., by Heusner and Enright (1966), Coombs (1972), Atkinson *et al.* (1974), Rebach (1977), and Cripe (1979).

D. Roberts (1975) showed experimentally that pesticides had an inhibitory effect on byssus formation in the mussel and suggested that this response might be used to screen seawater pollutants. Similarly, Conger *et al.* (1978) showed that the inhibition of shell growth in oysters by toxic substances could serve in bioassay. Morse *et al.* (1979b) pointed out that extremely low concentrations of pollutants interfered with the setting reaction of planktonic abalone larvae and that this reaction provided a much more sensitive assay than did survival *per se*. The spermatozoa of echinoderms are also very sensitive to pollutants and may serve as the basis of a rapid fertilization bioassay (Lönning and Hagström, 1975; Stober *et al.*, 1979). Bradley (1976) devised a temperature shock-recovery test with a species of copepod that made it possible to predict the animal's survival under different environmental conditions.

Theoretically, stress of any kind can serve as an early indicator of adverse environmental conditions, but the difficulty of detecting and quantifying the microscopic or biochemical changes that accompany stressful conditions is severely limiting. Studies by Bayne *et al.* (1976, 1979) and Widdows (1978) on stress in the mussel provide a good perspective of the problems. According to Moore and Stebbing (1976) and Ivanovici (1979), however, at least some sublethal biochemical changes are suitable as indicators. Sprague (1976) evaluates many aquatic sublethal tests.

Although the need for long-term bioassays, especially for testing sublethal levels of pollutants, is now well recognized, few such experiments have been undertaken, principally because it is difficult to maintain adequate controls over extended periods of time. Wilson (1968), Perkins (1972), Rand *et al.* (1976), and Lee (1977) have discussed long-term bioassays that utilize marine invertebrates. Similarly, the need for information on the effects of pollutants on various stages of the life cycle, particularly of species that have several different ontogenetic stages, has been emphasized (Buikema and Benfield, 1979). Nevertheless, relatively few tests have been designed to accomplish this (D'Agostino and Finney, 1974; Laughlin *et al.*, 1978; Nimmo *et al.*, 1978b; Tyler-Schroeder, 1978c, 1979; Connell and Airey, 1979).

Bioassays involving marine invertebrates have been devised to detect and measure specific substances: chlorine (Burton, 1977), aflatoxin and other mycotoxins (Harwig and Scott, 1971; Ďuračková *et al.*, 1977; Prior, 1979), red tide toxins (Trieff *et al.*, 1973), ciguatera toxins (Granade *et al.*, 1976), suspected carcinogens and anticancer drugs (Karnofsky and Simmel, 1963; Buu-Hoi and Chanh, 1970; Kinghorn *et al.*, 1977; Pesch and Pesch, 1980; Samoiloff *et al.*, 1980), and contaminants of culture media (Bernhard, 1977). The first-stage larva of the horseshoe crab is extremely sensitive to molting hormones and has been used to bioassay ecdysteroids and their synthetic analogs (Jegla and Costlow, 1979). Calton *et al.* (1978) developed a test for substances with cardiotoxic properties using an *in vitro* preparation of the heart of the blue crab. Couch and Courtney (1977) determined that the incidence of a virus disease in the pink shrimp, *Penaeus duorarum*, regularly increased when this crustacean was exposed to PCB and they devised a bioassay to measure the effects of low concentrations of pollutants on natural pathogen–host interactions. Tubiash (1971) found the soft-shell clam well suited for testing the pathogenicity of microorganisms to bivalves in general. According to Kimeldorf and Fortner (1971), the anemone, *Anthopleura xanthogrammica*, rapidly responds to the presence of ionizing radiation by withdrawing its tentacles.

There is an extensive literature on testing petroleum pollutants and related chemicals that has been critically reviewed in one of the FAO's *Reports and Studies* by C. H. Thompson *et al.* (1977). Recent studies that have featured marine invertebrates as test organisms for these chemicals include LaRoche *et al.* (1970), Allen (1971), Portman (1972b), Zillioux *et al.* (1973), Vanderhorst *et al.* (1976), Donahue *et al.* (1977), Lönning (1977), Lee and Nicol (1978), Le Roux and Lucas (1978), and Michael and Brown (1978).

Marine invertebrates regularly available as assay organisms in the 23 coastal states have been reviewed by Becker *et al.* (1973). Animals collected in the field for bioassays may vary seasonally in their susceptibility to a particular toxicant; life in captivity may also alter their responses. On the basis of tests conducted on wild-caught grass shrimp, Tatem *et al.* (1976) recommended that animals be tested as soon as possible after collection, that test conditions be made to resemble those in nature as nearly as possible (thus minimizing the acclimation period), and that the animals first be tested with a standard reference toxicant to ensure that they are healthy. Seldom have the effects of life in captivity on marine invertebrates been studied with reference to their use as test animals in bioassays. Sosnowski and Gentile (1978) could find no difference in the response to heavy metals of six successive, captive-bred generations of the copepod, *Acartia tonsa*. In contrast, Crow and Harrigan (1979) showed that certain behavioral responses of laboratory-reared nudibranchs differed from those of specimens collected in nature.

The factors that can influence bioassays are many and subtle; the problem of comparing results obtained under slightly different conditions is a serious one, as the investigations of Reish *et al.* (1978) have shown for the same test performed in different laboratories. A recently recognized factor is the influence of circadian rhythms on sensitivity to toxicants. Bellan-Santini *et al.* (1979) discovered that the toxicity of cadmium for two species of crustaceans and one polychaete depends on the time of day at which they are exposed to the heavy metal.

Descriptions of continuous-flow bioassay apparatus, designed for aquatic animals of moderate size and mobility, may be found in Connor and Wilson (1972), Bahner *et al.* (1975), Chandler and Partridge (1975), Harrison *et al.* (1975), Maciorowski (1975), Berg and Granmo (1976), and Auwarter (1977). Modifications for small organisms that might be affected by water currents were presented by Maki (1977). Apparatus to control and record pH was described by Lillie and Klaverkamp (1977) and salinity by Bahner and Nimmo (1975).

TISSUE AND ORGAN CULTURE

There is a vast literature on tissue culture of mammalian cells, a much smaller, but significant, number of papers on plant tissue culture, and some on invertebrates, most of which concerns insects. Papers on marine invertebrate tissue and organ culture are few and varied and do not constitute a coherent, well-developed area of research.

Papers prior to 1970 are reviewed in a two-volume work edited by Vago. In it methods of organ culture of the major invertebrate phyla and the Sipuncula are covered by Le Douarin (1971) and Gomot (1972). Peponnet and Quiot (1971) deal with cell cultures from *Limulus* and various crustaceans, and Rannou (1971) with those from sponges, coelenterates, echinoderms, *Bdelloura*, and *Sipunculus*. Another, more recent review by Maramorosch (1976) includes information on a few molluscs.

There is a modest amount of literature on crustacean organ and tissue culture. Holland and Skinner (1976) determined *in vitro* DNA synthetic activity in limb regenerates from the land crab, *Gecarcinus lateralis*. Lambert (1977) has been able to maintain chromatophores from the integument of the dorsolateral portions of the cephalothorax of the fiddler crab, *Uca pugilator*, in culture. The techniques employed by Lambert, as described in detail below, may be applicable to other tissues and organisms. All tissue culture supplies were obtained from GIBCO (Grant Island Biological Company).

In Lambert's study, an intact fiddler crab was washed in 70 percent ethyl alcohol for 1 min, then twice in sterile artificial seawater. Then, using autoclaved forceps and scissors, the dorsolateral portions of the carapace were removed and placed in 10 ml sterile crustacean saline containing 1 ml of 100 antibiotic/antimycotic mix. The saline used was that of Skinner, Marsh, and Cook (1965) adjusted to 1,000 mosmol with sodium chloride. The portions of the carapace were transferred to a fresh saline/antibiotic medium for 5 min, then transferred to an autoclaved Stendor dish containing 2 ml sterile saline with 1 percent trypsin. The dish was placed on a shaking device and the tissue was incubated, with moderate shaking, until most of the chromatophores had been released into the medium, a process requiring 60–75 min. At this point, the trypsin was inactivated with soybean trypsin inhibitor. Two different procedures were then followed.

In the first instance, freed cells were gently collected with a sterile Pasteur pipette, placed on collagen-coated coverslips in Falcon tissue culture dishes, and incubated at 25°C. The coverslips were coated with collagen as described by Ehrmann and Gey (1956) as follows: commercial collagen was dissolved in 0.1 percent acetic acid and spread on coverslips that were then placed into Petri dishes containing ammonium hydroxide-soaked cotton for 3 h. The coverslips were then washed several times with distilled water and stored in 95 percent ethyl alcohol. Before use, the coverslips were washed again several times in sterile water, then sterile crab saline, and finally in two washes of sterile tissue culture medium. The makeup of this medium (Table 9–1) was a modification of that of Holland and Skinner (1976). The required hemolymph from *Limulus* was prepared by removing 10 ml from each animal with a sterile syringe, allowing clotting to occur, centrifuging at 10,000 *g*, and collecting the supernatant. This was stored, frozen, until used.

The second procedure, a modification of that developed by Ide (1973), made it possible to purify the cell types by taking advantage of the different densities. After the soybean trypsin inhibitor had been added, the cells were applied to a discontinuous Ficoll density gradient consisting of 10, 20, 30, 40, and 50 percent autoclaved Ficoll/saline. This was centrifuged at 600 *g* for 20 min, as a result of which the melanophores and leukophores settled to the bottom of the centrifuge tube,

TABLE 9–1 Chromatophore Tissue Culture Medium[a]

Component	Amount
Powdered minimal essential medium (Eagle's)	9.51 g
100X Amino acids	30 ml
100X Vitamin	30 ml
100X Glutamine 200 m*M*	20 ml
10 Percent sodium citrate	40 ml
100X Pyruvate (0.1 *M*)	10 ml
100X Antibiotic mix	10 ml
4 *M* Potassium chloride (KC1)	2.5 ml
4 *M* Sodium chloride (NaC1)	100 ml
1 *M* Calcium chloride ($CaC1_2$)	12 ml
0.06 *M* Magnesium sulfate ($MgSO_4$)	100 ml
Glass-distilled water	400 ml
1 *M* Sodium hydroxide (NaOH) to adjust pH to 7.3	
Glass-distilled water	to 800 ml
Limulus hemolymph, added after filter sterilization to make 20 percent	200 ml

[a] From Lambert (1977), modified from Holland and Skinner (1976).

distinct from the erythrophores, xanthophores, and cellular debris. The melanophores and leukophores were resuspended in a culture medium that differed from the above in containing 20 percent commercial lobster—as opposed to *Limulus*—hemolymph. These cells were placed in Falcon tissue culture dishes and allowed to settle on the bottom of the dish, rather than collagen-coated coverslips.

To a limited extent, cells from echinoderm larvae can be cultured. Okazaki (1975) has been able to grow the micromeres of sea urchin larvae in seawater containing added horse serum. The cells will divide a number of times and produce spicules. This technique is, however, not suitable for long-term culture. Spiegel and Spiegel (1975) review the general procedures for the disaggregation and reaggregation of sea urchin larval cells and describe some of the results. Apparently, no one has been able to culture either larval or adult echinoderm cells for extended periods.

Pearson and Woodland (1979) review various attempts to culture the tissues of horseshoe crabs (e.g., Yamamichi and Sekiguchi, 1974) and describe in detail the *in vitro* cultivation of *Limulus* amoebocytes.

Over the years, the amoebocytes of bivalves have attracted attention because of their ability to remain alive and motile for several days after removal from the mollusc. Brewster and Nicholson (1979) were able to maintain these cells *in vitro* for as long as 6 mo even though no definite evidence for cell division could ever be obtained. Cultures of oyster larval tissue did yield actively dividing cells, but succumbed to microbial contamination before they were 2 wk old. Lubet *et al.* (1978) cultured various organs of the adult mussel, especially the gonads, in order to test the effects of DDT on them.

10

Some Commonly Used Species

The marine, multicellular, nonparasitic invertebrates that have been used as laboratory animals comprise but a tiny fraction of the more than one hundred thousand species know to inhabit the saltwaters of the earth. The number is increasing, however, partly because of their recognition as suitable models for biomedical research (Ray, 1958; Prosser *et al.*, 1973; Ruggieri, 1975a; Thomas, 1976; Bulla and Cheng, 1978; Hillman, 1978). In this chapter, we shall review the better-known experimental subjects among the marine invertebrates.

Table 10–1 is a summary of factors that are important in the care and maintenance of the invertebrates that may be appropriate for use in research. Listed temperatures may not always be optimum, but they are satisfactory. The foods should be regarded as suggestions and often represent just a few of the forms that can be used. Specimens will ordinarily survive at temperatures below those given, but very low temperatures (below 8°C) can be as lethal to some species as temperatures that are too high. Geographical abbreviations refer generally to the same areas discussed in Chapter 5.

MOLLUSCS

Molluscs are second only to arthropods in the number of described species and are both abundant and varied in the marine environment. There are five well-known classes, the Scaphopoda (tusk shells), Gastropoda (snails), Bivalvia (clams and mussels), Polyplacophora (chitons), and the Cephalopoda (octopus and squid). Two other classes, the Monoplacophora and the Aplacophora, are mostly known from fossils but are also found today in deeper waters.

SCAPHOPODA

Though they are well known as fossils, the scaphopods or tusk shells are poorly understood biologically. There is some older work on development but little contemporary research. These animals appear to burrow through the unconsolidated sediments and extract detritus as food. Specimens can be obtained from some biological supply companies.

GASTROPODA

The snails and their relatives are among the most-studied molluscs, particularily as regards their development and nervous system. The development, care, and maintenance of the aplysids is described in Chapter 14, and the mud snail, *Ilyanassa* (*Nassarius*), in Chapter 15.

Obviously, not all gastropods are equally available, and species differ considerably in their acceptance of captivity as well as morphology and size. Indeed, relatively few members of this diverse class have been studied in the laboratory aquarium.

A research project should involve, initially at least, species that are known to accept confinement and that can be fed in captivity. Not surprisingly, these generally include the nondiscriminating carnivores (some are scavengers) and herbivores. Some of the more commonly used types are indicated in Table 10–1. As a rule, unless they offer very special advantages for other reasons, grazing species that require a film of algae, and carnivores that require a specific live food (such as sponges, hydroids, or bryozoa) are poor candidates for the laboratory aquarium. Algal films are difficult to maintain, and live food is usually not readily available or, if obtained, may quickly die.

Carnivorous gastropods that have a nonspecific diet may be fed bits of clam, squid, shrimp, or fish; some prefer polychaete annelids, e.g., some species of *Conus*. Such nonspecific herbivores as the alpysids (sea hares) and the abalone will usually eat one or more species of large algae and will often eat fresh or frozen vegetables.

Some snails, such as *Navanax* and *Pleurobranchaea*, are vulnerable to specialized predators, either at a specific developmental stage or throughout their life-cycle. Episodes of cannibalism have been noted

TABLE 10–1 Routinely Used Species

Species	Geographic Source[a]	Holding Temperature, °C	Feeding Type[a]	Suggested Food	Comments
Phylum Cnidaria					
Class Hydrozoa					
Podocoryne carnea	A	16–22	C	*Artemia* nauplii	most conveniently grown attached to slides
Cladonema sp.	F, G	20–25	C	*Artemia* nauplii	as for *Podocoryne*
Cordylophora lacustris	Estuaries A, P	18–25	C	*Artemia* nauplii	does best in dilute sea water, 10–20 parts per thousand
Cl. Scyphozoa					
Cassiopeia frondosa	F, G	20–25	C	*Artemia* nauplii	polyps very hardy, medusae difficult
Cl. Anthozoa					
Aiptasia pallida	sA, F	20–25	C	*Artemia* nauplii, shrimp, fish	very tolerant of salinity and temperature (see Chapter 12)
Anthopleura elegentissima	P	15–22	C	shrimp, fish, clam	
Anthopleura xanthogrammica	P	15–22	C	shrimp, fish, clam	will not tolerate other anemones close by
Corynactis californica	P	15–22	C	*Artemia* nauplii, shrimp, fish	easily kept in aquaria
Metridium senile	nA, P	10–16	C	shrimp, fish, clam	
Pachycerianthus torreyi	P	15–22	C	*Artemia*, nauplii, shrimp, fish, clam	long, graceful tentacles demonstrate feeding method very well
Sagartia modesta	A	15–22	C		
Tealia lofotensis	P	15–22	C	shrimp, fish, clam	very omnivorous, hardy

Ph. Rhynchocoela					
Cerebratulus lacteus	A	15–22	C	small polychaetes	must have sandy-mud substrate in which to burrow
Ph. Annelida					
Amphitrite ornata	nA	15–20	D	detritus	
Chaetopterus variopedatus	A, P, G	15–22	S	bacteria, single-celled algae	can be kept in glass U-tube in aquarium
Eudistylia polymorpha	P	12–15	S	bacteria, single-celled algae	
Glycera americana	A, P	15–20	C	small polychaetes	need sandy-mud substrate
Glycera dibranchiata	A, P, G	15–20	C	small polychaetes	
Nereis spp.	A, P, G	15–22	D, C	shrimp, fish, *Ulva*	need sandy-mud substrate
Ph. Sipuncula					
Themiste pyroides	P	15	D	bacteria, etc., organic matter from mud or sand	see Chapter 10 for other species
Ph. Mollusca					
Cl. Gastropoda					
Anisodoris nobilis	P	15	C	sponges	
Aeolidia papillosa	nA, P	12–18	C	anemones (small)	
Aplysia californica	P	15	H	*Ulva*, spinach, lettuce	see Chapter 14
Aplysia wilcoxi	F, G	18–20	H	*Ulva*, spinach, lettuce	
Astraea sp.			H	diatoms, etc., sessile on rocks, aquarium walls	very hardy in aquarium
Busycon canaliculatum	A	16–18	C	clams, other gastropods	
Busycon carica	A	16–18	C	clams, other gastropods	
Conus californicus	P	15	C	annelids	
Cyprea spadicea	P	15–20	H	algal growth on rocks	hardy in aquaria
Fasciolaria tulipa	sA, F	15–22	C		
Hypselodoris californicus	P	15–20	C	sponges	will live long time even without food

TABLE 10–1 *Continued*

Species	Geographic Source[a]	Holding Temperature, °C	Feeding Type[a]	Suggested Food	Comments
Hermissenda crassicornis	P	14–16	C	hydroids	
Littorina littorea	A, G	15–20	H	various brown and green algae	tend to climb above water level of aquaria
Lunatia heros	nA	15	C	bivalves, other snails	produces lots of mucus
Megathura crenulata	P	15–20	C	bryozoans	
Ilyanassa obsoleta	A, P	16–22	C, D	squid, shrimp, worms	organic matter from mud (see Chapter 15)
Navanax inermis	P	14–18	C	opisthobranchs	*Bulla, Hermissenda* are eaten
Pleurobranchaea californica	P	10	C	squid, fish, opisthobranchs	will eat *Metridium* also
Polinices duplicata (south to Cape Cod)	A, F, G	18	C	bivalves, other snails	
Urosalpinx cinereus	A	16–18	C	mussels, barnacles	
Cl. Bivalvia					
Aequipecten irradians	A		S	single-celled algae	
Crassostrea gigas	P	15	S	single-celled algae	see Chapter 16
Crassostrea virginica	A	15	S	single-celled algae	for bivalve culture methods
Macoma balthica	nA, nP	12–18	S	single-celled algae	
Mercenaria mercenaria	A	15–18	S	bacteria, single-celled algae	will keep easily for weeks out of water in refrigerator
Mya arenaria	nA	12–18	S	bacteria, single-celled algae	
Mytilus californicus	A, P	15–18	S	bacteria, single-celled algae	
Ostrea lurida					
Spisula solidissmima	A	15–18	S	bacteria, single-celled algae	keeps best in sand bottom

Mytilus edulis	A	15–20	S	single-celled algae	
Cl. Polyplacophora					
Chaetopleura spp.		14–18	H	*Ulva*, various algae	
Cryptochiton stelleri	P	12–15	H	red algae	
Cl. Cephalopoda					
Octopus spp.	F, P	15–20	C	clams, crabs, shrimp, fish	preferably live crustacea (see Chapter 10)
Ph. Arthropoda					
Cl. Merostomata					
Limulus polyphemus	A, F	15–20	D	shrimp, clam, worms	very durable species
Cl. Crustacea					
Artemia salina		room temp.	H	*Dunaliella*, yeast	see Chapter 6
Balanus spp.	A, P, G, F	12–15	S	*Artemia* nauplii, phytoplankton	
Callinectes sapidus	A, G	12–16	C, D	fish, clams, worms, shrimp	
Cancer spp.	A, P	12–16	C, D	shrimp, clams, worms, fish	
Carcinus meanas	A	15–20	C, D	shrimp, clams, worms, fish	
Crangon septemspinosa	A	15–20	C, D	shrimp, clams, worms, fish	keep well on sand bottom
Homarus americanus	A	10–15	C, D	shrimp, clams, worms, fish	
Neopanope texana	A, F		C, D	shrimp, clams, worms	
Ovalipes ocellatus	A, F, G	15–20	C, D	shrimp, clams, worms	
Palaemonetes vulgaris	A, F, G	12–16	D	shrimp, clams, worms	
Panulirus spp.	F, P	14–16	C, D	shrimp, clams, worms	
Uca spp.	A, F, P	15–22	C, D	shrimp, clams, worms	also sift mud for organic particles
Ph. Echinodermata					
Cl. Asteroidea					
Asterias forbesii	A	15–20	C	bivalve molluscs	
Asterias vulgaris	A	10–15	C	bivalve molluscs	
Astropecten braziliensis	P	12–15	C	fish, squid, shrimp	need sand bottom small molluscs (eat whole)

TABLE 10–1 *Continued*

Species	Geographic Source[a]	Holding Temperature, °C	Feeding Type[a]	Suggested Food	Comments
Dermasterias imbricata	P	12–15	C	fish, squid, shrimp	
Patiria miniata	P	12–20	C, S, H	fish, algae, crab, detritus	very hardy
Pisaster spp.	P	12–15	C	bivalve molluscs	
Cl. Echinoidea					
Arbacia punctulata	A	16–22	H	*Ulva*, lettuce, spinach, see Chapter 19	
Dendraster excentricus	P	15	D		eaten by sea urchins
Echinarachnius parma	A	15	D		eaten by sea urchins
Lytechinus spp.	F, P	15–18	H	*Macrocystis, Ulva*, lettuce, spinach	see Chapter 19
Strongylocentrotus spp.	A, P	10–16	H	*Macrocystis, Ulva*, lettuce, spinach	
Cl. Ophiuroidea					
Ophioderma panamensis	P	14–18	C	squid, fish, crab, clams	
Ophiopholis aculeata	A	16–22	H	various algae, eel grass	
Cl. Holothuroidea					
Caudina arenicola	P	14–18	D		burrows in sand
Thyone briareus	P	14–18	D	diatoms, other unicells	
Parastichopus parvimensis	P	14–18	D		

Ph. Chordata					
Cl. Ascidiacea					
Ciona intestinalis	A, P, H	15–20	S	phytoplankton	breeds easily in aquaria
Botryllus schlosseri	A, P	14–18	S	phytoplankton	easy to clone culture
Molgula manhattensis	A, P	14–18	S	phytoplankton	breeds easily in aquaria
Pyura spp.	A, P	14–18	S	phytoplankton	
Styela spp.	A, P, H	14–18	S	phytoplankton	

[a]Key to Abbreviations

Geographic Source		Feeding Type
A—Atlantic Coast	P—Pacific Coast	C—Carnivore
F—Florida Coast	n—northern	D—Detritus, scavenger
G—Gulf Coast	s—southern	H—Herbivore
H—Hawaii		S—Suspension feeder

among some species of gastropods, e.g., *Megathura, Aplysia, Hermissenda*, and *Pleurobranchaea*.

Care must be taken to prevent some of the gastropods (e.g., *Littorina* and *Tegula*) from crawling out of the aquarium. Drains must be screened and protected.

A few of the opisthobranchs have been extensively used as experimental animals, particularly in neurophysiology (Willows, 1973; Kandel, 1979). This order includes hermaphroditic gastropods that usually do not have an exposed shell. The majority of species are carnivores that prey upon specific species of sponge, coelenterate, or bryozoan and are difficult to culture. A few are both predatory and scavengers, e.g., *Pleurobranchaea*. Some, such as *Aplysia*, are nonspecific herbivores. *Navanax*, another opisthobranch, differs from most others in that it swallows its prey without rasping or biting. In nature it preys upon other opisthobranchs and will do so in the laboratory. *Navanax* can fast for up to 3 wk (at 15°C), yet remain in good condition. It will feed on the bubble shell *Bulla* or the nudibranch *Hermissenda*, which can be raised in the laboratory. *Navanax* usually survives well in aquaria if given reasonable care.

BIVALVIA

A number of bivalve species have been used in research, particularly in studies of development, muscle physiology and biochemistry, and circulatory physiology. They can also be used in pollution studies. The culture of some of these species is discussed in Chapter 16.

All bivalves are suspension feeders; most feed on phytoplankton. As for laboratory husbandry, they can be separated into two groups: those that require a surrounding of sediment that will hold the valves shut and those that can live independently of sediment. Clams that require burial should be kept in sediment so arranged that water slowly passes through it. This prevents the lower portion from becoming anaerobic. The animals should be buried vertically at a depth that permits them to extend their siphons above the substrate. The siphons are at the end of the shell opposite the rounded bulbous portion of the hinge. As noted in Chapter 3, a rubber band can be used to hold the shells together until the animals are placed in sediment.

Bivalves that do not require burial include the clam *Mercenaria*, oysters, mussels, rock clams (*Chama*), scallops, and rock scallops. Some species, e.g., *Lima*, may survive better if conditions are provided in which they can form a nest.

Bivalves may be fed by periodically adding suitable phytoplankton to the aquarium while the filtration system is temporarily turned off, or by removing each animal from the aquarium and placing it in a phytoplankton suspension. Single-celled algae such as *Isochrysis*, *Monochrysis*, *Tetraselmis*, or *Dunaliella* are often cited as appropriate food.

POLYPLACOPHORA

Aside from studies of the unusual inclusion of magnetite in their radulae and the extraordinary enzymatic components in their digestive systems, relatively little work has been done with the chitons. Some types (*Cryptochiton, Stenoplax*) live well in the aquarium, where they browse on benthic algae. Chitons appear to be relatively undisturbed by predators other than sea stars. Some chitons will continuously rasp on the sides of aquaria and will penetrate protective coatings such as epoxy paint. They have not, however, been observed to carve their way through fiberglass (polyester resin).

CEPHALOPODA

Octopuses have been maintained in many laboratories for studies of learning, behavior, vision, and neurophysiology (Young, 1977; Wells, 1978). Squid have been used primarily as a source of the giant axon and in work on the biochemistry of vision (Rosenberg, 1973; Arnold *et al.*, 1974; Geduldig and Hoekman, 1979). Both are interesting and, probably, the most intelligent of the invertebrates (Nixon and Messenger, 1977).

Most species of squid and some species of octopus are poor candidates for the laboratory aquarium. Small squids have been raised in aquaria, but it is not easy to do so at a distance from the seacoast.

All species of cephalopods are carnivores. Octopuses generally feed on the bottom on clams, snails, crabs, some fish, shrimp, other octopuses and, occasionally, worms. Squid feed in the water column on fish, shrimp, and other squid. In nature cephalopods are preyed upon by fish, crabs, pinnipeds (seals), cetaceans (whales), and birds. They are also taken in recreational and commercial fisheries.

Octopuses

Octopuses (*Octopus bimaculatus, O. vulgaris*, and similar species) require a relatively large volume of very clean, well-oxygenated seawater. They prefer room in which to wander about and the security of a shelter such as a flower pot, a jar, or a plastic jug in which to hide. A ratio of about 100 liters per 500 g of octopus is recommended. Large aquaria, say 200 liters, will accommodate about 2 kg of octopuses. A temperature range of 10–20°C is tolerated, but about 15°C is optimal for *O. bimaculatus*. If more than one octopus is to share the aquarium, it is essential to provide an acceptable shelter for each and to house animals of similar size together.

Octopuses will enter anywhere they can. If given the opportunity, they will tear up subsand filters, pull out drain plugs from the bottom of the aquarium, disconnect air lines, and remove air stones. They will also crawl out of the aquarium unless it has a tight-fitting lid.

A diet of fresh fish or shellfish has been used successfully to maintain octopuses; frozen fish or shellfish are generally not preferred. If the animals will not accept flesh initially, it may be wise to provide live food and gradually acclimate them to bits of clam, shrimp, fish, or lobster. Feeding every day or every second day is recommended, the amount to be determined by the size of the animal. The aquarium should be kept free of debris and putrefying flesh.

During spring and early summer, males become aggressive toward other males and are best segregated; specimens of unequal size should always be segregated. Mating females may exhibit promiscuous behavior. If a female does spawn in captivity, she must have a place in which to nest. Once the female deposits eggs, she will guard them faithfully until they hatch, after which she dies.

Squid

Most species of squid are pelagic and spend their lives in an environment without solid barriers. When these animals are disturbed within the confines of an aquarium, they will often strike the side and incur minor abrasions that are slow to heal. In most cases these wounds, and any others received during capture, become infected and the individual eventually dies (Leibovitz *et al.*, 1977; Hulet *et al.*, 1979).

Two systems have been used with some success in maintaining *Loligo*. One employs a relatively large aquarium of about 1,000-liter capacity (1 m × 2.5 m × 1.5 m). The other requires a tank that is lined with a plastic curtain and specially designed for holding squid (Summers *et al.*, 1974). Both systems provide a relatively large area for swimming and seek to minimize the stress placed upon the highly excitable squid. An aquarium of 1,000 liters will accommodate up to 50 specimens. Even under the best of circumstances, there will be gradual attrition. An aquarium temperature of 6 to 20°C will be tolerated, but a temperature of about 10°C is preferable.

Another species, *Illex illecebrosus*, has also been maintained in a large pool (O'Dor *et al.*, 1977).

In captivity, squid will feed on small, swimming fish such as *Fundulus* (killifish), live shrimp, or, on occasion, goldfish. Goldfish do not live long in saltwater and should be removed as soon as they die.

Clean, well-aerated, chilled aquaria are essential for maintaining squid. The aquarium should be checked twice daily to remove dead or moribund animals. Bright lights, noise, excessive vibrations, or disturbances should be avoided.

Care must be taken to minimize damage to the animals during capture and transport, because there is a direct correlation between physical damage during capture and death in captivity. If possible, animals should be dip-netted, a night-light being used to attract the squid. They can also be captured on barbless lures, or in lampara nets (a large midwater purse seine). Trawling is generally the least desirable capture method, because it causes the most injury.

ARTHROPODS

This is the largest phylum of multicellular organisms, both as to number of species and numbers of individuals. It can be divided into nine classes (Fingerman, 1976), only two of which are exclusively marine, the Merostomata (horseshoe crabs) and the Pycnogonida (commonly called sea spiders). Four of the classes, Diplopoda (millipedes), Chilopoda (centipedes), Pauropoda, and Symphyla (centipede-like), have no marine representatives. There are a few marine insects (see Cheng, 1976). The class Arachnida (spiders and mites) consists almost exclusively of terrestrial species, but one family of mites lives in fresh water and a second is marine. Most members of the class Crustacea (crabs, lobsters, and barnacles) are marine.

Of the marine arthropods, only the North American horseshoe crab and the crustaceans are routinely used in research and teaching laboratories. The former is a very hardy organism and survives well in the laboratory (see Chapter 18). Many of the marine crustaceans can also be maintained in the laboratory with little difficulty (see Chapter 17). Highly active swimming forms such as the edible blue crab, *Callinectes sapidus*, must have a plentiful supply of oxygen.

MEROSTOMATA

This group has remained almost unchanged since the Mesozoic era. Today there are only five species. *Limulus polyphemus*, the North American horseshoe crab, occurs mainly along the Atlantic coast of the

United States, but is also present in the Gulf of Mexico, primarily in the northeast corner.

This species is best known for studies of its compound eyes and blood. Electrophysiological studies of visual receptors of the horseshoe crab led in large measure to H. K. Hartline being awarded the Nobel Prize in 1967 (Hartline, 1969). The blood of *Limulus* has been much studied because it is an excellent source of the copper-containing oxygen-transport pigment hemocyanin, which is blue when oxygenated but colorless when deoxygenated. More recently, a very sensitive *in vitro* test for bacterial endotoxins has been devised that utilizes horseshoe crab amoebocyte lysate (Cohen *et al.*, 1979; Pearson and Weary, 1980). This reactivity has been linked to the horseshoe crab's defense against microbes (Nachum *et al.*, 1979). There have also been a few endocrinological studies of *Limulus*. Ecdysones are highly effective in stimulating molting of larval horseshoe crabs (Jegla *et al.*, 1972). The central nervous system contains a substance that causes hypoglycemia in crayfish (Pezalla *et al.*, 1978). Chapter 18 describes the methods for culturing *Limulus*.

CRUSTACEA

The fiddler crab, *Uca pugilator*, is one of the most frequently used crustaceans. It occurs along the eastern coast of the United States and along the shore of the Gulf of Mexico and has been used in a variety of studies—hormonal control of color changes, hormonal control of carbohydrate metabolism, molting, limb regeneration, biochronometry, and sensitivity to auditory stimuli (Brown, 1950; Skinner and Graham, 1972; Crane, 1975; Kleinholz, 1976; Fingerman and Fingerman, 1977; Keller, 1977; Salmon *et al.*, 1977; Weis, 1977; Trinkaus-Randall and Mittenthal, 1978).

Larval and adult specimens of the mud crab, *Rhithropanopeus harrisii*, found along the coast from Maine to Texas, have been used to determine the effects of exposure to juvenile hormone mimics that might be used as insecticides (Christiansen *et al.*, 1977; Payen and Costlow, 1977). The larvae obtained from ripe females were cultured in the laboratory.

The blue crab, *Callinectes sapidus*, a commercially important species occurring along the Atlantic coast of the United States to the Gulf of Mexico, has been much studied. Recently, for example, the effects of several factors (weight, sex, temperature, salinity, stage of the molting cycle) on the rate of oxygen consumption have been determined (Laird and Haefner, 1976; Lewis and Haefner, 1976). Because it is relatively large, a small number of specimens will provide sufficient tissue for many biochemical procedures. The blue crab's hepatopancreas (midgut gland) stores calcium as calcium phosphate for the molting process, and after ecdysis releases much of the calcium to harden the new exoskeleton. Mitochondria in the hepatopancreatic cells appear to be involved in the sequestering and release of the calcium (Chen *et al.*, 1974).

The striped shore crab, *Pachygrapsus crassipes*, found intertidally along the coasts of California, Oregon, and Washington, has been used recently in studies of the control of molting and of yolk synthesis (Chang and O'Connor, 1977; Lui and O'Connor, 1977). This work provided the first conclusive evidence that the molting glands (Y-organs) of a crustacean secrete alpha-ecdysone, as do the insect molting glands (the prothoracic glands).

Metograpsus messor is a crab found in large numbers along shores of Oahu, Hawaii. It has been extensively investigated to elucidate its osmoregulatory capabilities and the endocrine mechanisms that control them. This crab can regulate both hyposmotically and hyperosmotically, depending upon the salt concentration in the aquarium (Kamemoto, 1976).

The American lobster, *Homarus americanus*, caught mostly off the coast of New England, can be kept easily in a marine aquarium and has been cultured (see "Literature Citations" at the end of this chapter). Partly because of its commercial importance, this species is being studied in several laboratories. Studies of the control of its growth, for example, have been published steadily (Hughes *et al.*, 1972; Gilgan and Burns, 1977).

The prawn, *Palaemonetes vulgaris*, which occurs along the coast of the United States from Massachusetts to Texas, has been and continues to be a popular animal for use in studies of the screening pigments found in its compound eyes (Sandeen and Brown, 1952; Fingerman *et al.*, 1971). The movements of the distal retinal pigment, also commonly called the iris pigment, can easily be observed in intact prawns.

Although they are relatively old and need revising in some areas, the volumes by Waterman (1960, 1961) and Lockwood (1967) are excellent sources of much worthwhile information on the physiology of crustaceans.

COELENTERATES

Coelenterates are commonly considered to include two phyla of radially symmetrical animals, the Cnidaria and the Ctenophora. Ctenophores, commonly known as comb-jellies, are delicate, pelagic, marine animals, little studied and seldom if ever maintained in seawater systems away from seaside laboratories. Cnidarians are, with a few exceptions, also marine. In contrast to ctenophores, cnidarians are popular demonstration and experimental animals and are among the most common invertebrates to be found in marine aquaria.

There are three classes of Cnidaria: Hydrozoa, Scyphozoa, and Anthozoa and two basic body types—polyp and medusa. The class Hydrozoa includes hydroids, hydromedusae, and a few types of coral; Scyphozoa, most of the large jellyfish; Anthozoa, the sea anemones, sea pens, and most corals. Most medusae actively swim by virtue of periodic pulsations, whereas most polyps are sessile. Most medusae are difficult to maintain in aquaria; they abrade themselves against aquarium walls while swimming and their tentacles tend to get caught in aquarium filters. Polyps, on the other hand, particularly sea anemones and hydroids, do very well in laboratory cultivation (see Chapter 12). Indeed, the floral beauty of sea anemones and the ease with which they can be maintained have made these animals very popular with amateurs who keep marine aquaria, and most retailers handling marine aquaria and supplies normally carry some species of anemones as well as tropical marine fish.

All cnidarians are carnivorous. Newly hatched brine shrimp are appropriate food for most small cnidarians and for larger forms that are planktonic feeders. Larger animals, such as anemones, can be fed pieces of fish or clam.

The only predators of cnidarians likely to cause difficulty in culture or in a community aquarium are nudibranchs. Many species of nudibranchs feed voraciously on hydroids, sea anemones, and scyphozoan polyps. Tiny nudibranchs introduced inadvertently into a culture can grow surprisingly rapidly at the expense of cnidarians.

HYDROZOA

Hydrozoans occur both as polyps and medusae. Polyps typically consist of a contractile hydranth borne on an elongate, immobile stem; they have a mouth, often forming part of a mobile proboscis, and one or more sets of tentacles. Individual hydranths are generally small, ranging from less than a tenth of a millimeter to several millimeters in diameter. Most are colonial, new members being added to a colony by budding. All individuals in a colony are descendants of a single polyp and are of the same sex. There is a single enteron throughout the colony, through which food particles can be moved by ciliary currents.

Medusae are the sexual stages of hydrozoa and arise from buds (termed gonophores) produced by polyps or modified polyps. In some species, the gonophores develop into free-swimming medusae, but in many species they form only rudimentary medusae, which are not released.

Many kinds of hydrozoan polyps have been raised in the laboratory. In 1971 Davis listed 36 species that had been successfully cultured; the number has increased since that time. Some more recent information can be found in Brenowitz (1975). Among the species of experimental interest that have been successfully maintained are: *Bougainvillia carolinensis* (Wyttenback, 1973), *Clava multicornia* (Kinne and Paffenhöfer, 1965), *Cordylophora lacustris* (Fulton, 1960), *Hydractinia echinata* (Hauenschild, 1954; Stokes, 1974), *Phialidium gregarium* (Roosen-Runge, 1970), *Podocoryne carnea* (Braverman, 1962), *Sarsia tubulosa* (Kukinuma, 1966), *Proboscidactyla flavicirrata* (Campbell, 1966), and several species of *Campanularia* and *Obelia* (Crowell, 1953; Crowell and Wyttenback, 1957; Miller, 1966; Morin and Cooke, 1971). The original papers should be consulted for details on the culture methods. Not all hydroid polyps, however, are easily cultured. Despite one claim of successful maintenance (Mackie, 1966), many workers have found *Tubularia* difficult to keep for extended periods, which is unfortunate since *Tubularia* is a large genus and a favorite experimental animal in developmental and neurobiological studies.

Hydrozoan polyps are conveniently grown on glass microscope slides (Crowell, 1957). A colony can be initiated by inserting a portion of an old colony or a single polyp beneath a string or rubber band tied around the slide. The edges of the slide are less likely to cut the string or rubber band if the edges are first rounded by rubbing them lightly with a file or with fine sandpaper. Several glass slides with adhering colonies can be conveniently kept in a glass rack of the sort used for histologic staining. The colonies can then be fed by transferring the rack to a dish containing brine shrimp and cleaned by transferring the rack into fresh seawater.

Many species of hydroid polyps will produce medusae in culture, but in general these medusae are difficult to raise. Such successes as have been obtained are mostly at marine laboratories where there is a continual source of good-quality seawater and fresh, natural food (e.g., Rees and Russell, 1937; Brinckmann, 1962, 1964; Roosen-Runge, 1970). A few hydromedusae are relatively easy to raise, in particular those that swim feebly, sporadically, or not at all. *Eleutheria* medusae creep about on their tentacles and do not swim; these have been reared to sexual maturity in the laboratory (Weiler-Stolt, 1960). *Cladonema* polyps readily produce medusae that cling to surfaces and only occasionally swim. *Cladonema* polyps and medusae are readily cultured (Josephson and

Schwab, 1979) using the Crowell method described in the previous paragraph. This genus should provide excellent material for teaching laboratories or for experiments in which a dependable source of hydromedusae is required. The solitary polyps of the limnomedusan *Vallentinia* produce medusae that are capable of swimming but rarely do so. Both polyps and medusae of this species are easily maintained (Foster, 1973).

SCYPHOZOA

The life cycle of a typical scyphozoan includes both a polyp and a medusa stage; the former is termed a scyphistoma. Scyphistomae reproduce asexually by various budding processes and periodically produce medusae by transverse fission. Medusa production is promoted by lowering the holding temperature and, at least in *Aurelia*, by the presence of iodine in the surrounding seawater (Spangenberg, 1967, 1969). The newly formed medusae grow, mature, and produce the gametes that give rise to the next polyp generation.

Scyphistomae of several common scyphozoan species (*Aurelia, Cassiopeia, Chrysaora, Cyanea*) are readily maintained in laboratory culture; in fact, the vigorous growth of scyphistomae sometimes make them a nuisance in community aquaria. Raising medusae from scyphistomae takes a measure of care, but has been achieved for *Aurelia aurita* (Spangenberg, 1965; Abe and Hisada, 1969) and for *Cyanea capillata* and *Chrysaora quinquecirrha* (Cargo, 1975). The original papers contain details on the rearing procedures used. Of the jellyfish likely to be collected around the shores of the continental United States, *Cassiopeia xamachana* from Florida is perhaps the easiest to maintain in the laboratory. This jellyfish lies on the bottom with its oral side uppermost, beating slowly.

ANTHOZOA

Anthozoa are exclusively polypoid. Anemones (order Actiniaria) and cerianthids (order Ceriantharia) are solitary, but the remaining nine or so orders of the anthozoa, including several kinds of soft and hard corals, are exclusively or primarily colonial. Anemones and cerianthids generally do well in laboratory aquaria if customary care is taken as to water temperature, salinity, and feeding. Corals, on the other hand, are difficult to maintain in healthy condition. One exception is the northern coral *Astrangia*, which has been found to survive well on brine shrimp and even grow in laboratory aquaria (R. L. Pardy, personal communication).

Among the anemones commonly maintained in aquaria are *Metridium senile, Diadumene leucolena,* and *Haliplanella luciae* from the Atlantic coast; *Aiptasia* sp. from Florida; and *Anthopleura elegantissima, A. xanthogrammica, Tealia felina, Metridium senile*, and *Stomphia coccinea* from the Pacific coast. Several species of sea anemones reproduce asexually as well as sexually (Chia, 1976). In animals with asexual reproduction, stocks can be increased without the bother of special care for delicate larval stages, and clones of genetically identical animals can be obtained. *Aiptasia, Anthopleura elegantissima, Haliplanella*, and *Metridium* will reproduce asexually in the laboratory. Under favorable conditions the doubling time of *Haliplanella* is less than 10 days (Minasian, 1976), making this a particularly favorable anemone for some laboratory cultures. Specific and detailed culture methods for several anemones are contained in Chapter 12.

ANNELIDS

The marine annelids that are used in experimental biology are all members of the class Polychaeta. This class is abundant and worldwide. The polychaetes range in size from almost microscopic to individuals that are many centimeters long, and they occupy a variety of habitats. Many live in sediments either freely moving about or in formed tubes. Many others live in and around communities of larger animals and large algae, hiding among the various interstices. The sabellariids can form extensive communities of their own and sometimes produce large outcroppings or reefs by cementing sand grains together in the formation of their tubes.

The polychaetes played a key role in early developmental studies; their development is similiar to that of the molluscs. They have currently become important in the evaluation of marine pollution and serve as good indicator species. Some of this work is covered in Chapter 13.

Table 10–1 includes several annelids that can be maintained in the laboratory, and Chapter 13 describes the methods for rearing a number of different polychaetes through many generations. The techniques are simple, and highly inbred strains have been raised. There is an unrealized opportunity here for combining genetics and modern experimental procedures in the study of a large and important group of animals.

ECHINODERMS

All five classes in the phylum Echinodermata are marine. These are the Echinoidea (sea urchins), Asteroidea (sea stars), Ophiuroidea (serpent stars), Holothuroidea (sea cucumbers), and the Crinoidea (sea lilies). Only the first two classes have been used extensively in the laboratory. The sea urchins, upon which the bulk of the research has been done, and the sea stars have been used primarily as sources of eggs and sperm for studies of developmental and cell biology, giving rise to an enormous literature. Adult echinoderms have been used to a limited extent in biomedical research, for example, to study immunology (Karp and Hildeman, 1976; Coffaro and Hinegardner, 1977). The coelomocytes have been used to study cell motility (Edds, 1977).

Unlike the eggs of other echinoderms, and of most other animals for that matter, sea urchin eggs are haploid when spawned. As a consequence they are easily fertilized and used, since there need be no concern with the timing of meiosis. By contrast, sea star eggs complete meiosis after spawning and must be allowed to mature prior to fertilization. Maturation is stimulated by 1-methyladenine (Kanatani, 1969, and subsequent papers by that author).

Both sea urchins and sea stars can be maintained in the laboratory. Sea urchins are omnivores and will eat a wide variety of foods. Most, but not all, sea stars are carnivores; some of those with long arms feed by extending their stomachs around prey or pieces of food and sometimes even through minute cracks between shells of bivalve molluscs. The husbandry of sea urchins is discussed in detail in Chapter 19.

General references on echinoderms include Boolootian (1966), Hinegardner (1967), Nichols (1969), Binyon (1972), the 1975 issue of *American Zoologist,* and Czihak (1975). Books by Horstadius (1973), Giudice (1973), and Stearns (1974) cover various aspects of echinoderm development.

ECHINOIDEA

A number of sea urchin species adapt well to captivity. *Strongylocentrotus droebachiensis* grows along the northeastern and northwestern coastlines of the United States and Canada. Specimens can also sometimes be purchased in fish markets, where they are sold as food. They spawn in late winter. The disadvantage of this species is that the temperature maximum is about 12°C for the adults and 9°C for good embryonic development.

Arbacia punctulata grows subtidally along the coast from New England to Florida. It spawns during the summer north of Cape Hatteras and for a longer season farther south. Animals can be maintained at about 22°C or less, 15 to 18°C being optimum. The embryos will develop at these temperatures also. Harvey's book (1956) is still a good reference for this, as well as for other species of sea urchins.

Lytechinus variegatus can be obtained along the Florida and Gulf coasts. Adults and embryos will survive at room temperature. The spawning season is during the winter.

Lytechinus pictus grows subtidally from Panama to Southern California. The spawning season is during the summer. Both adults and embryos will survive up to a maximum temperature of 24°C. Temperatures between 15 and 18°C are recommended for both embryos and adults. A temperature below 8°C is lethal.

The purple urchin, *Strongylocentrotus purpuratus,* is an intertidal and subtidal species found in abundance along the West Coast of North America. Adults and embryos should be kept at 17°C or less. Animals spawn from fall to spring, with a longer season in their more southerly range.

There are two species of sand dollar, *Echinarachnius parma,* on the East Coast, and *Dendraster excentricus,* on the West Coast. Both are used in research, but they are more difficult to maintain in the laboratory than are sea urchins. Sea urchins will readily prey upon them if given an opportunity. Sand dollars are essentially flattened sea urchins and their development is basically the same as that of the urchins. The embryos, however, seem to be more sensitive to various chemicals.

There are a number of readily available sea urchin species in Hawaii. *Tripneustes gratilla* is one that has been extensively used. It grows at room temperature, and fertile animals can be found throughout the year.

All species named above, except the last, can be obtained from professional collectors and all (except the sand dollars) can be raised in the laboratory by techniques described in the literature or by slight modifications of these techniques.

ASTEROIDEA

The sea stars that most often serve as sources of eggs and sperm are: *Asterias forbesii,* a species common along the East Coast, and *Pisaster ochraceus* and *Patiria miniata,* on the West Coast. Other species have also been used. In general, sea stars are more difficult to maintain in the laboratory than are sea urchins, but the task is not insurmountable when attention is given to the details of the animal's requirements.

ASCIDIANS

Ascidians are distantly related to the vertebrates; both groups belong to the phylum Chordata. The Ascidiacea are one of three classes in the subphylum Urochordata (=Tunicata), along with the Larvacea and Thaliacea. The last two are pelagic and planktonic throughout their lives, fragile, and generally do not survive well in closed aquarium systems. All three classes have but limited tolerance to reduced salinity.

The ascidians are commonly referred to as sea squirts and include several species that are classic objects of laboratory study. Many spawn (or the gametes can be removed by dissection), and the fertilized eggs have been used in embryological studies. More complex forms of reproduction occur widely in this class, particularly among the compound ascidians, and the group has a number of unique biochemical characteristics. In addition, the circulatory system is unusual in that periodically the direction of blood flow reverses. There have been some genetic studies (Milkman, 1967) and some recent work on immunology (Scofield and Weissman, 1981). The larvae have been used in a variety of studies, some of which are reviewed by Cloney (1978).

All ascidians are suspension feeders, generally relying on phytoplankton and other small bits of particulate matter. Much like the bivalve molluscs, they have a pair of openings (siphons) that draw water through a filter-feeding system. Many of the difficulties associated with feeding bivalve molluscs also apply to the ascidians. In addition, ascidians usually attach themselves to an available hard surface and cannot be moved about as readily as can a clam. Several species show a preference for vigorously moving, extremely clean, unpolluted seawater. These are characteristics not achievable in laboratory aquaria, where high rates of clean water movement can be achieved only with extensive filtration, which would, of course, remove most particulate food. As a consequence, most experimental work with ascidians has been done with species that occur naturally in calm waters.

There are three more or less natural ascidian groupings: solitary, social, and compound animals. Some solitary individuals reach considerable size, occasionally more than 15 cm in length. Solitary forms are most amenable to manipulation in the laboratory, and genera such as *Ascidia, Ciona, Corella, Molgula, Pyura,* and *Styela* have all been used in research.

Social ascidians grow as colonies. The individual zooids are distinct from one another but still remain joined together at the base by stolons or an equivalent communicating network of tissue. The genus *Perophora* has been studied in the laboratory.

Colonies of compound ascidians consist of several individual zooids that share a common tunic and, in some species, communal incurrent siphons. Reproduction is both sexual and by budding; one colony usually contains several generations simultaneously. It is difficult to study the individual zooid of compound ascidians; generally only the properties of the colony are observed. *Botryllus* and *Botrylloides* are two genera that have been used in research. Often, if one or more individual zooids dies, the situation may prove to be contagious and bring about the rapid decline of the whole colony. Thus it is important to inspect the colonial forms for deterioration and remove any colonies that show signs of decline.

Solitary ascidians found in protected waters, e.g., harbors and enclosed embayments or coves, usually survive reasonably well in the laboratory, provided suitable temperature and salinity are maintained. *Molgula* is particularly hardy. Any of the common species may be fed by temporarily turning off the filter system and adding such suitable phytoplankton as *Monochrysis, Isochrysis, Dunaliella,* or similar algal species. Alternatively, if the animals can be readily moved they can be placed in a container of seawater enriched with the alga and allowed to feed for 30 to 60 min. Feeding every other day should be sufficient. Social and compound ascidians can be maintained and fed in the same way.

Ascidians have few predators, either in nature or in the aquarium. Certain sea stars may attack them and certain fish, e.g., wrasses, feed on them. A few snails (e.g., *Lamellaria*) are specifically adapted to graze on certain ascidians. Finally, a few ascidian species harbor such symbiotic organisms as clams, shrimp, or amphipods that, if they succumb to the stress of the aquarium, may either attack the tissues of the host or die and threaten its health.

In general, if ascidians are to be maintained in the laboratory, they should be so arranged that all individuals can be observed during daily inspection. If specimens become discolored or tend to float, they are probably dead or dying and should be removed to another container or discarded. Some ascidians are remarkably acidic, and their death may bring about a sharp reduction in the pH of the seawater. As a precaution, the pH should be periodically determined to assure that the buffer capacity of the aquarium has not been exceeded.

There are very few published directions for husbandry of ascidians. Some techniques for maintaining particular species and for obtaining and culturing their larvae can be found in individual research papers such as those mentioned earlier.

SIPUNCULANS

Sipuncula is a phylum of wormlike animals that vary from a few millimeters to 30 or more centimeters in length. They are generally thigmotropic and, in the field, are found in protected situations. Many species, particularly tropical forms, occupy burrows of their own making in calcareous rock. Other species are found under rocks, in the interstices of rocks, or in coarse gravel. Still others inhabit burrows in sand and mud. Table 10–2 lists some common intertidal species and indicates typical habitats for each. Thus far, sipunculans have not been used extensively as research animals; however, they are fairly easy to maintain and may offer certain advantages.

Sipunculans commonly feed by ingesting the surrounding substrata or, in the case of those with well-developed tentacles, on suspended particulate matter in the water. In an aquarium, sipunculans will feed on sand mixed with the organic debris that frequently accumulates in the bottom of a tank. Temperature in the aquarium should approximate the mean temperature of the water from which the specimens were collected. Intertidal species from the coast of California and the Pacific Northwest should be maintained within a temperature range of 10–15°C. Species from the northeastern coast of the Atlantic should be kept at temperatures of 15–17°C, whereas southeastern and Caribbean species do well at ordinary room temperatures. For most sipunculans, the salinity in the aquarium should be full-strength seawater, since very few species are found in estuaries.

Sipunculans are relatively defenseless. If they are to be kept in a general aquarium with other animals, care must be taken to protect them from predators. Unless offered some protective cover, it is likely they will be devoured by crabs, urchins, anemones, certain gastropods, or fish. If sand or rocks are provided, the sipunculans will remain well hidden and can survive in a tank with a variety of other species.

For periods of a few weeks, most sipunculans will remain apparently healthy, will continue to spawn, and will behave in a generally normal manner without substratum or food. They may be kept in finger bowls, either submerged in a continuously flowing or recirculating seawater system, or in free-standing water that is changed daily. The sand-burrowing species, however, are the most difficult to maintain under these conditions: in the prolonged absence of substratum, they will suffer from blistering and rupturing of the cuticle. Under these adverse conditions, species of *Siphonosoma* will undergo a series of transverse constrictions that eventually rupture and kill the animal. Species associated with rocky habitats usually have thicker cuticles than do the sand-burrowers and remain healthy for longer periods in the laboratory in the absence of substratum. In fact, they may survive as long as a year, kept only in finger bowls in which the water is frequently changed. Under these conditions, however, the animals may be covered with overgrowths of diatoms, lose weight, and shrink. Experimental studies of biochemical

TABLE 10–2 Common Intertidal Sipunculans of the North American Coast

Species	Locality	Habitat
Aspidosiphon brocki	Florida, Caribbean	Burrows in calcareous rock
Golfingia hespera	Monterey Bay to Gulf of California	Mud flats among *Phyllospadix* roots; associated with *Mesochaetopterus* and *Cerianthus*
Golfingia pugettensis	San Juan Islands and Puget Sound	In muddy sand
Phascolion cryptus	Florida	Emply gastropod shells in sand
Phascolopsis gouldi	Northeastern Atlantic Coast	Muddy sand
Phascolosoma agassizii	Alaska to Baja California	Under rocks, in crevices, in gravel, among *Phyllospadix* roots and byssal threats of mussels
Phascolosoma antillarum	Florida, Caribbean, Gulf of California to Panama	Burrows in calcareous rock
Phascolosoma perlucens	Florida, Caribbean, Gulf of California to Panama	Burrows in calcareous rock
Siphonosoma cumanense	Florida, Caribbean	Sand
Sipunculus nudus	Southeastern Atlantic Coast, Caribbean, Monterey Bay to Panama	Sand
Themiste dyscrita	Oregon to Point Conception, California	Under rocks, among *Phyllospadix* roots, in pholad holes
Themiste pyroides	Washington to Baja California	Under rocks, rock crevices
Themiste zostericola	Point Conception to Baja California	In sand and mud, under boulders

SOURCES: Gerould, 1913; Fisher, 1952; Stephen and Edmonds, 1972; Smith and Carlton, 1975.

changes during starvation have been carried out by Wilber (1947) and Towle and Giese (1966). If removed from their burrows, many of the rock-boring species will in time undergo considerable swelling, but otherwise appear in good health.

General references on the systematics and biology of sipunculans include Hyman (1959), Stephen and Edmonds (1972), Rice (1967, 1976), and Rice and Todorovic (1975, 1976). There are no good long-term culture procedures.

LITERATURE CITATIONS TO MARINE INVERTEBRATES CULTURED IN THE LABORATORY

This is a selective list of references, almost all of which contain information specifically directed toward the maintenance and culture of marine, multicellular invertebrates. A helpful introduction to practically every group may be found in Kinne (1977).

GENERAL REFERENCES

Costello, D. P., and C. Henley, 1971
Hauenschild, C., 1962, 1970, 1972
Kinne, O,. 1977
Kinne, O., and H. P. Bulnheim, 1970
Needham, J. G., *et al.*, 1937
Paffenhöfer, G. A., and R. P. Harris, 1979
Reeve, M. R., 1977

PORIFERA (sponges)

Arndt, W., 1933
Fell, P. E., 1967, 1976, 1977
Wilson, H. V., 1937

CNIDARIA (coelenterates)

Chia, F., and L. R. Bickell, 1978
Crowell, S., 1967
Lenhoff, H. M., 1971

HYDROZOA (hydroids)

Brenowtiz, A. H., 1975
Davis, L. V., 1971
Rees, W. J., and F. S. Russell, 1937
Toth, S. E., 1965
Werner, B., 1968

Bougainvillia
Reed, S. A., 1971
Tusov, J., and L. V. Davis, 1971

Campanularia flexuosa
Brock, M. A., 1975
Crowell, S., 1953

Clytia attenuata
West, D. L., and R. W. Renshaw, 1970

Eirene viridula
Bierbach, M., and D. K. Hofmann, 1973

Eucheilota maculata
Werner, B., 1968

Eutonina indicans
Rees, J. T., 1978

Hydractinia echinata
Ivker, F. B., 1972

Microhydrula pontica
Spoon, D. M., and R. S. Blanquet, 1978

Pennaria tiarella
Pardy, R. L., 1971
Rees, W. J., 1971

Phialidium gregarium
Roosen-Runge, E. C., 1970
Worthman, S. G., 1974

Podocoryne carnea
Braverman, M. H., 1962, 1971, 1974

Siphonophora
Mackie, G. O., and D. A. Boag, 1963

Spirocodon saltatrix
Ikegami, S., *et al.*, 1978

Tubularia
Miller, R. L., 1976
West, D. L., and R. W. Renshaw, 1970

Vallentinia gabriellae
Foster, N. R., 1973

CUBOZOA (sea wasps)

Carybdea alata
Arneson, A. C., and C. E. Cutress, 1976

Tripedalia cystophora
Werner, B., 1975

SCYPHOZOA (true jellyfish)

Aurelia aurita
Abe, Y., and M. Hisada, 1969
Groat, C. S., *et al.*, 1980
Schwab, W. E., 1977
Silverstone, M., *et al.*, 1977
Spangenberg, D. B., 1965, 1969

Cassiopeia
Hill, S. D., and J. N. Cather, 1969

Chrysaora quinquecirrha
Cargo, D. G., 1975
Loeb, M. J., 1973

Cyanea capillata
Cargo, D. G., 1975

Dactylometra pacifica
Kakinuma, Y., 1967
Rhopilema verrilli
Calder, D. R., 1973
Cargo, D. G., 1975
ANTHOZOA (sea anemones and corals)
Actinia equina
Chia, F., and M. A. Rostron, 1970
Haliplanella luciae
Johnson, L. L., and J. M. Shick, 1977
Minasian, L. L., Jr., 1976
Pocillopora damicornis
Reed, S. A., 1971
Ptilosarcus guerneyi
Chia, F., and B. J. Crawford, 1973

CTENOPHORA (comb-jellies)
Greve, W., 1970
Walter, M. A., 1977
Ward, W. W., 1974
Mnemiopsis
Baker, L. D., and M. R. Reeve, 1974
Pleurobrachia
Greve, W., 1968, 1972
Hirota, J., 1972, 1974

TURBELLARIA (free-living flatworms)
Apelt, G., 1969
Convoluta roscoffensis
Provasoli, L., *et al.* 1968
Mesotoma productum
Heitkamp, U., 1972
Monocelis
Giesa, S., 1966
Parotocelis luteola
Kozloff, E. N., 1969

NEMERTEA (proboscis worms)
Coe, W. R., 1937
Ferraris, J. D., 1978
Gontcharoff, M., 1959

ROTIFERA
Solangi, M. A., and J. T. Ogle, 1977
Brachionus plicatilis
Droop, M. R., and J. M. Scott, 1978
Hino, A., and R. Hirano, 1976, 1977
Hirata, H., 1974
Lubzens, E., *et al.*, 1980
Theilacker, G. H., and M. F. MacMaster, 1971

NEMATODA (roundworms)
Lee, J. J., *et al.*, 1970
Lee, J. J., and W. A. Muller, 1975
Nicholas, W. L., 1975
Rothstein, M., and W. L. Nicholas, 1969
Thun, W. von, 1966
Tietjen, J. H., and J. J. Lee, 1977a
Aphelenchoides
Meyers, S. P., *et al.*, 1963
Chromadorina germanica
Tietjen, J. H., and J. J. Lee, 1977b
Chromadora macrolaimoides
Tietjen, J. H., and J. J. Lee, 1973
Deontostoma californicum
Viglierchio, D. R., and R. N. Johnson, 1971
Diplolaimella schneideri
Chitwood, B. G., and D. G. Murphy, 1964
Enoplus paralittoralis
Hopper, B. E., and R. C. Cefalu, 1973
Metoncholaimus
Hopper, B. E., and S. P. Meyers, 1966
Monhystera
Chitwood, B. G., and D. G. Murphy, 1964
Gerlach, S. A., and M. Schrage, 1971
Tietjen, J. H., 1967
Tietjen, J. H., and J. J. Lee, 1972
Oncholaimus oxyuris
Heip, C., *et al.*, 1978
Prochromadora orleji
Bergholz, E., and U. Brenning, 1978
Rhabditis marina
Bergholz, E., and U. Brenning, 1978
Tietjen, J. H., *et al.*, 1970
Theristus pertenuis
Gerlach, S. A., and M. Schrage, 1971

PRIAPULIDA
Priapulus caudatus
Lang, K., 1948

MOLLUSCA
POLYPLACOPHORA (chitons)
Pearse, J. S., 1979
Watanabe, J. M., and L. R. Cox, 1975

GASTROPODA
  Mason, C. F., 1977
Prosobranchia (snails and limpets)
  D'Asaro, C. N., 1970
  McKillup, S. C., 1979
  Moore, D., 1960
Anachis avara
  Scheltema, R. S., and A. H. Scheltema, 1963
Batillaria minima
  Wagner, A., 1960
Bedeva paivae
  Black, J. H., 1976
Charonia lampas
  Latigan, M. J., 1976
Concholepas concholepas
  Castilla, J. C., and J. Cancino, 1976
Conus
  Cruz, L. J., *et al.*, 1978
  Perron, F. E., 1980
Crepidula fornicata
  Pilkington, M. C., and V. Fretter, 1970
  Silberzahn, N., 1977
Haliotis
  Kan-no, H., 1976
  Kikuchi, S., and N. Uki, 1974a,b
  Koike, Y., 1978
  Leighton, D. L., 1972
  Morse, D. E., *et al.*, 1977, 1978, 1979a,b
Haliotus (Sulculus) diversicolor
  Nishimura, K., *et al.*, 1969
Littorina
  Chu, G. W. T. C., and E. P. Ryan, 1960
  Struhsaker, J. W., and J. D. Costlow, Jr., 1969
Nassarius
  Pilkington, M. C., and V. Fretter, 1970
  Scheltema, R. S., 1962
  Vernberg, W. B., and F. J. Vernberg, 1975
Patella
  Dodd, J. M., 1957
Strombus
  Brownell, W. N., 1978
Urosalpinx cinerea
  Federighi, H., 1937
Opisthobranchia (sea slugs and sea hares)
  Franz, D. R., 1975
  Poizat, C., 1972
Alderia modesta
  Seelemann, U., 1967
Aplysia
  Audesirk, T. E., 1977
  Capo, T. R., *et al.*, 1979
  Carefoot, T. H., 1979, 1980
  Kandel, E. R., 1979
  Kriegstein, A. R., *et al.*, 1974
  Stephens, L. L., and J. E. Blankenship, 1974
  Strenth, N. E., and J. E. Blankenship, 1978
  Switzer-Dunlap, M., and M. G. Hadfield, 1977
Clione limacina
  Conover, R. J., and C. M. Lalli, 1972
Doridella obscura
  Perron, F. E., and R. D. Turner, 1977
Elysia chlorotica
  Harrigan, J. F., and D. L. Alkon, 1978a
Haminoea solitaria
  Harrigan, J. F., and D. L. Alkon, 1978a
Hermissenda crassicornis
  Harrigan, J. F., and D. L. Alkon, 1978b
Paedocline doliiformis
  Lalli, C. M., and R. J. Conover, 1973
Phestilla
  Harris, L. G., 1975
Phyllaplysia taylori
  Bridges, C. B., 1975
Rostanga pulchra
  Chia, F. S., and R. Koss, 1978
Tornatina (Acteocina) caniculata
  Franz, D. R., 1971
Tritonia diomedea
  Kempf, S. C., and A. O. D. Willows, 1977
SCAPHOPODA (tusk shells)
  Davis, J. D., 1977
PELECYPODA (bivalves)
  Calabrese, A., and H. C. Davis, 1970
  Chanley, P., 1975
  Culliney, J. L., *et al.*, 1975
  Loosanoff, V. L., 1954
  Loosanoff, V. L., and H. C. Davis, 1963
  Morse, D. E., *et al.*, 1978
  Rhodes, E. W., *et al.*, 1975
  Sagara, J., 1958
  Walne, P. R., 1964, 1974a
Arctica islandica
  Landers, W. S., 1976
Argopecten gibbus
  Costello, T. J., *et al.*, 1973

Bankia
Culliney, J. L., 1975
Townsley, P. M., *et al.*, 1966
Crassostrea
Epifanio, C. E., and C. A. Mootz, 1976
Hidu, H., 1975
Stiles, S. S., 1978
Veitch, F. P., and H. Hidu, 1971
Lyrodus pedicellatus
Board, P. A., and M. J. Feaver, 1973
Pechenik, J. A., *et al.*, 1979
Mercenaria mercenaria
Carriker, M. R., 1961
Hartman, M. C., *et al.*, 1974
Walne, P. R., 1974b
Mulinia lateralis
Calabrese, A., 1969
Rhodes, E. W., *et al.*, 1975
Mytilus
Bayne, B. L., 1965, 1976
Bayne, B. L., and R. J. Thompson, 1970
Brenko, M. H., 1973
Brenko, M. H., and A. Calabrese, 1969
Morse, D. E., *et al.*, 1977
Riley, J. D., 1973
Ostrea
Dix, T. G., 1976
Helm, M. M., 1977
Helm, M. M., and B. E. Spencer, 1972
Hidu, H., 1975
Millar, R. H., and J. M. Scott, 1968
Walne, P. R., 1970, 1974b
Panope generosa
Goodwin, L., *et al.*, 1979
Pecten maximus
Comley, C. A., 1972
Gruffydd, Ll. D., and A. R. Beaumont, 1972
Saxidomus giganteus
Breese, W. P., and F. D. Phibbs, 1970
Spisula solidissima
Cable, W. D., and W. S. Landers, 1974
Rhodes, E. W., *et al.*, 1975
Tagelus plebeius
Chanley, P., and M. Castagna, 1971
Teredo
Culliney, J. L., 1975
Karande, A. A., and S. S. Pendsey, 1969
Morton, B., 1978
Tridacna
La Barbera, M., 1975

CEPHALOPODA
Nautiloidea
Nautilus macromphalus
Hamada, T., and S. Mikami, 1977
Kanie, *et al.*, 1979
Kawamoto, N., 1978
Mikami, S., and T. Okutani, 1977
Coleoidea
Boletzky, S. von, 1974a,b
Grimpe, G., 1933
Decapoda (squid and cuttlefish)
Choe, S., 1966
Choe, S., and Y. Ohshima, 1963
Euprymna scolopes
Arnold, J. M., *et al.*, 1974
Illex
Boletzky, S. von, *et al.*, 1973
O'Dor, R. K., 1978
O'Dor, R. K., *et al.*, 1977
Loligo
Hanlon, R. T., *et al.*, 1978
Loligo opalescens
Hanlon, R. T., *et al.*, 1979
Hurley, A. C., 1976
Loligo pealei
Arnold, J. M., *et al.*, 1972
Summers, W. C., *et al.*, 1974
Summers, W. C., and J. J. McMahon, 1974
Loligo plei
Hanlon, R. T., 1979
Ommastrephes sloani
Hamabe, M., 1963
Sepia
Boletzky, S. von, 1975
Ohshima, Y., and S. Choe, 1961
Pascual, E., 1978
Richard, A., 1976
Sepiola
Boletzky, S. von, *et al.*, 1971
Sepioteuthis
LaRoe, E. T., 1971, 1973
Ohshima, Y., and S. Choe, 1961
Todarodes pacificus
Flores, E. E. C., *et al.*, 1976, 1977
Mikulich, L. V., and L. P. Kozak, 1972
Octopoda (octopuses)
Hapalochlaena maculosa
Tranter, D. J., and O. Augustine, 1973
Octopus
Taki, I., 1941

Octopus briareus
Hanlon, R. T., 1977
Octopus dofleini
Gabe, S. H., 1975
Octopus joubini
Boletzky, S. von, and M. V. von Boletzky, 1969
Bradley, E. A., 1974
Forsythe, J. W., and R. T. Hanlon, 1980
Mather, J. A., 1978
Opresko, L., and R. F. Thomas, 1975
Thomas, R. F., and L. Opresko, 1973
Octopus maya
Van Heukelem, W. F., 1977
Walker, J. J., *et al.*, 1970
Octopus tetricus
Joll, L. M., 1976
Octopus vulgaris
Hirayama, K., 1966
Itami, K., *et al.*, 1963
Wells, M. J., and J. Wells, 1972
Wodinsky, J., 1971

ECHIURA
Balzer, F., 1933
Pilger, J., 1978
Urechis caupo
Gould, M. C., 1967
Gould-Somero, M., 1975

SIPUNCULA
Rice, M.E., 1978

POLYCHAETA (bristle worms)
Dean, D., 1977
Dean, D., and M. Mazurkiewicz, 1975
Fauchald, K., and P. A. Jumars, 1979
Hauenschild, C., 1970
Reish, D. J., 1974
Reish, D. J., and T. L. Richards, 1966
Arenicola marina
Farke, H., and E. M. Berghuis, 1979
Autolytus
Schiedges, K. L., 1979
Capitella
George, J. D., 1976
Reish, D. J., 1974, 1977a
Dinophilus gyrociliatus
Traut, W., 1968
Hydroides
Grave, B. H., 1937b
Vuillemin, S., 1968
Laeonereis culveri
Mazurkiewicz, M., 1975
Nereis (Neanthes)
Goerke, H., 1971a,b, 1979
Guberlet, J. E., 1937
Kay, D. G., and A. E. Brafield, 1973
Reish, D. J., 1953, 1954
Ophryotrocha labronica
Åkesson, B., 1970
Perinereis nuntia
Yoshida, S., 1976
Polydora
Blake, J. A., 1969
Day, R., *et al.*, 1979
Polyophthalmus pictus
Guérin, J. P., 1971
Pomatoceros triquerter
Fφyn, B., and I. Gjφen, 1954
Segrove, F., 1941
Proceraea cornuta
Hamond, R., 1974
Sabellaria
Eckelbarger, K. J., 1975
Wilson, D. P., 1970
Scolelepis
Guérin, J. P., 1973
Guérin, J. P., and J. P. Reys, 1978
Stauronereis rudolphi
Richards, T. L., 1967

XIPHOSURIDA
LIMULIDAE (horseshoe crabs)
Limulus polyphemus
French, K. A., 1979
Neff, J. M., and C. S. Giam, 1977

CRUSTACEA
BRANCHIOPODA
Onbé, T., 1974, 1977
Alona taraporevalae
Shirgur, G. A., and A. A. Naik, 1977
Artemia
Bowen, S. T., 1962
Dempster, R. P., 1953

Person Le-Ruyet, J., 1976
Persoone, G., and P. Sorgeloos, 1972
Provasoli, L., and A. D'Agostino, 1969
Seki, H., 1966
Sorgeloos, P., *et al.*, 1976, 1977b
Sorgeloos, P., and G. Persoone, 1973, 1975
Tomey, W. A., 1978

Penilia avirostris
Takami, A., and H. Iwasaki, 1978
Takami, A., *et al.*, 1978

COPEPODA
Battaglia, B., 1970
Gonzalez, J. G., *et al.*, 1975
Nassogne, A., 1970
Omori, M., 1973
Paffenhöfer, G. A., and R. P. Harris, 1979
Zillioux, E. J., 1969
Zillioux, E. J., and N. F. Lackie, 1970

Asellopsis intermedia
Hardy, B. L. S., 1978

Acartia
Corkett, C. J., 1968
Gentile, J. H., *et al.*, 1974
Gentile, J. H., and S. L. Sosnowski, 1978
Heinle, D. R., 1969, 1970
Nassogne, A., 1970
Parrish, K. K., and D. F. Wilson, 1978
Person Le-Ruyet, J., 1975
Zillioux, E. J., and D. F. Wilson, 1966

Calanus
Corkett, C. J., 1970
Mullin, M. M., and E. R. Brooks, 1973
Paffenhöfer, G. A., 1970

Centropages
Lawson, T. J., and G. D. Grice, 1970
Person Le-Ruyet, J., 1975

Eurytemora
Heinle, D. R., 1969
Heinle, D. R., *et al.*, 1977
Katona, S. K., 1970
McLaren, I. A., 1976

Euterpina
Haq, S. M., 1972
Nassogne, A., 1969
Neunes, H. W., and G. F. Pongolini, 1965
Zurlini, G., *et al.*, 1978

Gladioferens imparipes
Takano, H., 1971b

Labidocera
Barnett, A. M., 1974
Gibson, V. R., and G. D. Grice, 1977

Laophonte setosa
Goswami, S. C., 1977

Lerneaenicus radiatus
Shields, R., 1977

Nitocra typica
Lee, J. J., *et al.*, 1976

Oithona nana
Murphy, H. E., 1923

Pseudocalanus
Corkett, C. J., 1967
Corkett, C. J., and D. L. Urry, 1968
Katona, S. K., and C. F. Moodie, 1969
Paffenhöfer, G. A., and R. P. Harris, 1976

Rhincalanus nasutus
Mullin, M. M., and E. R. Brooks, 1967

Scottolana canadensis
Harris, R. P., 1977
Heinle, D. R., *et al.*, 1977

Temora longicornis
Corkett, C. J., 1967
Harris, R. P., and G. A. Paffenhöfer, 1976
Person Le-Ruyet, J., 1975

Tigriopus
Nassogne, A., 1969
Takano, H., 1971a

Thompsonula hyaenae
Sellner, B. W., 1976

Tisbe
Barr, M. W., 1969
Battaglia, B., 1970
Betouhim-El, T., and D. Kahan, 1972
Hoppenheit, M., 1975
Rieper, M., 1978

CIRRIPEDIA (barnacles)

Balanus
Costlow, J. D., Jr., and C. G. Bookhout, 1957
Hirano, R., 1962
Kalyanasundaram, N., and S. S. Ganti, 1975, 1976
Karande, A. A., and M. K. Thomas, 1971

Landau, M., and A. D'Agostino, 1978
Landau, M., *et al.*, 1979
Molenock, J., and E. D. Gomez, 1972

Elminius modestus
Crisp, D. J., and P. A. Davies, 1955
Moyse, J., 1960
Tighe-Ford, D. J., *et al.*, 1970
Wisely, B., 1960

Octolasmis
Colón-Urban, R., *et al.*, 1979

Pollicipes polymerus
Lewis, C. A., 1975

Tetraclita serrata
Griffiths, R. J. I., 1979

MYSIDACEA

Metamysidopsis elongata
Clutter, R. I., and G. H. Theilacker, 1971

Mysidopsis bahia
Nimmo, D. R., *et al.*, 1978a

ISOPODA

Crinoniscus equitans
Bocquet-Védrine, J., 1974

Limnoria tripunctata
Menzies, R. J., 1972
Morton, B., 1978
Parrish, K. M., and J. D. Bultman, 1978

Sphaeroma quadridentatum
Lee, W. Y., 1977

AMPHIPODA
Yayanos, A. A., 1978

Amphithoe valida
Lee, W. Y, 1977

Calliopius laeviusculus
Dagg, M. J., 1976

Corophium insidiosum
Nair, K. K. C., and K. Anger, 1979a

Gammarus
Borowsky, B., 1980
Bulnheim, H. P., 1978
Hartnoll, R. G., and S. M. Smith, 1978

Hyperoche medusarum
Westernhagen, H. von, 1976

Jassa falcata
Nair, K. K. C., and K. Anger, 1979a

Microdeutopus gryllotalpa
Myers, A. A., 1971

Parathemisto gaudichaudi
Sheader, M., 1977

EUPHAUSIACEA (krill)
Komaki, Y., 1966
Lasker, R., and G. H. Theilacker, 1965
Murano, M., *et al.*, 1979

Nematoscelis difficilis
Gopalakrishnan, K., 1973

Nyctiphanes couchii
Le Roux, A., 1973

DECAPODA
Provenzano, A. J., Jr., 1967
Rice, A. L., and D. I. Williamson, 1970
Roberts, M. H., Jr., 1975
Sandifer, P. A., *et al.*, 1974b

NATANTIA (shrimps and prawns)
Dobkin, S., 1969
Sandifer, P.A., and T. I. J. Smith, 1979
Wickins, J. F., 1976b

Alpheus heterochaelis
Knowlton, R. E., 1973

Crangon crangon
Meixner, R., 1966

Leander
Albrechtsen, K., 1979
Schulte, E. H., 1976

Palaemon
Campillo, A., 1976
Fincham, A. A., 1977
Reeve, M. R., 1969
San Feliu, J. M., *et al.*, 1976
Wickins, J. F., 1972

Palaemonetes
Hall, L. W., Jr., and A. L. Buikema, Jr., 1977
Sandifer, P. A., *et al.*, 1975
Tyler-Schroeder, D. B., 1978a

Pandalus
Modin, J. C., and K. W. Cox, 1967
Paul, A. J., *et al.*, 1979

Penaeus
Beard, T. W., *et al.*, 1977
Brown, A., *et al.*, 1979
Cook, H. L., 1969
Ewald, J. J., 1965
Hirata, H., 1975
Hirata, H., *et al.*, 1978
Hudinaga, M., 1942
Lumare, F., 1976
Missler, S. R., 1980
San Feliu, J. M., *et al.*, 1973, 1976
Venkataramiah, A., *et al.*, 1976
Wear, R. G., and J. A. Santiago, 1976

Thor amboinensis
Sarver, D., 1979
REPTANTIA (crabs and lobsters)
Buchanan, D. V., *et al.,* 1975
Costlow, D. J., Jr., and C. G. Bookhout, 1960a
Ingle, R. W., and P. F. Clark, 1977
Sastry, A.N., 1970
Callinectes sapidus
Cook, D.W., 1972
Miller, M. R., 1978
Sulkin, S. D., *et al.,* 1976
Sulkin, S. D., and C. E. Epifanio, 1975
Winget, R. R., *et al.,* 1973
Cancer
Fisher, W. S., and R. T. Nelson, 1978
Hartman, M. C., 1977
Reed, P. H., 1969
Sastry, A. N., 1970
Carcinus
Dries, M., and D. Adelung, 1976
Williams, B. G., 1968
Homarus americanus
Boghen, A. D., and J. D. Castell, 1979
Chanley, M. H., and O. W. Terry, 1974
Gallagher, M. L., and W. D. Brown, 1976
Gilgan, M. W., and B. G. Burns, 1976
Hughes, J. T., *et al.,* 1974, 1975
Sastry, A. N., 1976
Serfling, S. A., *et al.,* 1974
Trider, D. J., *et al.,* 1979
Hyas araneus
Anger, K., and K. K. C. Nair, 1979
Mictyris longicarpus
Quinn, R. H., and D. R. Fielder, 1978
Nephrops norvegicus
Figueiredo, M. J. de, and M. H. Vilela, 1972
Pagurus samuelis
Coffin, H. G., 1958
Panulirus
Chittleborough, R. G., 1976
Dexter, D. M., 1972
Serfling, S. A., and R. F. Ford, 1975
Uchida, T., and Y. Dotsu, 1973
Rhithropanopeus harrisii
Frank, J. R., *et al.,* 1975
Sulkin, S. D., and L. L. Minasian, 1973
Sulkin, S. D., and K. Norman, 1976
Sulkin, S. D., and D. L. Pickett, 1973
Scylla serrata
Brick, R. W., 1974
Sesarma cinereum
Costlow, J. D., Jr., *et al.,* 1960
INSECTA
HEMIPTERA
Gerridae (water-striders)
Halobates
Andersen, N. M., and J. T. Polhemus, 1976
Herring, J. L., 1961
BRYOZOA
Abbot, M. B., 1975
Jebram, D., 1977a,b, 1980
Winston, J. E., 1977
Bugula
Grave, B. H., 1937a
Conopeum tenvissimum
Winston, J. E., 1976
Selenaria
Cook, P. L., and P. J. Chimonides, 1978

BRACHIOPODA
McCammon, H. M., 1972, 1975

POGONOPHORA
Southward, E. C., and A. J. Southward, 1977

CHAETOGNATHA (arrow worms)
Sagitta
Dallot, S., 1968
Greve, W., 1968
Reeve, M. R., 1970
Reeve, M. R., and M. A. Walter, 1972

ECHINODERMATA
Ruggieri, G. D., 1975b
Strathmann, R. R., 1975
ASTEROIDEA (sea stars)
Kanatani, H., 1969
Kanatani, H., and H. Shirai, 1971
Acanthaster planci
Henderson, J. A., and J. S. Lucas, 1971
Lucas, J. S., and M. M. Jones, 1976
Asterias forbesii
Fuseler, J. W., 1973
Mediaster aequalis
Birkeland, C., *et al.,* 1971
Stichaster australis
Barker, M. F., 1979

ECHINOIDEA (sea urchins)
de Angelis, E., 1977a,b
Hinegardner, R. T., 1969, 1975a,b
Osanai, K., 1975
Tyler, A., and B. S. Tyler, 1966
Arbacia
Bernhard, M., 1957
Cameron, R. A., and R. T. Hinegardner, 1974
Evechinus chloroticus
Dix, T. G., 1969, 1970
Lytechinus
Cameron, R. A., and R. T. Hinegardner, 1974
Fuseler, J. W., 1973
Mazur, J. E., and J. W. Miller, 1971
Strongylocentrotus purpuratus
Leahy, P. S., *et al.*, 1978

TUNICATA
ASCIDIACEA (sea squirts)
Berrill, N. J., 1937
Goodbody, I., 1977
Grave, C., 1937
Botryllus
Milkman, R., 1967
Halocynthia roretzi
Numakunai, T., 1965
Symplegma reptans
Nakauchi, M., *et al.*, 1979
THALIACEA (salps)
Thalia democratica
Braconnot, J. C., 1963
Heron, A. C., 1972
LARVACEA (appendicularians)
Oikopleura
Fenaux, R., 1976
Fenaux, R., and G. Gorsky, 1979
Paffenhöfer, G. A., 1973

11

Larvae

For nearly a century embryos and larvae of marine invertebrates have been widely used by research biologists as experimental material. Indeed, many modern concepts of developmental biology have emerged from studies of these early developmental stages. Recent research on larvae and early development of selected groups of invertebrates, particularly the coelenterates, echinoderms, and spiralians (primarily the mollusca), has been reported in a series of symposia sponsored and published by the American Society of Zoologists (1974, 1975, 1976). General information on the biology and morphology of larvae can be found in such comprehensive references as Vannucci (1959), Kumé and Dan (1968), Reverberi (1971), Anderson (1973), and in a series of volumes by Giese and Pearse (1974, 1975a,b, 1977, 1979).

The development of some marine invertebrates is direct, the embryo gradually taking on adult features; more commonly, however, there are one or more larval stages. Larvae of certain species are brooded by the parent, whereas others develop independently within an egg capsule and hatch out in a juvenile form. However, for the majority of bottom-dwelling marine invertebrates there is, at some time during the life-cycle, a free-swimming larval stage that may differ markedly from the adult in morphology, habitat, nutrition, and behavioral and physiological responses. The duration of the free-swimming stage varies, from species to species, from a few hours or days to several months for larvae that are capable of feeding. The latter ingest microscopic plankton.

The larval stage ceases with the onset of metamorphosis, which is often abrupt. In the course of metamorphosis, larval organs are lost, and the larva settles on the substrate, assuming the morphological features and mode of existence of the adult. Settlement is often the most critical stage in the life cycle and may well depend on the presence of a suitable substrate and on a variety of endogenous and exogenous interacting factors that are highly specific for individual species. In the absence of appropriate conditions, metamorphosis may be delayed for some time or not occur at all.

For species that are confined during their adult existence to the floor of the sea or to various intertidal or subtidal substrata, a pelagic larval stage contributes to dispersal and geographic relocation and facilitates genetic interchange among populations. The pelagic larva is subject to many environmental hazards, and, out of the huge number usually produced, few survive to become adults. Because they are sensitive to environmental pollutants and can often be cultured in large numbers, larvae have become increasingly used in monitoring environmental quality and for testing the effects of various toxicants and pollutants. Such groups as polychaete worms, bivalve molluscs, and certain crustaceans have received the most attention, the last two because of their commercial importance. Studies of marine invertebrate larvae have also figured prominently in advances in various aspects of experimental biology. For example, the sea urchin larva has been widely used in elucidating fundamental processes in development and molecular biology and the ascidian larva has figured prominently in studies of cell contraction.

Techniques for culturing the larvae of a number of marine invertebrates are now available. Many can be reared on a diet of single-celled algae. Some of the culturing techniques are described in Chapters 12–19.

POLYCHAETE LARVAE

The typical larva of polychaete worms is known as the trochophore. Usually oval or biconical in shape, it varies widely. An equatorial band of prominent locomotory cilia, the prototroch, passes just above the mouth and divides the larva into anterior and posterior spheres (Figure 11–1). Frequently there are additional ciliary bands: a telotroch, which encircles the posterior end of the larva in the region of the anus; and a midventral band, the neurotroch, which runs anteriorly from the telotroch to the mouth. Some larvae may also have a metatroch that encircles the larva just posterior to the mouth. In the anterior hemisphere (episphere), there is a pair of eye spots and at the apex an apical tuft

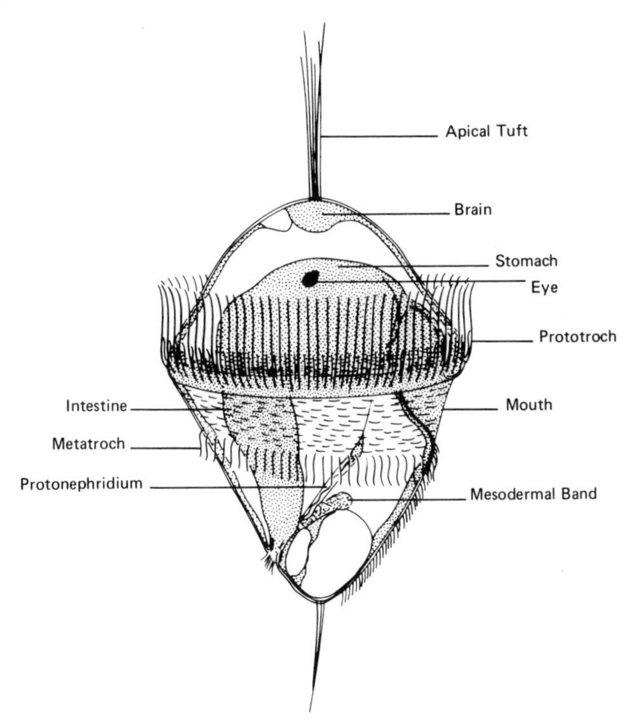

FIGURE 11–1 Trachophore larva of polychaetes. (From Segrove, 1941.)

of long cilia emerging from an underlying sensory plate. Nerves radiate posteriorly from the apical plate to join a nerve ring on the inside of the prototroch. The ventral mouth, posterior to the prototroch, opens to the digestive tract, which consists of an esophagus, expanded stomach, and narrow intestine, terminating at the posterior anus. Located within the blastocoelic body cavity on either side of the anus is a pair of ventral mesodermal bands that later split to form the coelomic cavity. A pair of larval excretory organs, the protonephridia, which open by ventral pores to the exterior, may be present. Before undergoing metamorphosis, the trochophore often passes through several intermediate stages characterized by additional bands and rows of cilia and an elongation of the posterior hemisphere to form several segments.

In recent years polychaete larvae of numerous species have been reared through the metamorphosis and settlement stages and there is considerable information on mass culturing (Reish, 1974; Dean and Mazurkiewicz, 1975; and Chapter 13). Many polychaete larvae procure their nutrients from yolk previously stored in the egg. Thus they have a short pelagic existence, no external food requirement, and are relatively easy to rear. For planktotrophic or feeding larvae, on the other hand, an appropriate food source must be provided.

A species in which a nonfeeding larva is brooded by the adult, *Capitella capitata*, has been cultured in the laboratory for 30 to 40 generations (Reish, 1974). As an established indicator of pollution, this species has been useful for bioassays. Recently it has been shown that sublethal concentrations of detergents, copper, zinc, and organic compounds will produce abnormalities in the larvae of *Capitella capitata*, sometimes appearing in the second generation of larvae (Foret, 1972; Reish, 1974). Another species that broods its larvae, *Ophryotrocha labornica*, has been maintained in the laboratory for 50 generations and has been used in studies of pollution and genetics (Åkesson, 1970, 1972).

Methods for large-scale culturing of polychaete worms as a commercial source of fish bait have been reported (D'Asaro, 1973). The worm used is the common lugworm, *Arenicola*. The larvae undergo most of their development within jelly masses deposited by the female, hatching out as swimming stages that continue only about 24 h before settling.

MOLLUSCAN LARVAE

The veliger is the larval form characteristic of gastropod and bivalve molluscs. Frequently the veliger is preceded by a trochophore stage, from which the larval velum develops as an expansion of the prototroch.

In gastropod molluscs, the trochophore stage is usually passed while the organism is still within the egg capsule and the larva hatches as a veliger (Figure 11–2). The gastropod veliger has an asymmetric larval shell and a bilaterally symmetrical velum, the latter with two—in some species as many as four to six—large velar lobes that can be retracted within the shell. The foot and operculum are well developed, and the digestive system includes mouth, radula, esophagus, stomach, digestive gland, intestine, and anus. A pair of dorsal pulsating ectodermal sacs function as the larval heart, and two large cells on either side of the esophagus serve as excretory organs. The nervous system consists of an apical plate and, late in development, also a pair of tentacles, one or more pairs of eyes, and statocysts. Torsion occurs in the young veliger, the shell and visceral mass rotating 180°. At metamorphosis the larval heart and excretory cells are shed, the velar lobes are lost, and the young snail drops to the bottom where it enters the crawling phase.

In the last few years attention has focused on the rearing of opis-

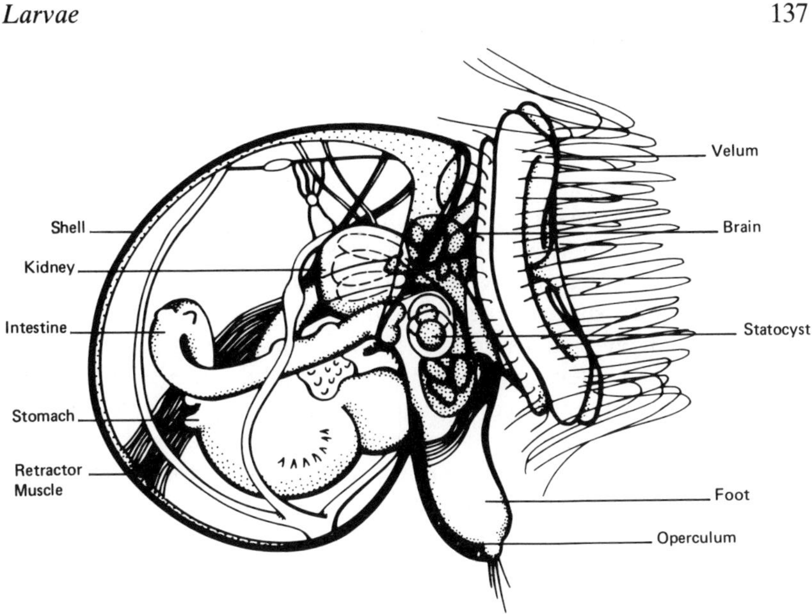

FIGURE 11–2 Veliger larva of molluscs. (From Kriegstein, 1977a.)

thobranch gastropods, particularly *Aplysia*. Development of husbandry techniques has been stimulated by the demand for adult specimens in neurophysiological research. There are large identifiable neurons in the central nervous system of these species that have been used in a variety of investigations. The veligers of some aplysiids have an extended pelagic existence, making it inconvenient to rear them in the laboratory. However, techniques have been devised for rearing several species (Kriegstein *et al.*, 1974, for *Aplysia californica*, and Chapter 14). Larval morphology and metamorphosis are now fairly well understood. The abalone, *Haliotus*, has also been cultured (Bardach *et al.*, 1972).

The veliger larva of bivalves differs from that of gastropods in that it is entirely symmetrical, does not undergo torsion, has a circular rather than lobed velum, and has a laterally compressed larval body enclosed by a bivalved larval shell. At metamorphosis the velum is cast off and the juvenile clam becomes benthic.

Bivalves generally have a short generation time and thus could be useful in genetic studies. At least two dozen species have been reared through an entire life cycle, some through several generations. There is a vast literature on culturing bivalve veliger larvae, particularly those of commercial importance such as the oyster and several species of clams

(Joyce, 1972). The classic work on rearing bivalve molluscs is that of Loosanoff and Davis (1963); updated reviews are to be found in the proceedings of a symposium on culture of marine invertebrates (Rhodes *et al.*, 1975; Chanley, 1975; Culliney *et al.*, 1975; see also Chapter 16). The larva's high sensitivity to pollutants has been utilized in studies of pollution effects (Calabrese and Rhodes, 1974; Renzoni, 1974).

As is true for other larval forms, bivalve larvae are highly sensitive to variations in water quality, and growth may be inhibited or survival reduced by undue silt content, metabolites of certain bacteria and algae, and traces of heavy metals or zinc (Walne, 1964). Much information is available on the effects of temperature, salinity, and pH on bivalve larvae, and the optimum combination of factors has been determined for several species (Loosanoff and Davis, 1963; Calabrese and Davis, 1970; Lough, 1974). Foods used most successfully in rearing larval bivalves are unialgal cultures of *Isochrysis galbana* and *Monochrysis lutheri* (Rhodes *et al.*, 1975). Specific techniques are described in Chapter 16.

CRUSTACEAN LARVAE

The nauplius is the basic larval type in crustaceans, occurring as the first free-living stage in some groups, within the egg capsule in others. The characteristic feature of the nauplius is the three pairs of appendages: a uniramous first antenna, a biramous second antenna, and a biramous mandible (Figure 11–3). The antennae, adapted for swimming, serve as the locomotor organs. There is a single median eye associated with a simple neryous system, and a pair of excretory glands. The digestive tract may or may not be complete.

Whereas most species hatch either as a nauplius or, in the more advanced groups, as a later larval stage, there are a few that hatch as a miniature adult. The stages subsequent to the nauplius are usually distinctive for different groups and are designated as specific larval types. Each new stage is preceded by a molt; however, there may be several molts in one stage. The number of larval stages varies from group to group. For instance, the brine shrimp, *Artemia salina*, hatches as a nauplius. Later the body becomes segmented and more than three pairs of appendages appear; this stage is designated a metanauplius. With each molt, additional body segments and appendages develop. After approximately 14 instars it reaches sexual maturity.

A barnacle, such as *Balanus*, passes through six naupliar stages and a nonfeeding, bivalve cypris before molting into a sessile barnacle. A copepod develops through six naupliar and four metanaupliar stages

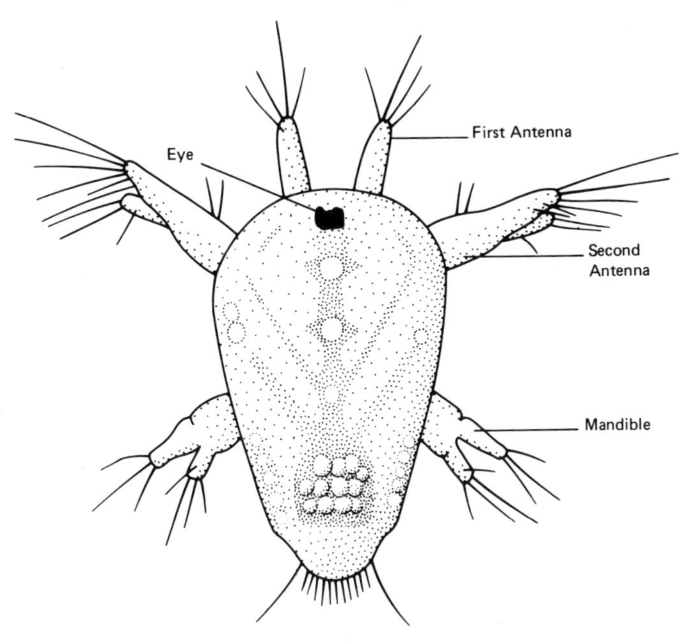

FIGURE 11–3 Nauplius larva of crustaceans. (From Reverberi, 1971.)

before molting into a copepodid larva. The copepodid resembles an adult copepod in shape and head appendages but has fewer thoracic segments and appendages and fewer abdominal segments. There are a total of six copepodid stages before sexual maturity is reached.

Foods suitable for rearing decapod crustacean larvae, as well as various culturing methods, have been reviewed by Roberts (1975). Some can be fed cultured algae. When reared in the laboratory, decapod crustacean larvae usually feed on such small zooplankton as newly hatched larvae of *Artemia* (see Chapter 6). Smaller food, such as cultured algae, rotifers, or larvae of bivalves or sea urchins, may be required for small crustacean larvae (see Chapter 6).

Selection of an appropriate technique from the many culture methods that have been reported will depend upon the number of animals required for the study under consideration. Methods are available for commercial culturing of certain edible crustaceans such as penaeid shrimp (Tabb *et al.*, 1972) and lobsters (Hughes *et al.*, 1975). For experimental purposes, culturing can be on a smaller scale, although the same basic requirements must be met. Chapter 17 describes methods for rearing crabs.

Crustacean larvae have been much studied in work on chemoreception and settlement phenomena (Chia and Rice, 1978). Studies on barnacle larvae first showed that larval settlement could be chemically induced by a factor isolated from adult tissues (Crisp, 1974). Studies of various crustacean larvae have provided much information on control of larval development and metamorphosis, especially hormonal influence (Costlow, 1968). Crustacean larvae have been widely used in research on temperature and salinity tolerance and are being increasingly utilized in investigations of pollutant effects. For example, it has been demonstrated that larvae of the fiddler crab, *Uca pugilator*, show marked behavioral changes and decreased metabolic rates in response to very low concentrations of mercury that have no apparent effect on the adult (DeCoursey and Vernberg, 1972). Tests of the pesticide Mirex on several species of crab larvae showed that the death rate was greater in later than in earlier larval stages and that larvae exposed to juvenile insect hormone mimics experienced a higher percentage of abnormalities at the higher salinities of seawater (Bookhout and Costlow, 1974a).

ECHINODERM LARVAE

Free-swimming larval forms are found in all classes of echinoderms, including crinoids (feather stars or sea lilies), holothurians (sea cucumbers), ophiuroids (brittle stars), asteroids (starfish), and echinoids (sea urchins and sand dollars). Although the larvae of each group are distinctive, all have certain basic features in common, including external bilateral symmetry. In feeding larvae, which are present in all groups except crinoids, the mouth and anus both open on the midventral line. The mouth is located in the center of a depression bordered by a ciliated band that varies in shape and extent in different species. Internally, the larvae are characterized by three pairs of coelomic sacs and a tripartite gut.

The larvae of sea urchins (echinoids) have been the most widely studied of the echinoderms. The earliest larval stage is known as the pluteus. The fully formed larva is cone shaped, the apex pointed posteriorly with four to six pairs of arms directed upward and outward (Figure 11–4). The arms are supported by calcareous skeletal rods and bear ciliated bands. Movement is by means of ciliary activity of these bands. In some species, portions of the bands (called epaulettes) are thickened, arched, and develop long cilia.

The pluteus-type larva also occurs in the ophiuroids or brittle stars. One of the most obvious differences between the two groups is the greater horizontal extension of the arms of the "ophiopluteus."

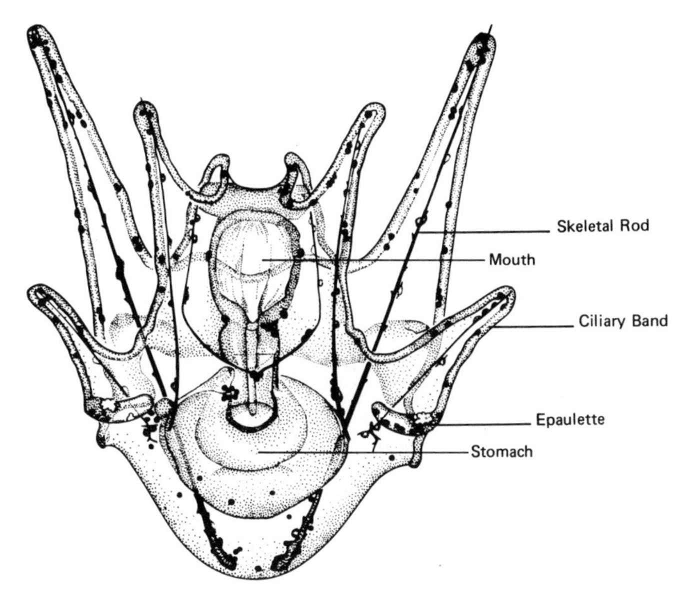

FIGURE 11–4 Pluteus larva of an echinoid. (From Reverberi, 1971.)

The larva of the crinoids is a barrel-shaped doliolaria with four to five horizontal ciliated bands and an apical tuft of long cilia arising from a sensory plate (Figure 11–5). This is usually a nonfeeding form with incomplete gut.

Holothurians, or sea cucumbers, often have a larva known as the auricularia (Figure 11–6). In this form the ventral depression or vestibule is flanked by four lobes, two directed anteriorly and two posteriorly. A continuous ciliated band, variously shaped, surrounds the vestibule. In some holothurians the auricularia develops into a doliolaria by reformation of the ciliary band into several latitudinal bands. In other species, the holothurian embryo may transform directly into the doliolaria, bypassing the auricularia stage.

The bipinnaria larva of sea stars (asteroids) resembles the auricularia, but the lobes are elongated into a number of arms and there are two continuous ciliated bands. This larva develops into a later larval stage, the brachiolaria, by forming three additional anterior arms at the base of which is an adhesive area or sucker by which the larva attaches to the substratum at the time of metamorphosis (Figure 11–7).

Techniques for culturing some echinoderms are described in Chapter 19. They have been summarized in a review by Ruggieri (1975). Gametes can be obtained from adult sea urchins in great quantities by electrical stimulation, injection of isotonic potassium chloride, or by removal of

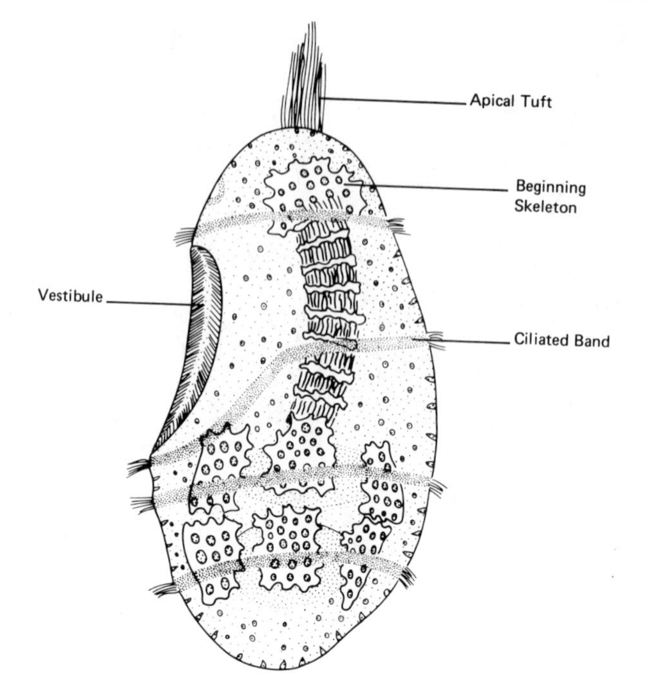

FIGURE 11–5 Doliolaria larva of crinoid echinoderms. (From Hyman, 1955.)

gonads. Large numbers of synchronously developing embryos and larvae can be produced by appropriate mixtures of eggs and sperm. The foods most commonly used in rearing the larvae are the algae *Rhodomonas lens* and *Dunaliella tertiolecta*. The ease with which large numbers of gametes can be secured and the minimum care required in maintaining synchronous cultures of the early larvae account for the popularity of these animals in experimental biology.

Over the past hundred years the gametes and developmental stages of echinoderms, especially sea urchins, have provided a major source of research material for embryologists and developmental biologists. Echinoderms, more than most organisms, have contributed to solutions of fundamental problems in development from gametogenesis and fertilization to protein synthesis, differentiation, and larval morphogenesis. Literally thousands of papers have been written about early sea urchin development. For bibliographies and reviews the reader is referred to the books by Giudice (1973), Stearns (1974), and Czihak (1975). Now that techniques for rearing sea urchins through their entire life cycle are

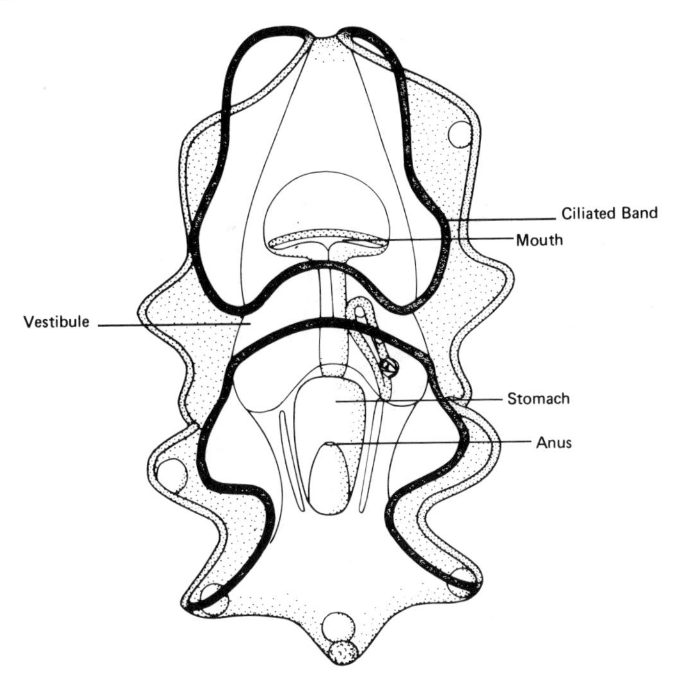

FIGURE 11–6 Auricularia larva of holothurian echinoderms. (From Mortensen, 1938.)

available, it should be possible to expand the developmental studies to include investigations of genetics (Hinegardner, 1969, 1975a).

As in the case of many other groups, echinoderm larvae are becoming increasingly recognized in studies of pollution effects and in bioassay. A variety of larval abnormalities are produced by exposures to environmental fluctuations and contaminants in seawater. Examples of such studies are those of Rio *et al.* (1965), Timourian (1968), and Bougis *et al.* (1979).

OTHER LARVAE

AMPHIBLASTULA

This is the larval type characteristic of many sponges. It is an oval or rounded larva, one half comprised of narrow flagellated cells and the other half of nonflagellated large cells (Figure 11–8). The larva swims

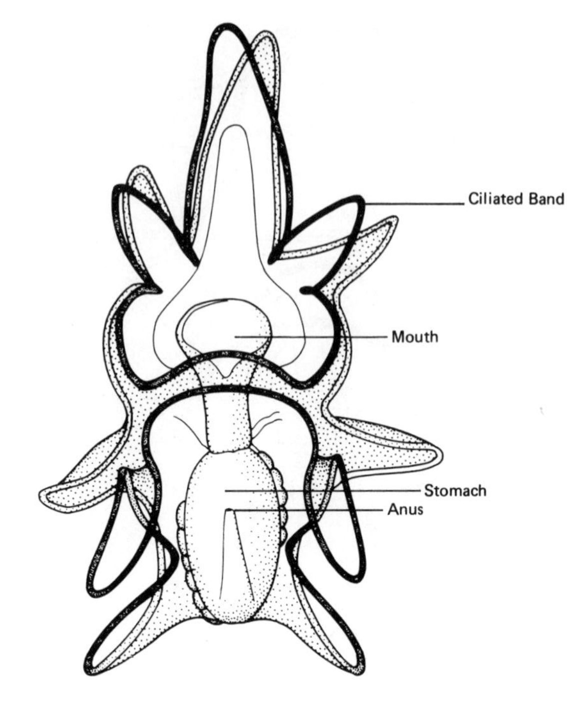

FIGURE 11–7 Bipinnaria larva of asteroid echinoderms. (From Mortensen, 1938.)

for a short time before it attaches to a substratum. Upon settling, cells from the flagellated half move into the interior, either by invagination or epiboly, and form the characteristic choanocytes of the adult sponge (Minchin, 1896).

PLANULA

The cnidarians, including hydroids, jellyfishes, gorgonians, anemones, and corals, exhibit a myriad of reproductive patterns, both sexual and asexual, often with alternation of sexual and asexual generations. One common developmental feature is the planula, a ciliated, nonfeeding larval stage; comprised of two cell layers, and commonly elongated and bipolar (Figure 11–9). Larval shape, patterns of ciliation, and internal structures vary among different classes. Planulae usually swim near the bottom for a short time, from a few hours to a few days. In some species they may become large, reaching several millimeters in length. The planulae of certain groups within the Cnidaria undergo further devel-

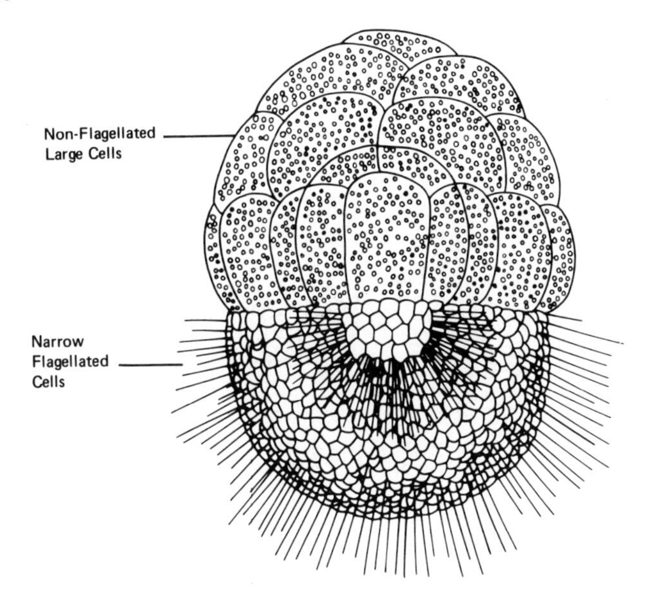

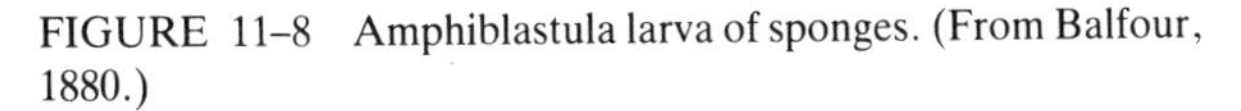
FIGURE 11–8 Amphiblastula larva of sponges. (From Balfour, 1880.)

opment into other characteristic larval types before settlement. Metamorphosis of the planula usually results in a polyp, an attached form with mouth, gut, and tentacles. This is the asexual phase of hydroids and some jellyfish, which produces additional individuals by budding. Polyps also give rise to the swimming medusae of the sexual phase, which in turn forms the gametes that produce the planula larva. Although there are many variations, often rather complex, of this basic pattern, in hydroids the polyp is generally the dominant phase, whereas in the jellyfishes or scyphozoans the medusa is dominant and the polyp either reduced or absent. The anemones and corals completely lack a medusoid phase. For a general reference in this area, see Hyman (1940). Chapter 12 describes methods for rearing sea anemones and the literature on coelenterate culture is reviewed in Chapter 10.

MÜLLER'S LARVA

Although most turbellarian flatworms develop directly, such forms as marine polyclads have a free-swimming larval stage known as Müller's larva, after its discoverer. The shape of the larva resembles a rounded triangle with eight posteriorly directed ciliated arms or lobes (Figure

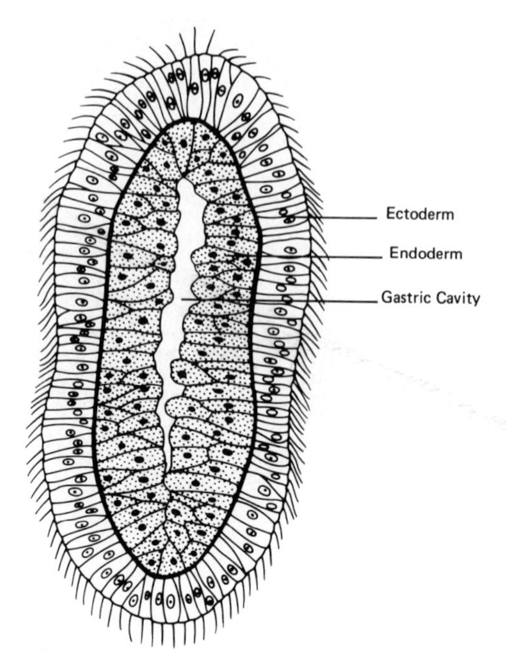

FIGURE 11–9 Planula larva of cnidarians. (From Reverberi, 1971.)

11–10). There are also long anterior and posterior sensory tufts, one or more eyespots, a brain, and a frontal organ. A mouth is present, but the larva does not feed until after settlement and metamorphosis into a young worm. At metamorphosis the lobes degenerate and are resorbed, the typical pharynx of the adult is formed, and the body becomes first ovoid, then flattened (Kato, 1940).

CYPHONAUTES

The free-swimming larval stage of many bryozoans, or moss animals, is known as the cyphonautes (Figure 11–11). It is flat, triangular in side view, enclosed by two chitinoid valves, and characterized by a pyriform organ, used to test the substratum before settlement, and an adhesive sac by which the larva attaches to the substratum at metamorphosis. There is an apical sensory organ bearing a tuft of long, stiff cilia and a basal ring of cilia by which the larva swims. A continuous current of water flows from an anterior inhalant chamber to a posterior exhalant chamber. This current directs food, consisting of small phytoplankton,

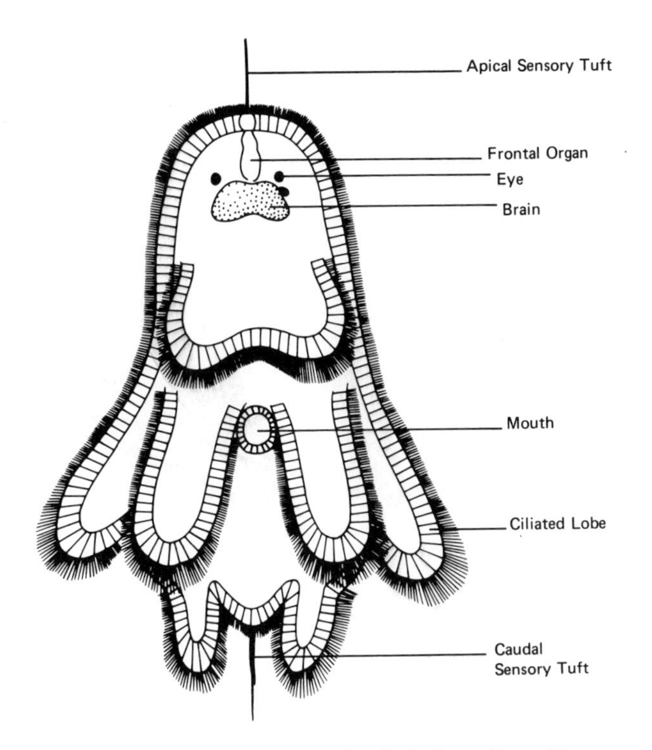

FIGURE 11–10 Müller's larva of turbellarians. (From Hyman, 1951.)

into the pharynx and removes fecal material from the anal region (Atkins, 1955a,b).

At settlement, the adhesive sac is everted, attaching the larva to the substratum. The larval organs regress, leaving only an undifferentiated mass of cells. From these cells a young polypide develops and by a process of asexual budding a new colony of bryozoans is formed.

Bryozoans are a widely distributed group, numbering about 4,000 species, and an important component of the fouling community. They commonly attach to ship hulls and are considered pests in commercial shellfish beds, where they compete for space and food. The study of larval behavior and settlement therefore assumes economic significance. As experimental subjects, the larvae have been studied for behavioral responses to light and gravity and substrate selectivity (Ryland, 1974). Methods for laboratory cultivation have been reviewed by Abbott (1975) and Jebram (1977a).

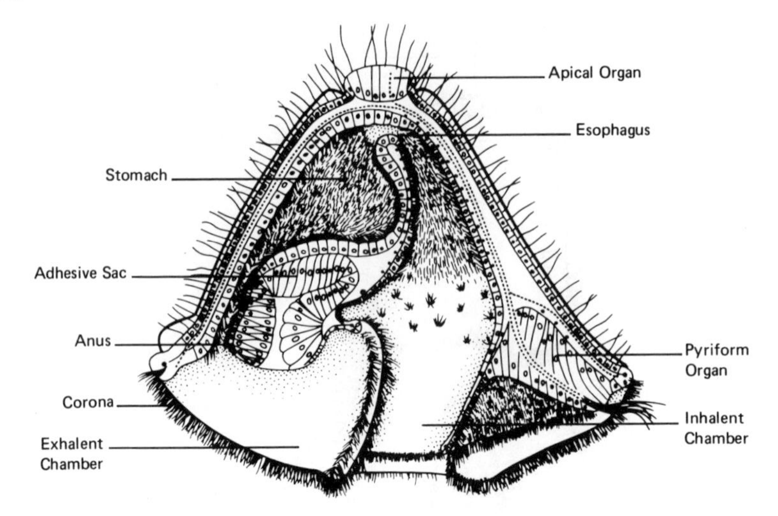

FIGURE 11–11 Cyphonautes larva of bryozoans. (From Nielsen, 1971.)

TADPOLE OF ASCIDIANS

The swimming larva of ascidians is notably different from the sessile adult and is known as a "tadpole" because of its shape (Figure 11–12). The larval tail is several times the length of the trunk and encloses a hollow dorsal nerve cord and notochord. These latter features, although lost in the adult, and the presence of gill slits, are the basis for classifying the ascidians with the Chordata. The tail serves as a swimming organ, the dorsal and median flaps acting as fins. Mouth and rudimentary gut are present, but the larva is enclosed in a gelatinous cuticle and does

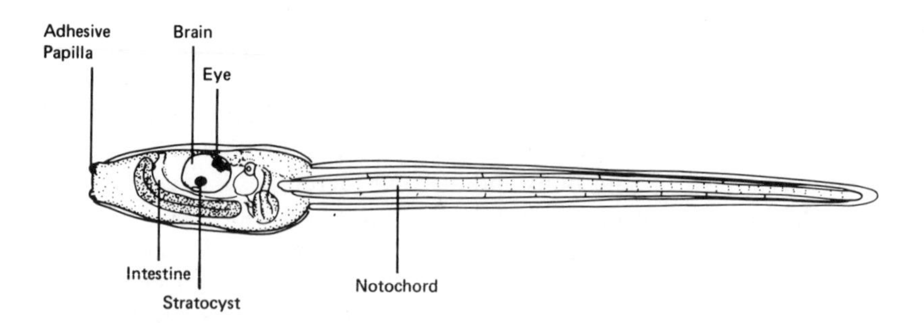

FIGURE 11–12 Tadpole of ascidians. (From Berrill, 1947.)

not feed until after metamorphosis. Characteristic larval organs include a dorsal eye, statocyst, and anterior adhesive papillae (Berrill, 1950).

After swimming for a short time, for some species only a few hours, the larva attaches to the substratum by a secretion from its adhesive papillae. After attachment, the larva undergoes a rapid metamorphosis in which the tail is resorbed, larval organs degenerate, and rudiments of adult organs differentiate to transform the motile larva into a sessile adult.

The mechanism of tail resorption has attracted the attention of developmental biologists and cytologists as a model for cell contractility. Studies of ultrastructure have shown that contraction of tail epidermis during metamorphosis is due to contraction and alignment of microfilaments in the epidermal cells (Cloney, 1961).

PART II

12

Methods for Rearing Sea Anemones in the Laboratory

DAVID A. HESSINGER *and* JUDITH A. HESSINGER

INTRODUCTION

GENERAL DESCRIPTION

The group of organisms that includes the jellyfish, corals, sea anemones, and hydra (the cnidarians) is among the simplest metazoan groups, possessing only an ectoderm and endoderm separated by an acellular mesoglea. Cnidarians are primarily marine; only a few species occupy freshwater habitats. The name comes from the presence of nematocysts or cnidae in all species. Nematocysts are highly structured, intracellular secretory bodies used in food-gathering and defense, although in some special cases they are adapted for substrate attachment during locomotion of normally sessile forms. The phylum is divided into three classes—the Hydrozoa (hydra, for example), the Anthozoa, and the Scyphozoa (jellyfish). Sea anemones are included among the anthozoans, along with the corals. Anthozoans are characterized as sessile polyps that possess no swimming medusoid stage during any part of the life cycle (Hyman, 1940). Anemones are especially suited for laboratory culturing and study because they are sessile and solitary. *Aiptasia pallida*, for instance, grows on vertical, hard surfaces just below the low-tide mark and can be grown in the laboratory in high density on racks of vertically supported Plexiglass plates. *A. pallida* is subtropical and adapted to normal ambient laboratory temperatures. *A. pallida* is big enough to be individually manipulated, yet not so big that it cannot be handled or moved without disruption. Anemones, in general, and *A. pallida*, in particular, are suitable models for studying several biological processes:

1. *Symbiosis*. Many anemones, and most corals, in addition to ingesting prey, maintain their caloric requirements through a symbiotic association with a group of dinoflagellate algae called zooxanthellae. Along with certain species of symbiotic *Hydra* (Muscatine and Lenhoff, 1963), both anemones (Muscatine, 1961) and corals (Goreau, 1961) have been much used in studying algal–animal symbiotic relationships.

2. *Asexual reproduction*. Both sexual and asexual modes of reproduction occur among most sessile cnidarians, whether solitary or colonial. Asexual reproduction normally occurs when conditions for survival are stable and optimal. Sexual reproduction assures species dispersal and survival in periods of adverse environmental conditions. Among sea anemones budding and pedal laceration are the most common types of asexual reproduction. Pedal lacerates begin as undifferentiated, syncytial masses that undergo complete regeneration of the adult form in miniature within a few days. Because pedal lacerates can be collected in fairly large numbers and at different stages of development corresponding to the time of their release from the parent, they are well suited for morphogenetic studies (Smith and Lenhoff, 1976), but as yet they are little studied. In addition, frequent pedal laceration and the ease of collecting them enables the investigator to "clone" individual sea anemones for studies in which large numbers of very uniform animals are needed.

3. *Behavior*. Anemones possess a simple nervous system and exhibit simple behavior patterns that are suited for quantifying data. Anemones show both positive and negative photoresponses. *Aiptasia pallida* will extend its tentacles and body column in the presence of light and will quickly retract if the light is interrupted even briefly. They lose their responsiveness, however, during and after feeding. Feeding behavior is believed to be primarily under the control of soluble effector molecules and specific chemoreceptors. The chemoreceptors can detect extremely low concentrations of the effector, usually specific amino acids and small peptides (Lenhoff and Heagy, 1977). Such "feeding hormones" are believed to originate from the puncture wounds inflicted on the prey by the stinging nematocysts. The triggering of nematocyst discharge is itself a response of the nematocytes to both chemical and mechanical stimuli. Thus nematocytes, the cells that produce and deploy nematocysts, behave as independent effectors (Pantin, 1942).

4. *Nematocysts*. Many features of nematocysts merit study: Nemat-

ocysts are the most complex secretory structures known and the assembly of nematocysts and the packaging of their contents are posttranslational feats of impressive dimensions. Although tentacle nematocysts, particularly those of sea anemones, behave as independent effectors, virtually nothing is known about the structures or the processes involved in the intracellular transduction and conduction of the external stimuli that trigger discharge. Whereas the physical mechanism of nematocyst eversion has been well described (Picken and Skaer, 1966), it is not known what initiates or drives the eversion of the nematocyst tubule during nematocyst discharge. The ultrastructural and functional relationships of the elements in nematocyst-forming cells are almost unkown (Cormier and Hessinger, 1980; Slautterback, 1967; Westfall, 1970). Nematocyst toxins have been studied for the last 70 yr, beginning with Richet's (1902) classical work. Only recently have methods been available for obtaining toxins directly from functionally mature nematocysts (Barnes, 1967; Blanquet, 1968). This was done much to help resolve the criticism and ambiguity of earlier work that used whole tissue extracts.

In this regard, the acontia nematocysts from certain sea anemones have proved extremely useful. Acontia are threadlike extensions of the gastric filaments that occur in certain anemones. In *Aiptasia pallida* the acontia carry two types of different-sized nematocysts. Because of their intercellular location, it is possible to isolate nematocysts that are completely free of tissue debris by using a modification (Blanquet, 1968) of the original method of Yanagita (1959). The two types of nematocysts can then be separated. When prepared in this way they can be stimulated to discharge upon exposure to low ionic strength buffered media, and nematocysts and nematocyst venom can be prepared in pure form for *in vitro* physiological and biochemical studies. The venom is composed of several proteins, some of which are unique, potent, specific toxins (Hessinger *et al.*, 1973). Since every species of cnidarian possesses a unique complement of nematocysts, each with its own venom, cnidarians constitute the taxonomically most extensive group of venomous animals. For instance, the microbasic mastigophores from the acontia of *Aiptasia pallida* possess two different kinds of protein neurotoxins (Hessinger *et al.*, 1973) plus a complex system of three synergistically acting cytolytic proteins (Hessinger and Lenhoff, 1976). Since there are 24 distinct morphological types of nematocysts and about 10,000 species of cnidarians, it seems highly likely that many more of these unique and potentially useful nematocyst toxins exist with activities affecting various aspects of membrane structure and function. *Aiptasia* toxins and those recently reported in the Portuguese Man-of-War (Lin and Hessinger, 1979) are unique. Venom from closely related species usually represents variations on a general theme of action, thus providing a useful arsenal of related, yet distinctly varied, physiological and biochemical tools for cell membrane research.

ADVANTAGES OF CULTURING ANEMONES IN THE LABORATORY

Anemones can be raised that are genetically, developmentally, and environmentally virtually alike, thus facilitating carefully controlled experiments.

Genetic Control

Whenever genetic cloning is not possible, restriction or elimination of genetic differences among experimental animals can, at times, be partially accomplished by employing statistically large numbers of standardized animals reared under strictly controlled conditions. This approach has been applied to a few asexually reproducing species (e.g., *Hydra*, flatworms) whose generation times are short. In the case of pedal lacerating anemones, the generation time of an individual from pedal lacerate to lacerating adult is on the order of weeks. Yet once an anemone commences pedal laceration, it will do so at the rate of one or more per day under optimal conditions, making it possible to "clone" sizable populations of anemones from a single progenitor within a month or two. Such progeny should be genetically identical, except for mutations that may have occurred.

Environmental Control

Advantages of genetic cloning and genetic selection, combined with the capacity to rear anemones under strictly standardized and controlled laboratory conditions, give the potential for biological uniformity that has rarely been realized for metazoans (Lenhoff and Brown, 1970). There are several environmental variables that must be taken into account in standardizing sea anemone culture conditions: (i) Light is especially important for the optimal growth of symbiotic anemones, since photosynthetic products released from the algae nourish the host. The intensity, duration, and even the wavelength of the lighting can be used to optimize growth conditions for the anemones. The use of external lighting may necessitate removing the algae that grow on the sides of the aquaria. Nonsymbiotic anemones, on the other hand, seem to grow reasonably well even in total darkness. (ii) Salinity should be controlled. Whereas anemones can tolerate fairly wide ranges of salinity, *Aiptasia* is fairly sensitive to slight but abrupt changes, particularly from lower to higher concentrations. (iii) pH fluctuations, particularly increases in acidity, are apt to occur in laboratory aquaria. Sea anemones are sensitive to pH values that are more than a few tenths of a unit below 7.8. (iv) Culture temperatures near those of ambient native marine habitats are usually ideal. Sea anemones do not react adversely to mild temperature fluctuations, but slight long-term changes in temperature will affect growth rates. It should also be noted that seasonal changes in water temperature often trigger gonadogenesis in anemones. (v) In most instances, it is impossible or impractical to duplicate natural sources of food. It will, therefore, be necessary to find a suitable, nutritionally complete food source. The amount and frequency of feeding must be ascertained. In their native habitat they feed opportunistically and continually, but it will be necessary in most laboratory-sized systems to feed the anemones periodically, and regularly.

These five environmental factors are very important in establishing a laboratory culture. Subsequently, it may be necessary to consider other factors (e.g., the increase of certain metabolic by-products to toxic levels or the depletion of certain essential seawater constituents).

HISTORICAL PERSPECTIVE

Individual and small numbers of sea anemones may be maintained in laboratories and hobbyist aquaria for many years relatively easily and without special equipment or technology. Historically, the first systematically controlled mass culture of sessile cnidarians was developed by Loomis (1954) and refined by Lenhoff and Brown (1970), using the freshwater *Hydra*. The procedures described as dish culture methods are based upon those of Lenhoff and Brown. Because of the larger size of *Aiptasia pallida*, the numbers of animals required and the need to conserve seawater and time, an aquarium method was developed that maximizes the numbers of anemones cultured per gallon of seawater and enhances the efficiency of the day-to-day feeding and cleaning. This culture method is particularly appropriate for *Aiptasia pallida*.

GENERAL DESCRIPTION

The method involves maintaining mature, full-sized *Aiptasia* on Plexiglass plates hung vertically in seawater aquaria from overhead racks. Anemones are fed to repletion daily by transfering the anemone racks to separate smaller feeding tanks teeming with freshly hatched and harvested *Artemia* nauplii. The quality of seawater in the aquaria is maintained by circulating it through a series of external particulate and bacterial filters. Anemones thus cultured have been maintained for over 5 yr.

MATERIALS

ANEMONES

For a variety of reasons, we have chosen to work with *Aiptasia.* They occur from North Carolina to the Florida Keys, on the eastern coast of Baja California, and in the Hawaiian Islands, or can be purchased from Carolina Biological Supply Co., Burlington, N.C. Specimens can be fed exclusively on *Artemia* nauplii without apparent loss of vitality. The anemones will adhere strongly to flat surfaces and can be moved on these attachment sites for daily feeding and cleaning. Pedal laceration provides a continual supply of new individuals.

SEAWATER

We attempted to use artificial seawater (Seven Seas Mix) and found that the anemones slowly deteriorated. No additional attempts to use artificial seawater mixes were made, but it is possible that this method could be used successfully. We routinely use natural seawater that is transported in 5-gal plastic carboys or 110-gal horizontal Nalgene storage tanks (#6319–00, Cole-Parmer Instrument Co., Chicago, Ill.). Freshly collected seawater must be filtered before it is introduced into the culture system. Two Supreme Aquamaster Power Filters with a generous amount of fiberglass and top-mounted centrifugal pumps (Model PL, Danner Corp., Isilip, N.Y.) will effectively clear the water of most particulate matter within 24 h.

DISTILLED WATER

A supply of good-quality distilled or deionized water is needed for large cultures of anemones. Distilled water is added to the aquarium system daily to compensate for evaporative losses.

To avoid troublesome osmotic fluctuations, water lost through evaporation is replaced through the bacterial filter rather than directly to the aquaria.

ARTEMIA ENCYSTED EMBRYOS

Encysted *Artemia* embryos are available commercially (San Francisco Bay Brand, Newark, Calif.), in 15-oz (425-g) vacuum-packed cans.

Normally, we purchase two cases of 15-oz cans each year, taking care to test them for viability as soon as received. Embryos from the Great Salt Lake are of low viability, probably owing to a recent ecological upset there. In addition, the Salt Lake embryos have been found to contain high levels of DDT residues.

ROCK SALT

Rock salt is an inexpensive source of salt for making the brine solution used to hatch *Artemia* nauplii. Any commercial source of rock salt such as that used in water softeners is suitable.

FLUORESCENT LIGHTS

We use 4-ft, grounded, single and double bulb fixtures combining 35- and 40-watt plant lights (e.g., Gro-Lux, Sylvania) with soft-white bulbs (e.g., Watt-Miser or Cool White, General Electric).

ELECTRICAL TIMER

A 24-h electrical timer is used to supply both the aquaria and the dish cultures with a 12-h lighting schedule (7 a.m. to 7 p.m., yearround).

SPECIAL EQUIPMENT

Several pieces of special equipment have been developed. Some of them are constructed by us or for us; others have been modified from commercially manufactured items. These items are described in sufficient detail here to make duplication possible. Further modifications and improvements are doubtless possible.

Culture Dishes

Flat Pyrex baking dishes (Corning Glass Co., Corning, N.Y.) are well suited for countertop dish culture of anemones. They will stack on top of each other (as many as eight to a stack), thus conserving counter space and retarding evaporation. The dishes are available in a variety of configurations and sizes to accommodate anemones of different sizes and to optimize use of available counter space. For *Aiptasia pallida,* we prefer the square dishes (8″ × 8″ × 2″; #222), although we have used larger rectangular dishes (11¾″ × 7½″ × 1¾″; #232) and the larger 3-qt (13½″ × 8½″ × 1¾″; #233). For exceptionally large *A. pallida,* such as those from Marathon, Fla., we use a much deeper loaf dish (9″ × 5″ × 3″; #215–B). Topmost dishes on stacks are covered with ⅛″ Plexiglass.

Culture Dish Scrapers

Both rubber and teflon ("policemen"-styled) scrapers are used to clean the inner dish surfaces of adherent anemone and *Artemia* debris and algal growth. We also use small pieces of plastic netting (available in fabric stores as colored veiling) folded over the index finger to scrape and clean in the rounded corners of the dishes.

Wash Bottles

Nalgene wash bottles containing seawater are used to loosen debris from around anemones and from the culture dish bottoms during cleaning operations following feeding. The jets from these bottles should be wide enough to provide adequate amounts of water for effective flushing of the areas and yet fine enough so that the jet of water will not injure the anemones. The tip of the plastic spout can be trimmed back with a scalpel or razor blade until the desired stream of water is attained.

Anemone Detachers

When anemones must be detached from the surface to be moved or transferred, a single-edged razor blade or sharp-edged teflon "policeman" should be used. The razor blades will not tear or traumatize the anemone. Because razor blades will corrode quickly after exposure to seawater, it is best to coat them with a very light film of high-vacuum silicone stopcock lubricant (Dow Corning Corp., Midland, Mich.) before use. To ensure that no traces of the grease are deposited on the culture dish or plate, it is advisable to remove as much as possible with tissue paper before using.

Anemone Pipettes

Whenever it is necessary to transfer individual anemones, we use pipettes made from different-sized single bore glass tubing. One end is fitted with a rubber bulb and the other end, after fire polishing, is used to pick up the anemones. Anemones should always be picked up from their pedal discs (or bases) to protect them from evisceration, and care should be taken to avoid drawing them into the tube past the tentacles.

Aquaria

Our aquaria measure 90 × 41 × 35 cm (36″ long × 16″ high × 14″ wide) and have a capacity of approximately 114 liters (30 gal).

Aquarium Filters

Maintaining the quality of seawater is the key to successful mass culture. Anemones will raise the level of soluble nitrates in the aquaria to toxic levels in a matter of weeks if the water is not replaced or reconditioned. Additional difficulties arise from accumulations of particulate debris and wastes. Two types of filtering systems are employed to maintain the condition of the seawater in the light of these waste problems.

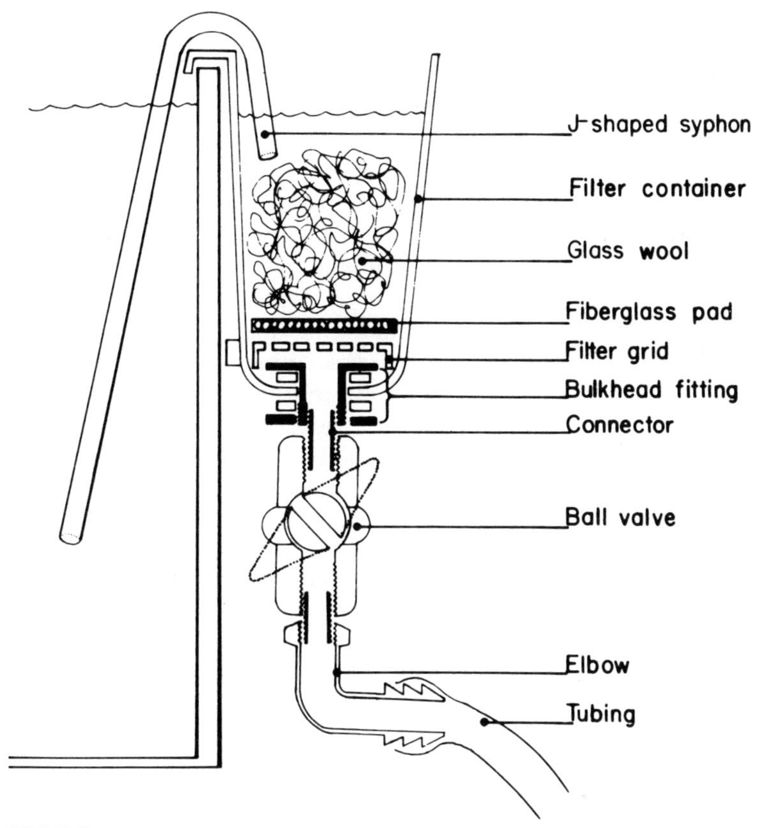

FIGURE 12–1 Particle filter.

Particulate Filters A filter assembly is located at both ends of each aquarium (Figure 12–1) and consists of a Model PL, Supreme Aquamaster Power Filter (Danner Corp., Isilip, N.Y.) containing a plastic filter grid with fiberglass pad. Above this pad the assembly holds a generous amount of loosely packed fiberglass (0.81 oz box of Soft Spun Glass, Metaframe Corp., East Paterson, N.J.). Because this filter assembly will remain functional for several months, it is advisable not to use Dacron wool, which would be biodegraded in that time. The filter assembly is connected to the aquarium via two or three J-shaped plastic syphons (5/8″ ID) that are included with the purchased Aquamaster assembly. The chimney of the filter grid is not used, and a size #1 rubber stopper is inserted in its top. A circular hole of 1¼″ diameter is made in the bottom of the filter tank and a ¾″ (MPT) PVC bulkhead fitting (#6445–30, Cole-Parmer Inst. Co., Chicago, Ill.) is fitted into the hole. A double female ¾″ threaded PVC ball valve (R and G Sloane Manuf. Co., Inc., Sun Valley, Calif.) is connected to the bulkhead fitting with a threaded PVC male connector. At the other end of the valve a ¾″ linear polyethylene elbow (¾″ MIPT) joint with tapered hose fitting (fits ½″ ID tubing) is threaded (#6451–85 Cole-Parmer). A length (3–4 ft) of thick-walled (1/8″) Tygon tubing (½″ ID) is fitted to the tapered hose fitting. The other end of the tubing extends to the external bacterial filter on the floor beneath the aquaria. Thus, the water in the aquarium will be passively connected to the bacterial filter via syphons, an external particulate filter, and a length of tubing.

Bacterial Filter The bacterial filter (Figure 12–2) consists of a large reservoir that contains layers of crushed oyster shell and white quartz gravel supported over a collecting cistern from which the filtered water is pumped back to the aquaria. The reservoir is constructed from ¾″ marine plywood that has been internally coated with epoxy resin, reinforced with cloth fiberglass. The fiberglass extends over the top edges and down the other sides for about 4″–6″. In the bottom of the reservoir there are four epoxy-coated 2″ × 4″ (cross section) pieces of wood, plus two additional half-width pieces. Two of the full-width and the two half-width pieces have slots cut out of one of their narrow sides (1½″ long × 1″ high) about 6″ apart. The 2″ × 4″s are arranged so that they span the width of the reservoir, are parallel to each other, and stand on their narrow, slotted edges. They are evenly spaced with the two nonslotted, full-width pieces against the ends of the reservoir and the two half pieces arranged in the center about 6½″ apart so that the end of each is against an opposite side of the reservoir. Centered between the two shorter 2″ × 4″s and standing upright is an 18½″ length of PVC conduit (6½″ OD;

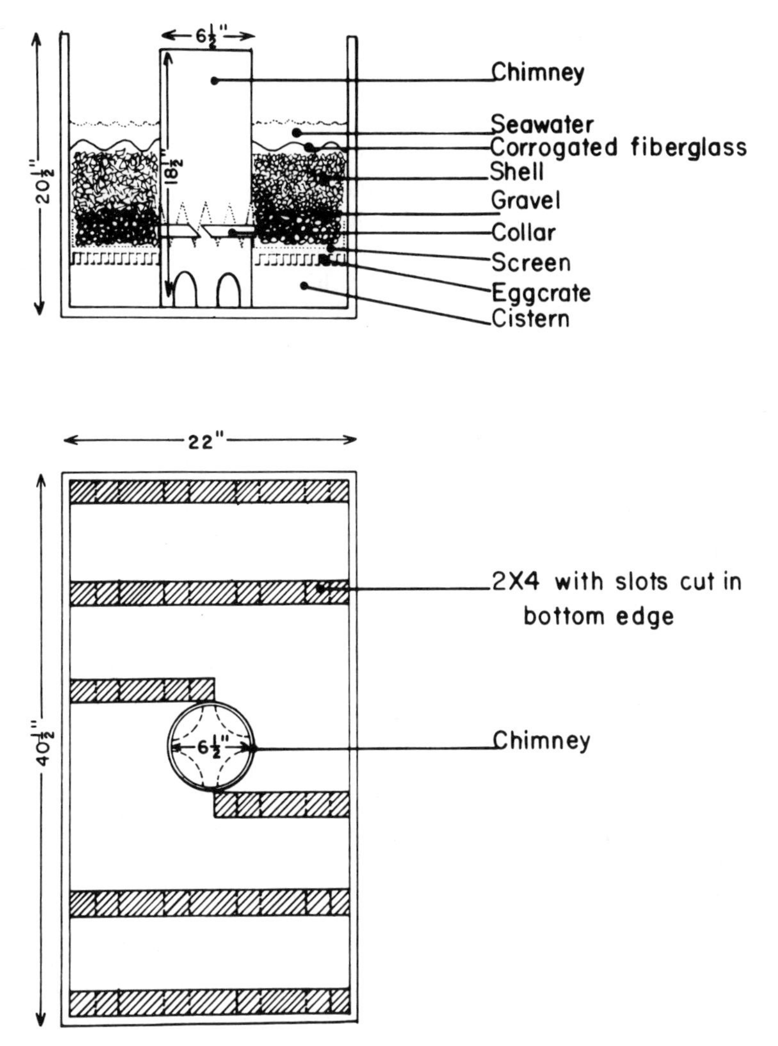

FIGURE 12–2 Bacterial filter.

6″ ID) that had four equally spaced, semicircular pieces (3″ diameter) cut from the bottom end. This conduit forms the filter's chimney into which a submersible pump will be set. The pump will take water that has passed through the layers of gravel and shell and has collected in the underlying cistern and pass it back into the overhead aquaria. The 2″ × 4″s that are standing in the reservoir on their narrow, slotted edges support the filtering media.

Supported immediately atop the 2″ × 4″s is a piece of plastic ceiling light "eggcrate" (diffuser for fluorescent lights) that is cut to fit over the chimney as well as against all four inner sides of the reservoir. On top of the eggcrating will be fitted a double layer of nylon window screening cut 4″ longer and wider than the inside dimensions of the reservoir. Eight 3¼″ pie-shaped cuts radiating from the center allow the screening to fit over the chimney, leaving the pie-shaped flaps of screening pointing upward. The outer edges of the screening should be molded and smoothed to follow the surface of the eggcrating and extend part way up the walls of the reservoir. The corner edges of the screen should be buttfolded and then cemented with silastic sealer (RTV 108, General Electric Co., Waterford, N.Y.). The upward pointing center flaps are clasped tightly around the chimney by a ¾″ annulus of the PVC conduit. This ring has been cut obliquely through at one point to allow the annulus to fit around both the chimney and the screen flaps, thereby enabling it to hold the flaps against the chimney at the level of the eggcrating. This entire assembly constitutes the supporting structure for the filtering media.

Next, 50 lb of unpainted quartz aquarium gravel, available from local wholesale aquarium suppliers, are washed repeatedly under running tap water until the washwater is completely clear, then rinsed with distilled water and allowed to drain. The cleaned gravel is placed on top of the nylon screening of the support assembly and distributed evenly across the entire surface. Care must be taken to keep the gravel out of the chimney and cistern, and in particular to keep it from getting between the wall of the reservoir and the screening.

On top of the gravel is placed 100 lb of washed, crushed oyster shell, available in 50-lb bags at local feed stores. Finally, a piece of corregated fiberglass cut to fit around the chimney and into the reservoir is placed on top of the shell layer. The fiberglass should have ⅜″ holes spaced 1½″ to 2″ apart in each lower trough or "valley." The purpose of this fiberglass is to protect the underlying layers of filtering media from being disturbed or channeled by incoming water, at the same time not interfering with the passage of water into the filter.

Aquarium water that has first passed through the particulate filters flows passively to the top of the bacterial filter and then percolates through the shell and gravel layers and drains into the lower collecting cistern. There it is pumped out of the cistern and returned to the aquarium by an immersible pump (Model 1P681 Teel pump, Dayton Electric Mfg., Chicago, Ill.) that is set in the cistern through the chimney.

The removal of particulate wastes from the water before it reaches the bacterial filter assures that the bacterial filter will remain free flowing. The crushed shell and gravel provide the microniche for microorganisms that can fix nitrates and detoxify other soluble biological wastes. We have kept our bacterial filter functioning continuously for over 3 yr without any evidence of the nitrate build-up that previously required us to replace the aquarium water every 6–8 wk. In addition, pH values of the aquarium water are stable at 8.0–8.1 in comparison with 8.2 for the fresh seawater.

It seems likely that anemones are very sensitive to the levels of free Ca^{2+}, for we have noted in the past that when anemones began to drop off the plates in large numbers, this phenomenon can sometimes be arrested by adding crushed oyster shell or crushed coral to the aquarium. Normally, fresh seawater is saturated with $CaCO_3$, but as the pH drops in the culture system the solubility of $CaCO_3$ increases, making the aquarium water subsaturated. The presence of saturating levels of $CaCo_3$ and excess solid $CaCO_3$ acts as a buffering system to keep the pH from dropping as much as it otherwise might. Thus the crushed shell layer serves as both a calcium "buffer" and pH buffer in addition to providing a physical substrate for detoxifying microorganisms.

In a reservoir measuring 40 ½″ long × 22″ wide × 20 ½″ high (outside dimensions) we keep between 25–30 gal of seawater, connected to two 30-gal aquaria, each of which is housing 15–18 plates of sea anemones. We do not know how much larger a culturing system this sized bacterial filter will sustain for extended periods.

Water Flow Rate Control

In an open circulating system such as the aquarium culture system here described, with passive outflow and active inflow, the regulation of water levels within the aquaria is critical and difficult: Flooding will occur if there is too much inflow or too little outflow; if there is too little inflow or too much outflow, the aquaria will empty and the anemones die. Since the return pump can be assumed to pump water into the aquaria at a fixed rate (~180 gal/h), it becomes necessary to match or balance this active inflow rate to the passive outflow rate. This can be done by regulating the flow of the water through the particulate filters (and into the bacterial filter) by adjusting the PVC ball valves. When the flow rate through each of the particulate filters is approximately equal and when the level of seawater in the aquaria is at the desired level and constant, the system is said to be balanced and at equilibrium.

Water Level Control

Maintenance of the water in the aquaria to predetermined levels is complicated by several factors, including fluctuations in pumping rates

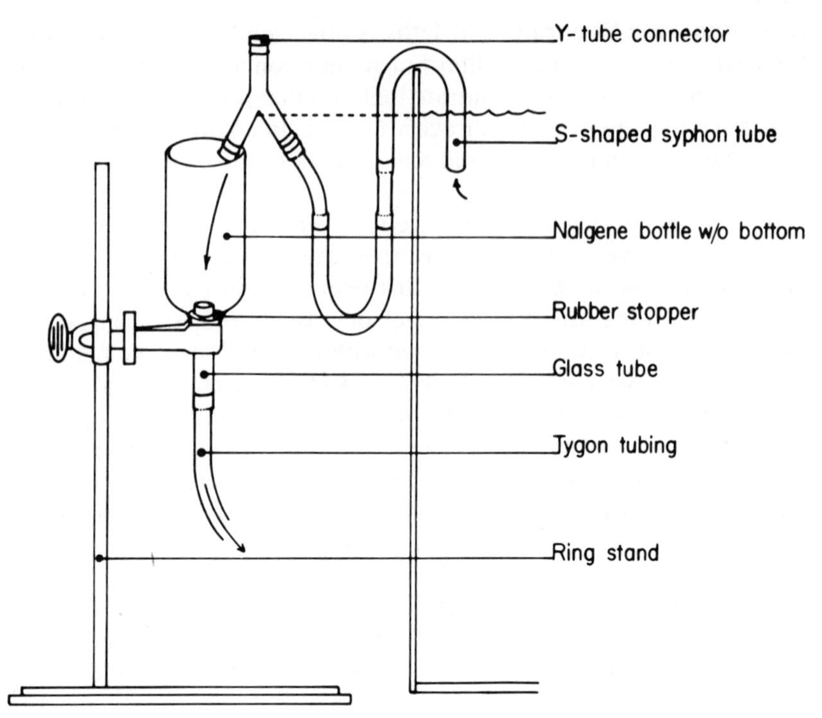

FIGURE 12–3 Syphon overflow assembly.

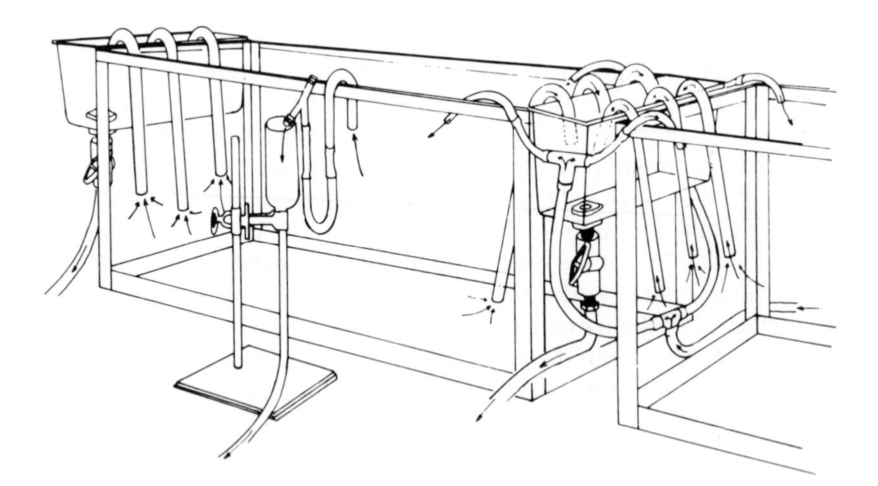

FIGURE 12–4 Water flow and level control.

(e.g., due to variations in electrical service) and by changes in flow resistance in the particulate filters (e.g., as they accumulate more particulate matter). With these fluctuations in mind, we decided that it would be best to have the inflow rate slightly exceed the passive outflow rate through the particulate filters. To compensate for this necessitates a second, but minor, outflow route. We achieved this with a syphon overflow assembly (Figures 12–3 and 12–4) that is designed to let a small portion of the water bypass the particulate filters and flow directly into the bacterial filter in response to any increase of water level above the desired preset level.

The overflow assembly (Figure 12–3) consists of two separate pieces: a syphon tube and a collecting funnel. The syphon tube is made in part by connecting two U-shaped (3½″ high) pieces of glass tubing (½″ ID) with a section of Tygon tubing to form an S-shaped syphon. One of the two arms of a Nalgene Y-tube connector (#6295–60, Cole-Parmer) is joined to one end of the S-tube via a short section of Tygon tubing. The free, downward pointing arm of the Y forms a spout that can be hooked over the edge of the collecting funnel while the vertical body of the Y points upward to serve as an antisyphon device. The lower part of the S functions to keep the syphon in the S-tube from breaking when water levels are at or below the preset level and no water is flowing through the syphon tube. The upper part of the S, of course, hangs over the edge of the aquarium and extends far enough down into the aquarium to be at least 1½–2″ below the desired water level.

The collecting funnel is essentially a funnel with attached piece of tubing that extends into the bacterial filter as do the tubings that drain the particulate filters. In our case, we have made funnels by cutting out the bottoms of 250-ml Nalgene reagent bottles and inserting a one-holed rubber stopper with glass tubing connector into the neck of the bottle from the inside. A length of Tygon tubing was joined to the funnel via the glass connector. The collecting funnel is fastened to a ring stand so that its height can be adjusted. The desired water level in the aquarium is set by adjusting the height of the top edge of the collecting funnel, which supports the spout (outflow end) of the syphon tube. In principle, the collecting funnel could be replaced with a piece of tubing connected to the outflow end of the syphon tube, thus giving a one-piece overflow assembly. Although simpler in design, it is troublesome when the assembly must be dismantled to remove trapped air bubbles or because it must be cleaned. In these cases, it is much easier to remedy the problem by simply lifting off the separate syphon tube rather than a much larger and cumbersome one-piece overflow assembly.

Review of Essential Water Control Features (Figure 12–4)

There are several features of the already-described water level and flow rate control systems that assure that the water in the aquaria will not go below the level needed for anemone survival and also protect, in most cases, against possible flooding. The low-level control systems are supposed to be "fail-safe" and consist of two features already noted: (i) All syphons will break when the water level drops below their inlet openings. Hence, the inlet openings of all syphons should never extend more than a few inches below the preset water level. (ii) The openings of the inflow tubes from the electric pump should also never extend more than a few inches below the preset water level. This is to protect the aquarium against back-syphoning through the inflow tubes should there be a failure in the pump or in electrical service.

The overflow or flood control systems should protect against most potential difficulties and consist of two related design features: (i) The overflow syphon assembly will handle small net increases in inflow. If the rate of water flow from the syphon tube spout of the overflow assembly is observed casually, on a day-to-day basis, even small increases in net inflow rates will be reflected and remedial action can be taken. (ii) The syphons on the particulate filters will also accommodate modest increases in flow in response to increases in the height of the aquarium water level. Nevertheless, if the flow of several syphons were to be interrupted, the inflow may exceed the maximum flow capacities of the remaining syphons together with the overflow assembly and flooding occur. Flooding can also occur if there is an intermittent power outage in which the pump stops long enough for the syphons to break before resuming pumping. Although flooding should not harm the anemones, it would be wise to have a functional floor drain.

Anemone Racks

Large anemones are maintained in the aquaria on vertical Plexiglass plates (Figure 12–5A 10½″ high × 8½″ wide × ⅛″) to which they adhere by their basal discs. The plates are hung from a horizontal Plexiglass rack (Figure 12–5B; 10½″ × 8½″) with 5 or 6 plates to a rack. The plates have small holes (⅛″ diameter) in the four corners of the plates, which serve to connect the plate to the rack through either two loops of 15 lb test monofilament line (1″ loop length) or two plastic curtain clips from each side of which one of the two gripping teeth has been removed. The general design and dimension of the racks are given in Figure 12–5. There are three such racks in each aquarium.

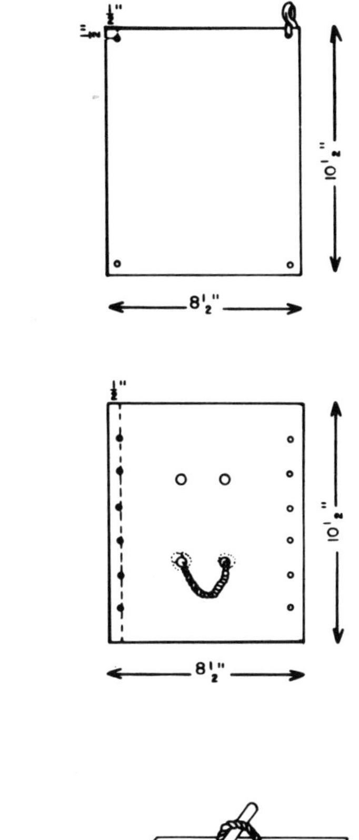

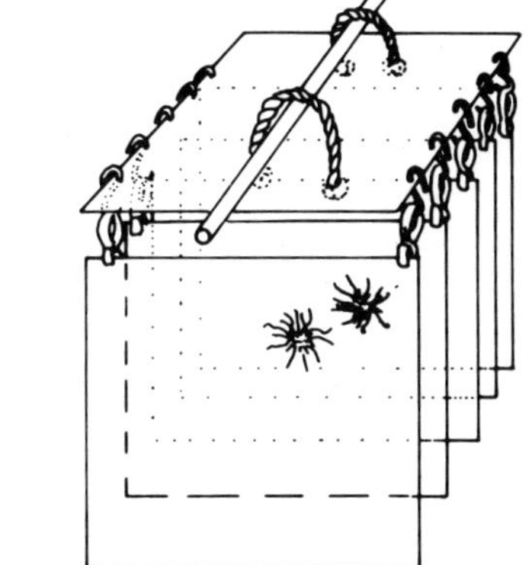

FIGURE 12–5 Aquarium plates.

HATCHING AND HARVESTING ARTEMIA NAUPLII

Hatching Tank

Hatching tanks are made by cutting the bottoms out of 5-gal, narrow-necked plastic bottles (Figure 12–6A) such as Radiac-wash bottles (Atomic Products Corp., Center Moriches, N.Y.). A rubber stopper fitted with a 1½″ long piece of glass tubing and a 2- or 3-ft length of flexible plastic tubing is inserted into the neck of the bottle from the inside and secured with a silastic adhesive. A pinch clamp is put on the loose end of the plastic tubing and is used to drain the tank upon completion of the hatching period. A small stopper with a length of monofilament line tied to it is inserted into the open end of the glass tube on the inside neck of the hatching tank (Figure 12–6B); the glass tube should not extend beyond the top surface of the rubber stopper. The small stopper is kept in place during the incubation to prevent the eggs from settling into the plastic tubing. The other end of the string can be tied to the stand that supports the hatching tank. The metal supporting stand can be obtained from either Atomic Products Corp. or a local supplier of distilled or spring water. At the present we have two hatching tanks on 48-h incubations that are harvested on alternating days in order to supply our daily need of fresh nauplii. Depending on the particular lot of *Artemia* cysts, the temperature, the rate of aeration, and the salinity, the incubation periods for optimal hatches may vary.

Harvesting Container

When the incubation period is complete, the hatching suspension, containing freshly hatched nauplii, cysts, cyst shells and debris, is drained into a 3-qt plastic refrigerator tray equipped with a spigot at one end (Pacific Plastic Products, San Francisco, Calif.) to allow the unwanted shells and debris to separate from the live nauplii.

Nets

Harvesting is completed by collecting the nauplii in a 125-mesh nylon dip net (General Biological Supply, Chicago, Ill.).

FEEDING AND RINSING TANKS

The anemones are fed outside of the aquaria in smaller feeding tanks. At the present, we use two rectangular 11-gal Nalgene tanks (18″ long

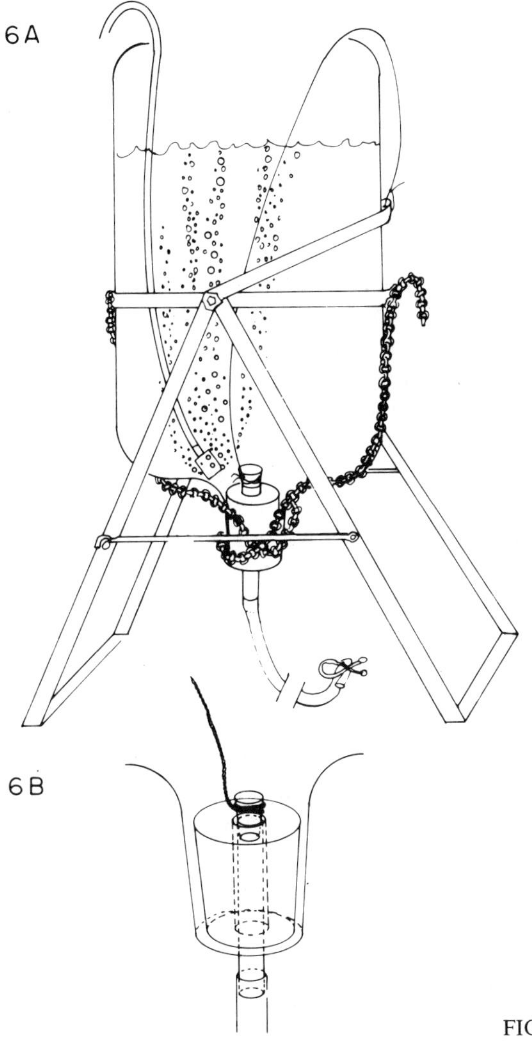

FIGURE 12–6 Hatching tank.

× 12″ × 12″; #6323–21; Cole-Parmer), one each for feeding and rinsing. Each is fitted with a faucet (#6078; Cole-Parmer) and a ¾″ PVC bulkhead fitting (#6445–30; Cole-Parmer) for draining after each use. Each tank can comfortably accommodate two full racks (6 plates each) of anemones. In the past, we have also used hard rubber, battery jar-sized developing tanks (Kodak, Rochester, N.Y.) which will accommodate one rack of 5 plates.

METHODS

PREPARATION OF FRESH ARTEMIA NAUPLII AS FOOD FOR ANEMONES

Hatching

Unopened cans of encysted *Artemia* embryos ("eggs") are stored at −20°C. Two 15-oz cans are always kept at 4°C, one opened for use. After using the contents of the opened can, the second can is opened and another brought from −20°C storage to 4°C. The viability of the eggs remains high and uniform over long periods (at least 2 yr) from can to can of the same lot if stored at −20°C. We find, however, that such eggs require an adjustment period of from 1 to 2 wk at 4°C before they will give optimum hatches.

The amount of eggs needed to supply the feeding needs of the culture should be determined empirically. Eggs should be washed well in a 125-mesh net under running tap water to remove organic matter that will otherwise support bacterial and fungal growth in the hatching tank and thereby decrease the viable yield of nauplii. The eggs are transferred to the hatching tank containing 0.25 *M* NaCl. Aerate with a weighted air stone in the neck of the tank. Adjust the aeration rate to a gentle rolling "boil," sufficient to keep the eggs from settling and suffocating on the bottom. The oxygen requirements for developing embryos and particularly for swimming nauplii are very high (Clegg, 1964), and asphyxiation can occur within a few minutes. Incubate at room temperature (70–75°F) for 48 h (some batches of eggs may require as little as 24 h).

Toward the end of the incubation period examine an aliquot of the hatching mixture to assess the condition of the hatch. An optimal hatch will have very few unhatched eggs that sink to the bottom, and proportionally many empty shells will tend to float to the surface. The nauplii will be actively swimming, and there should be no immature, nonswimming newly emerged nauplii ("parachute" forms) that are still attached to the cyst shells and encased within a membrane sac.

Harvesting

The purpose of the harvesting procedure is to (i) separate the live nauplii from the unhatched eggs and shells, (ii) to concentrate the living nauplii into a dense suspension, and (iii) to rinse the newly hatched nauplii and transfer them into seawater.

Remove the air-stone and stopper from the neck of the hatching tank and let the nauplii settle toward the bottom for a period of about 3 min. This will also allow most of the egg shells to float toward the surface. Drain the nauplii suspension into the harvester (a 3-qt plastic refrigerator tray). Allow the nauplii suspension to set for a few more minutes while the remaining shells float to the surface and the unhatched eggs settle to the bottom. The nauplii will accumulate toward the bottom of the harvester but should not settle out completely.

Collect the nauplii suspension in a 125-mesh net and immediately transfer into seawater and aerate well. Repeat this procedure if more nauplii are found in the hatching tank.

The freshly harvested nauplii should be fed to the anemones as soon as possible. Normally they will keep in seawater for only about 24 h, and if attempts are to be made to keep them that long they should first be diluted to a larger volume with fresh seawater. Care should be taken to avoid underaerating, which will cause suffocation, as well as over-aerating, which will physically disrupt the nauplii.

Be sure to clean and rinse the hatching tank well before setting it up for the next nauplii hatch. Likewise, the net must be scrupulously rinsed from both sides and hung up to dry to prevent it from becoming clogged. If properly cared for, a net will give 6–12 mo service.

DISH CULTURES

Purpose

Raising anemones in flat, Pyrex baking dishes serves several purposes: (i) This method is by far the least complicated and is best suited for the maintenance of small numbers of anemones. Since laboratory cultures generally begin with relatively few animals, this method will be described before the aquaria method that is designed for en masse culturing. (ii) The dish cultures serve as a reserve or "bank account" in the event the larger aquarium cultures are inadvertently lost. (iii) The dish cultures are a convenient means of separately maintaining small cultures of several subspecies that would otherwise interbreed or be difficult to separate and identify in larger mixed cultures. (iv) The dish cultures provide a convenient and accessible source of pedal lacerates that can be collected regularly. The lacerates can be reared in separate "nursery" dishes until they are ready for experimentation or until they reach a size suitable for aquarium culture. (v) Dish cultures also serve as a repository for anemones that have dropped off their plates in the aquaria.

Feeding

The anemones are fed freshly hatched *Artemia* nauplii daily (six times each week). Each dish receives sufficient nauplii from a plastic baster such that the anemones will ingest ⅔–¾ of them after 20 min. Care should be taken to ensure that the nauplii are initially evenly distributed among the anemones and that the anemones do not ingest more than half their capacity. If the anemones overfeed, they will regurgitate during the manipulations of the subsequent cleaning procedure. During the feeding period, the noningested nauplii will accumulate in a corner or side of the dish that receives the brightest light.

Cleaning

Culture dishes should be cleaned 20–30 min after the initiation of feeding. Hold the dish with both hands and rock it from left to right in order to resuspend killed but noningested nauplii. Decant most of the water into a nearby sink or receptacle leaving about ¼″ of water in the bottom. Set the dish on a flat surface and use the policeman to loosen anemone secretions, wastes, and algal growth from the inner surface and sides of the dish. Take care not to disturb the anemones or disrupt their bases, as this may cause the anemones to regurgitate and/or detach from the glass surface. The netting scraper is particularly suited for cleaning the sides, the corners, and the large areas on the bottom that are devoid of anemones. Hold the dish upright (vertically) over the sink, allowing the remaining water to drain out. Use a wash bottle filled with seawater (if aquarium cultures are also being used, use filtered seawater from the feeding tank) and apply a stream of water to loosen and wash debris from around the anemones. Begin rinsing at the uppermost edge and work systematically toward the draining edge or lowermost corner of the dish. If anemones detach from the dish and have to be retrieved, use an appropriate-sized anemone pipette. When cleaning and rinsing is completed, replace the dish to its original place and fill to ½–⅔ full with prefiltered seawater.

It should be noted that selective breeding of anemones is best done in dish cultures. The alert culturist can enhance the degree of expression of certain traits in the anemone colony by selecting those animals that markedly express the desired traits and/or discarding those that lack the degree of certain traits desired. Such traits as size, phototaxis, ability to adhere to plates, and ability to produce frequent (or few) pedal lacerates, among many other possible traits, can be selected for and the anemones possessing them cloned.

Removing, Transferring, and Reseeding Anemones

Anemones that have become detached from their plates and reattached to the bottom or sides of the aquarium can be removed carefully by separating them from the substrate surface with a single-edged razor blade. The animals can be picked up with an anemone pipette. When care is taken to avoid cutting the animal or eviscerating it or otherwise traumatizing it, the anemone will release very little mucous and may not even contract reflexively. Such animals will generally adhere quickly to a clean culture dish surface and will be ready for feeding and cleaning the next day. Slightly traumatized or wounded animals appear tightly retracted, often within a clot of clear, viscous mucous. Severely traumatized animals appear withered, develop a gray-brown cast, and fail to withdraw their acontia into the gastrovascular cavity and do not recover. All traumatized animals seem to depress the general state of healthy animals, and care should always be taken to avoid placing more than a few traumatized anemones in a dish with well animals.

AQUARIUM CULTURE

Purpose

The advantage of culturing sea anemones in aquaria is the ability to culture large numbers. This method can also be performed with relative ease, minimal disruption of the anemones, and under uniform and controlled laboratory conditions simulating the native habitat of the anemones.

Seeding Culture Plates

Individual aquarium culture plates are laid horizontally in the bottom of a plastic dish pan. Specimens of *A. pallida* (or other suitable species) should be selected that are healthy, large, and have clean pedal discs (or bases). They should also have been starved for the last 24 h. Anemones are carefully transferred to the plate using anemone pipettes, so that as much of their bases touch the plate as possible. Care should be taken to keep adjacent anemones from touching each other, particularly at their bases, as they will preferentially adhere to each other. Several plates can be seeded in the same dish by stacking them one on top of the other using small (60 × 20 mm) plastic Petri dish bottoms as spacers at each corner. The plates and the anemones should be left undisturbed for 24 h in order to let the anemones attach securely. At the end of this period, the plates can be connected to the racks in the aquaria. Newly seeded anemones should be acclimated to their vertical position in the aquarium for several hours before being fed.

Feeding

Avoid contaminating the aquaria with *Artemia* nauplii by transferring culture plates to smaller feeding tanks. Add enough nauplii to the feeding tank so that the feeding anemones become half-full in 20 min. Circulate water gently to prevent settling of the nauplii by aerating from an air stone. If anemones overfeed they will regurgitate in the aquaria, thereby causing a cleaning problem and also causing the anemones to be underfed. After feeding, transfer the racks to similar-sized rinse tanks to allow the anemones to relax and to finish ingesting captured nauplii (15–20 min). Before returning plates to the aquaria carefully raise and lower each rack 2–3 times to dislodge adhering nauplii. The feeding and rinsing tanks should be washed and their water filtered through 125-mesh nets and returned to these tanks.

Cleaning

Once a month the Plexiglass plates should be cleaned to remove the accumulation of algae, shrimp eggs, anemone pedal disc secretions, and pedal lacerates. In addition, algae should be removed from the walls of the aquaria.

To clean a plate, first remove it from the rack and place it in a plastic dish pan, in about 4 in. of seawater. To protect anemones from being squashed between the bottom of the pan and the plate, the plate may be supported above the bottom of the pan by placing a small object under each corner (e.g., small plastic Petri dishes). Cleaning is accomplished by scraping the areas around the large anemones with a single-edged razor blade and a rubber policeman. This is done to both sides of the plate before the cleaned plate is returned to its position on its rack in the aquarium.

The inner walls of the aquarium are most effectively cleaned with a handful of plastic veil netting. To prevent the removed algae from being released into the water, use only single lengthwise wiping strokes, taking care to raise the netting out of the water at the end of each stroke. Before returning the netting to the aquarium, rinse it well in tap water followed by distilled water. Repeat this procedure until the sides are cleaned of algal plaques.

Control of Environmental Factors

To maintain the anemones under optional and standardized growing conditions, it will be necessary to monitor and control several environmental factors, including water quality, lighting, and temperature.

Water Quality Salinity changes of more than 10 percent should be prevented by replacing water lost by evaporation each day. Salinity can be assessed either by a refractometer (American Optical Co., Buffalo, N.Y.) or a conductivity meter. The pH of the aquarium system should be kept stable and slightly alkaline. The bacterial filter described earlier is designed to regulate and control the pH of the system. If it is suspected that the water is acidic as evidenced by a grayish cast to the tentacle tips and a shortening of the tentacles, then immediate confirmation with a pH meter is necessary. If the pH is more than 0.5 pH units below fresh, prefiltered seawater then it must be assumed that either the seawater is not being replaced frequently enough (see below, p. 178) or that the bacterial filter is not meeting the waste demands of the culture system. The first remedial step is to replace the seawater in the system. If this cannot be done, slowly add 100 ml increments 0.1 *N* NaOH to the top of the bacterial filter until the pH of the system is within 0.2 pH units on the acid side of fresh seawater. If subsequent replacement of the seawater does not solve the problem, then the bacterial filter must either be too small for the size of the culture system or not functioning as it should. If the filter is too small, then either it should be replaced with a larger one or augmented with a second. If the filter is not functioning properly, the filter bed material will have a foul or hydrogen sulfidelike odor. In this case, the filter bed media must be discarded, the structural parts cleaned, and the filter reassembled using new supplies of gravel and crushed shell.

Aside from the above situations and remedies, it is necessary to follow routine water maintenance and recycling. Firstly, it is important that only prefiltered seawater be used in the aquaria. If raw seawater is collected, then it must be filtered at the laboratory as described (see the section on "Management of Water Quality," p. 19).

Secondly, it is important to have an on-going method for replenishing seawater in the aquarium system. This is necessary to ensure that some unknown essential trace elements are not depleted from the seawater or that possible nonbiodegradable materials do not accumulate to toxic levels. In addition, we attempt to conserve water by recycling it. The procedure is diagrammed in Figure 12–7. Basically, incoming raw seawater from the ocean is prefiltered and the salinity adjusted to match

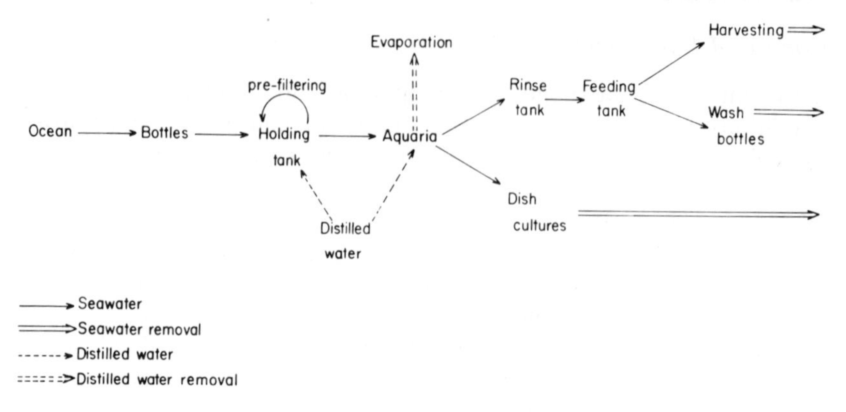

FIGURE 12–7 Seawater recycling and replenishment.

that desired for the aquarium culture. Prefiltered seawater is added to the aquarium system at a rate designed to match the rate of removal from the aquaria. About half of the removed water amounts to 5 gal once a week when the dish cultures are placed in aquarium seawater after cleaning rather than the usual fresh prefiltered seawater. The remaining amount of removed water occurs as the result of using feeding tank water each day to harvest the *Artemia* nauplii (1½ liters/day) and to fill the wash bottles (2 liters/day). The water removed in this way from the feeding tank is replaced with cleaner water from the rinse tank, and this in turn is replaced with water from the aquarium. The amount of water removed from the aquaria through the rinse and feeding tanks is about 5 gal/wk. Thus, the total amount of water removed and replaced in the aquarium cultures is about 10 gal/wk, or a little more than 10 percent of the seawater in the aquaria cultures.

Lighting Three 40-watt fluorescent bulbs supply most of the light needed by the symbiotic algae in each aquaria. The bulbs are 4 ft long and are stacked vertically about 2″ apart and about 4″ from the back wall of the aquaria. As mentioned, they are on a 12-h on/12-h off schedule. Other light is provided via ceiling fluorescent light fixtures. Natural sunlight has not been found essential. As long as the anemones maintain their natural brown pigmentation as imparted by the symbiotic zooxanthellae and orient their oral discs and tentacles toward the fluorescent light fixtures at the side of the aquarium, it can be assumed that their lighting needs are being adequately supplied. Anemones reared under darkened conditions will lose much of the brown pigmentation in a relatively short period.

Care should be taken to remove algal growths routinely from the sides of the aquaria so that the light is not obstructed from reaching the anemones. Care should also be taken to remove salt spray deposits periodically from the fluorescent light bulbs and lighting fixtures to prevent corrosion.

Temperature Subtropical anemones such as *A. pallida* do very well at ambient room temperature and are also very tolerant of normal diurnal and seasonal fluctuations within the range of 70–80°F. If the laboratory is suspected of being too cold for the anemones, an immersion-type aquarium heater can be used. If, however, ambient temperatures are too high for the particular anemones then a refrigeration system will have to be installed in the aquarium cultures or the room will have to be cooled.

Aeration Anemones do not seem to require any supplemental aeration in order to meet their respiratory needs. Aeration of the aquarium water is recommended, however, to facilitate vertical circulation of the water and to prevent the accumulation of a scum layer at the air–water surface. Care should be taken to position the air stones so that air bubbles do not interrupt the syphons and so that the seawater spray does not come in contact with corrodible equipment (e.g., fluorescent light fixtures).

13

Culture Methods for Rearing Polychaetous Annelids Through Sexual Maturity

DONALD J. REISH

INTRODUCTION

Polychaetes are segmented annelids that are present in all types of marine environments. They are ecologically important in the benthos where they generally constitute at least one-half the total number of macroscopic species and specimens. In the past many of the studies with polychaetes have been concerned with the systematics, ecology, and developmental aspects. Because of the limitations of seawater facilities, much of the experimental work had been done at marine biological stations. With the advent of artificial sea salts, greater perfection of cultural techniques, rapid transportation, etc., it is now possible to conduct experiments with polychaetes far from the ocean and at all times of the year. However, in spite of these advances, experimental research with polychaetes has not reached its full potential. Many of the species that have been successfully cultured, such as those described below, are small and are not conducive to certain types of research. The routine use of living polychaetes in educational institutions is minimal, and therefore students have not been exposed to this group.

The advantages of polychaetous annelids as research organisms are many. The unique position of the annelids on the phylogenetic tree with the appearance of true body segmentation, the appearance of well-developed organ systems, and the occurrence of different circulatory systems and blood pigments offer not only opportunities along academic lines, but also along fundamental medical research lines.

Papers dealing with developmental life history studies began about 140 yr ago and have continued to appear to the present time. In spite of the nearly 1,000 papers written on some aspects of reproduction and development of polychaetes, it was not until recent times that these data were synthesized (Schroeder and Hermans, 1975). No cultural technique manual has been written previously for the polychaetes, although some procedures are given in Needham *et al.* (1937) and Reish (1974). While some of the procedures included in these two papers plus what is given herein can be utilized for many other species of polychaetes, each species has its own set of needs and requirements which must be met before any degree of success can be expected.

While there are over 10,000 species of polychaetes, the complete life cycle (egg to egg) has only been worked out for less than 20 species (Reish, 1974), several of which belong to the single genus *Ophryotrocha* (Åkesson, 1973). Routine laboratory cultures of even a few species of polychaetes are a development within the past 20 yr. Shipments of *Ophryotrocha* from Sweden to other parts of Europe and the United States have been made by Åkesson. Subcultures of *Neanthes arenaceodentata* have been made in the states of California, Washington, Texas, Mississippi, Maryland, and Rhode Island from the original isogenetic strain established by the author from six specimens collected from Los Angeles Harbor in 1964. *Capitella capitata* was shipped to Rhode Island and Marseille, France, on two occasions in connection with intercalibration laboratory bioassay experiments (Reish *et al.*, 1978). The convenience of laboratory cultures of an isogenetic strain frees the laboratory researcher from the uncertainties of field collections. Experiments can be scheduled ahead of time with the knowledge that the organism will be available. As these species of polychaetes are used in research, undoubtedly the cultural procedures for additional ones will be worked out that will open additional opportunities for research.

MATERIALS AND METHODS

SEAWATER

A good supply of seawater is essential in order to be successful in culturing polychaetous annelids. This does not necessarily preclude that such work can only be done at a marine biological station equipped with a flowing seawater system. Seawater can be transported to the laboratory where it is stored and used as required, or commercially available sea salts can be purchased and mixed as needed. Whatever the source, it may be necessary to filter the water additionally to remove sediments, organisms, or impurities. In some instances it is helpful to filter the water additionally through a Millipore filter to remove some of the bacteria. Ultraviolet light treatment of seawater has proven to be useful in killing bacteria. Depending upon the nature of the research, periodic water quality analysis for salinity, pH, nutrients, heavy metals, and organic materials may be important not only in the interpretation of results, but equally important in interpreting unexplained failures.

For best results it is important to culture the animals at a more or less constant temperature using either a cold bath system, running seawater, or a constant temperature room. The particular temperature depends upon the geographical source of the living material. Polychaetes collected from a cold climate logically require a colder laboratory temperature. The three species described here are cultured at 19.5 ± 1°C. It is inconvenient to culture these polychaetes below 17°C because of the greatly reduced growth rate.

AQUARIA

Various-sized aquaria have been used to culture polychaetes. The size depends, in part, on the size of the worm and whether or not it exhibits cannabilistic behavior. Petri dishes have proved to be adequate for minute *Ctenodrilus serratus*. One-gallon jars are readily available and make good aquaria for culturing *Capitella capitata*. Mature pairs of *Neanthes arenaceodentata* are maintained in gallon jars from the reproductive period through incubation of the young. When the juvenile worms begin to feed, they are transferred to 15-gal aquaria that provide adequate space for this cannabilistic species.

AERATION

Aeration is provided by aquaria stones connected by tubing to a compressed air system. A filtration system may be provided (Figure 13–1) which helps to maintain water quality.

Hinegardner (1969) described an aeration–circulation device that was adapted for culturing developing eggs and embryos. This is described on p. 294. This device has been successful in rearing *Neanthes arenaceodentata* embryos that had been abandoned by the father (see below) and species that have planktonic larvae.

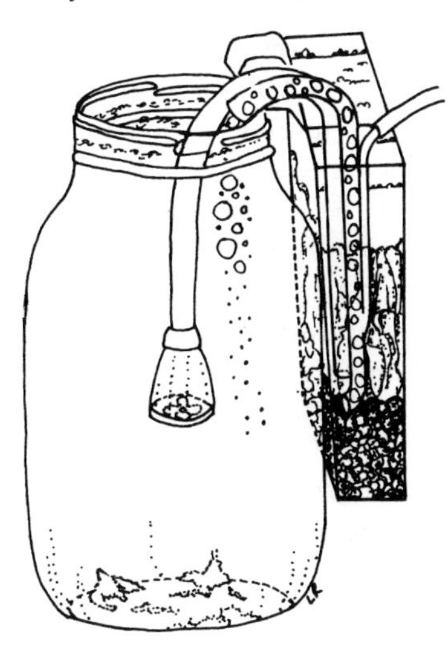

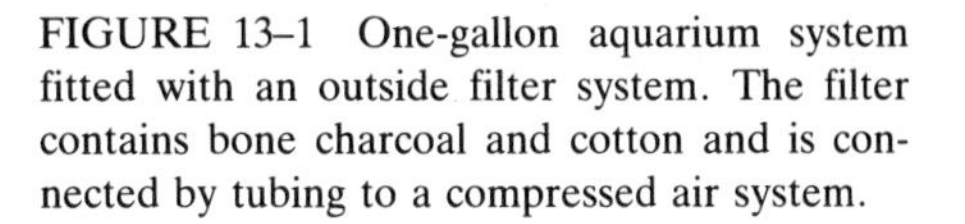

FIGURE 13–1 One-gallon aquarium system fitted with an outside filter system. The filter contains bone charcoal and cotton and is connected by tubing to a compressed air system.

FOOD

Many types of food have been utilized in culturing polychaetes, including phytoplankton, diatoms, frozen brine shrimp, frozen spinach, alfalfa flour, *Enteromorpha* (Figure 6–2), and the commercially available fish food Tetramin and Biorell. The emphasis herein will be to make use of nonliving food to feed both the embryos and adults. However, in some species, which are not covered here, it has been necessary to feed the planktonic larval stages living phytoplankton. Several different foods can be used in culturing *Neanthes arenaceodentata, Capitella capitata,* and *Ctenodrilus serratus.* Dried foods are favored since they will be available whenever needed. Furthermore, time and space need not be used to culture one organism to feed another.

Many species of *Enteromorpha,* such as *E. crinita,* grow in large quantities along the intertidal zone of estuaries. Peaks in abundance occur that will vary according to the locality, but generally they occur in the spring and fall. Collections are more conveniently made during the high-tide periods because the seaweed will be floating upon the water surface and will be more easily collected. The more vigorous growths of this alga are selected, which are grass-green in color. The fronds are washed at the locality, squeezed, and placed in a container. After a sufficient quantity has been collected, the alga is transported to the laboratory where it is washed several times in seawater. The alga is dried either in an oven or in sunlight. Generally, the latter method is more satisfactory. The alga should be spread out on screens that speed the drying process. After drying, the alga can be stored indefinitely.

Enteromorpha is used as polychaete food either as resoaked fronds or as a ground-up flour. In the former case, the dried *Enteromorpha* is placed in a dish containing seawater and kneaded until it is well soaked. Dried *Enteromorpha* should not be placed directly into an aquarium containing worms, because the fronds generally do not separate, which leads to a formation of a fungal growth. *Enteromorpha* flour is used to feed small worms such as *Ctenodrilus*. It is prepared by grinding the dried alga in a blender. The ground alga is then passed through a series of sieves, with the finest sieve having an opening of 0.06 mm. The *Enteromorpha* powder is collected on a pan beneath the finest screen. This powder can be stored for an indefinite period of time.

Tetramin is a commercially available fish food that has proved to be a useful polychaete food. Since this product consists of flakes, its direct use leads to fungal growth, foul water, and death of the organism. Tetramin powder is prepared in the same manner as described for *Enteromorpha*.

Alfalfa flour is a commercially available product that can be used as a prepared food source for *Neanthes arenaceodentata*. It has been used with *Capitella capitata*, but the tetramin powder is a better food supply for that species. The alfalfa flour is mixed with seawater at a ratio per weight of about 0.01 g alfalfa to 1,000 ml seawater. The mixture is stirred and this amount of food is sufficient for a population of *N. arenaceodentata* in one 15-gal aquarium for 1 wk.

PROBLEMS IN CULTURING POLYCHAETOUS ANNELIDS

Most of the difficulties and failures in culturing species can be attributed to the lack of regular care. Since many aquaria of different sizes are involved and since there may be different stages involved in the life history, it is convenient to maintain records on the sides of the aquaria and to arrange aquaria according to the stage of development. Feeding should be done on a specified day(s) of the week.

Two primary feeding problems are the result of either feeding too much food or not soaking and kneading *Enteromorpha* sufficiently prior to use. Both of these situations lead to the appearance of a fungal growth over the surface of the food resulting in the inability of the worm to penetrate to feed. As a general rule, it is better to underfeed than overfeed.

Microorganisms, especially bacteria, protozoans, nematodes, and copepods, are almost always present and almost impossible to eliminate. Frequent changes of water controls the growth of these microorganisms. Sterilized glassware, the use of millipore-filtered seawater, ultraviolet light, and antibiotics can help to minimize the problems caused by microorganisms. If periodic care is provided, these additional measures are not necessary in the culturing of *N. arenaceodentata, C. capitata,* and *C. serratus.*

Occasionally, the male of *N. arenaceodentata* and the female of *C. capitata* will abandon the embryos within their tube. The cause of abandonment is unknown, but it seems to be the result of foul water or if the worm or container is handled too much. Embryos within the tube that have been abandoned by the parent generally die within a day. The use of the magnetic stirring device of the type described on p. 294 has been the only successful method in culturing these abandoned embryos.

Gregarine protozoans occur in polychaetes, but no evidence exists that this or any other parasite, if indeed there are any, cause any problems in culturing the three species of worms discussed here.

NEANTHES ARENACEODENTATA

OCCURRENCE

Neanthes arenaceodentata is widely distributed throughout the world; it has been reported from California, Baja California, New England, Florida, Europe, Central Pacific, New Zealand, Australia, Philippine Islands, India, and South Africa. Typically, it is found in intertidal to subtidal estuarine environments living within the sediments. It may be present in numbers up to 1,000/m^2. Specimens burrow into the sediment and construct a tube that is lined with mucous. Specimens measure up to about 4 cm in length and are tan to yellow-tan in color. It will be necessary to make positive identification of this species in the laboratory. Consult Pettibone (1963) for description and figures of the worm. Since this species lives with sediment, it is difficult to collect them individually in the field. They can be collected most conveniently by washing sediment through a 1.0 mm mesh sieve in the field. Specimens are often injured during this procedure, but many of them will regenerate the lost posterior end under laboratory conditions.

SYSTEMATIC NOTES

The name of this species is not stabilized in the scientific literature. It was known as *Neanthes caudata* (delle Chiaje) until 1963, and it is still referred to under this name in the European literature. Pettibone (1963) renamed it *N. arenaceodentata* (Moore) because *N. caudatus* was a homonym of an earlier species. Later, Day (1973) changed the name to *Neanthes acuminata* (Ehlers) because this rather than *N. arenaceodentata* was the next available name after *N. caudata.* Nevertheless, the name that is applied to this species will probably be one of these three names for some time.

LIFE HISTORY

The early stages in the life history of this species were studied by Herpin (1926) at Cherbourg, France, and the complete life cycle was followed by Reish (1957) in Los Angeles, California.

It is impossible to distinguish immature males from females on the basis of morphology. However, a behavioral difference occurs that can be utilized to distinguish the sexes. Males will fight males and females will fight females. This fighting behavior consists of extending the jaws to grasp the opposing worm (Figure 13–2). They can be cannibalistic. When placed together, a male and female will come together and lie side by side. If one specimen has constructed a tube, the two will lie within the tube until eggs are laid. It is possible to detect this behavior difference in young specimens possessing 21–25 segments. In order to

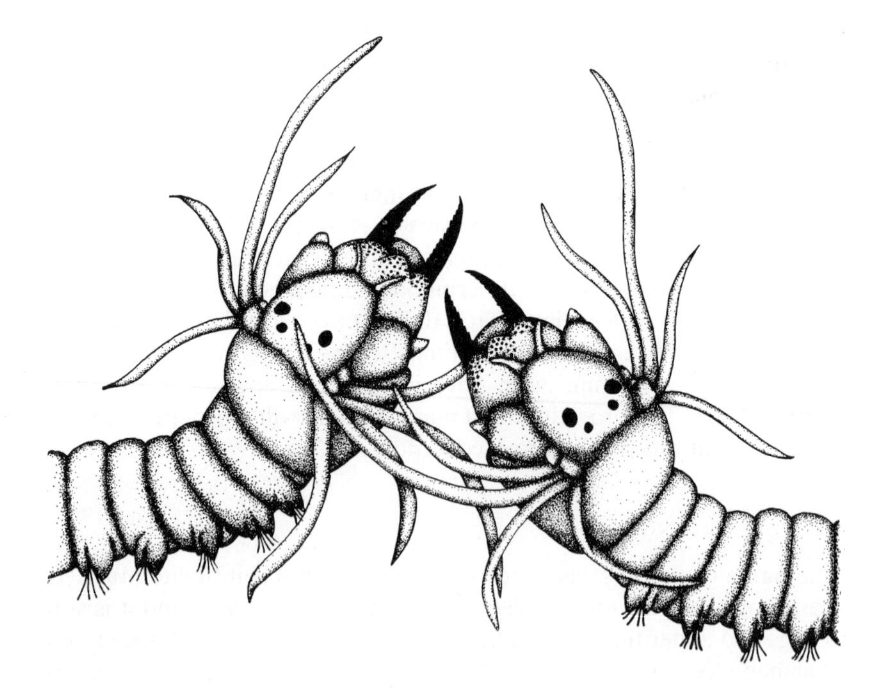

FIGURE 13–2 Two adult *Neanthes arenaceodentata* of the same sex in fighting position.

determine the sex of an immature worm, place it in a Petri dish with a female containing developing eggs within her coelom and observe the behavior of the two when they come in contact with each other. If they fight, the immature worm will be a female. If they come to lie side by side, then the immature worm is a male.

The eggs of *N. arenaceodentata* are formed within the walls of the parapodia. Shortly thereafter they break free and mature within the coelom. The muscle walls, especially the longitudinal muscles, slough off and are taken in by the eleocytes, which transfer material to the maturing eggs. The mature eggs of *N. arenaceodentata* are large, measuring up to 650 μ in diameter. Spawning has never been observed in this species, but presumably it occurs within the tube with external fertilization. Presumably the eggs are laid through bilateral breaks in the body wall between successive parapodial lobes. During maturation of the egg, about 75 percent of the body weight of the female is transferred to the eggs. The female either dies within a day following spawning or is eaten by the male. The number of eggs laid vary from about 700 in field-collected specimens to about 150–250 in laboratory-reared specimens. The embryos are clumped in the central part of the tube around the mid-body region of the male.

Development proceeds with the tube from fertilization until the 18–21 setigerous segment stage (Figures 13–3a,b,c). The developing eggs are incubated by the male, who presumably circulates water through the tube with his body undulations. The fertilized eggs are yellow in color as the result of the large amount of yolk material, but as the yolk supply is utilized during development, the young become tan in color. *Neanthes arenaceodentata* lacks the free-swimming trochophore larval stage characteristic of most polychaetes. The shape of the body is distorted as a result of the large quantities of yolk material; in fact, the yolk bodies may be observed in the future digestive tract up to nearly the 18 setigerous (i.e., setae bearing) segment stage (Figures 13–3a,b,c). The young worms leave the tube of the parent when they attain 18–21 setigerous segments (Figure 13–3d). Shortly after leaving the tube, they construct a tube and commence feeding. Growth, as measured by number of segments, is rapid during the prefeeding period, with about one segment added per day. No new segments are added for about a week during the period in which the young worm leaves the tube and commences feeding. Thereafter the growth rate is again one new segment a day until the adult number of segments (65) is reached (Reish, 1957). Growth, as measured by body weight, occurs at a regular rate in the male throughout its life. However, the female no longer increases in weight after the ova appear in the coelom. Presumably feeding is pre-

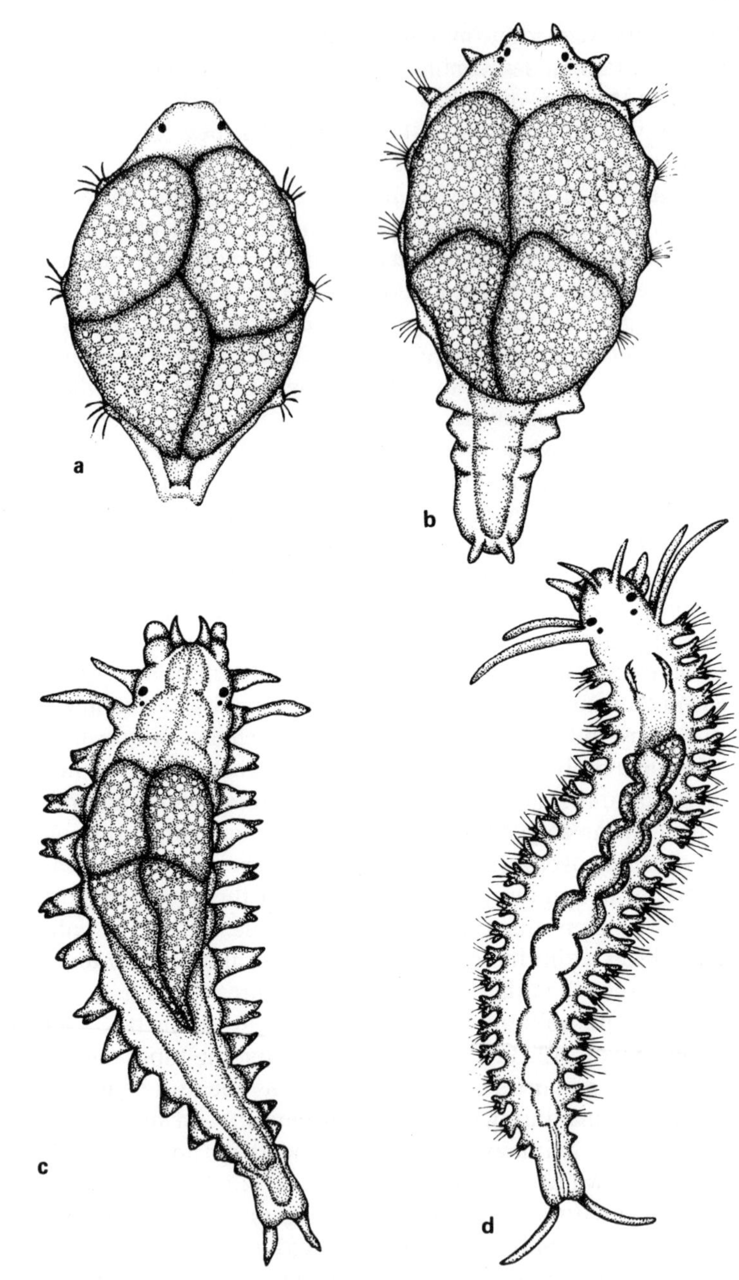

FIGURE 13–3 Larval stages of *Neanthes arenaceodentata:* a. three-segmented stage; b. four-segmented stage; c. twelve-segmented stage; d. juvenile stage.

vented by pressure of the developing ova upon the digestive tract. The female loses weight during the maturation of the ova and only weighs about 25 percent of her former weight after spawning (Reish and Stephens, 1969). Under laboratory conditions it takes 3 to 4 mo for *N. arenaceodentata* to complete its life cycle at about 20–22°C. Presumably a longer time is required in the field.

Though the female dies within a day following spawning, under laboratory conditions the male is capable of reproducing at least a second time after the completion of the initial incubation period.

TECHNIQUES OF HANDLING *NEANTHES ARENACEODENTATA*

Adults

Stock colonies of adults can be maintained at room temperature with a minimum of time by utilizing aquaria of about 10–15 gal size with a maximum population concentration of approximately 75–100 worms. The aquarium must be provided with at least two aquaria stones. Add about 0.5 g of alfalfa flour per 5 gal of water per week. The alfalfa flour should be mixed with seawater prior to use.

If all specimens are going to be used for experimental studies within a month, it will not be necessary to change the water in the aquarium. However, the water level should be checked periodically and distilled water added to keep the level constant. If an outside filter system is used, the air stem must be checked weekly and cleared of accumulated salts.

If the specimens are not being utilized within a month, then at about 4–6 wk the water should be changed. Since sexual maturity will be reached in 2–3 mo after aquaria are established, at about 10 wk all specimens should be removed and paired. Single males and females are used to establish additional aquaria (see below).

Larvae

No special care is required for specimens having less than 18–21 setigerous segments. These nonfeeding larvae are cared for by the male parent within his tube.

Immature Worms

After the young worms leave the tube of the parent, the young worms, which are just beginning to feed, should be removed as soon as possible to prevent cannibalism by the parent. The young worms should be placed in 10–15-gal aquaria at a maximum concentration of about 10 specimens per gallon; however, if this concentration is used, it will be necessary to divide the population into two similar sized aquaria within 4 wk. Add about 0.5 g alfalfa flour per 5 gal per week. Prior to feeding each week, examine the aquarium to make certain food is required.

Reproductive Specimens

Females with maturing eggs appear yellow-orange in color. During routine care and examination of specimens, set aside in a separate Petri dish any female containing eggs. Separate from the stock colony a similar sized worm whose sex is unknown and test for its sex as described above. Repeat the procedure until a male is found. Place both specimens in a 1-gal aquarium filled with seawater and provided with an outside filter system (Figure 13–1). Add about 0.2 g dried *Enteromorpha* that has been soaked and kneaded in seawater. Each succeeding week feed about 0.2 g dried *Enteromorpha;* however, do not add any food if the worms have not utilized all the food from the previous feeding. After the initial week, use a hand lens to see if eggs have been laid. Check every 2–3 days thereafter until they are laid. Note the approximate date when eggs were laid and continue to feed on schedule. Examine the developing eggs every other day beginning at 2 wk for emergence from the parent's tube. Since the male can become cannibalistic following emergence of his offspring, it is important to remove him. The recently emerged young worms should be handled as outlined above. The parent may be utilized either for a second reproduction period or discarded.

USE OF *NEANTHES* IN EXPERIMENTAL RESEARCH

Neanthes arenaceodentata has been used in many experimental studies, most of which have been concerned with environmental effects. These studies have involved the effect of some parameter on survival in *Neanthes;* they include the effect of experimentally reduced dissolved oxygen concentrations, reduced salinities, increased concentratons of nutrients, petrochemicals, and heavy metals (Reish, 1970; Reish *et al.*, 1974a,b; Oshida and Reish, 1975; Oshida *et al.*, 1976; Mearns *et al.*, 1976; Rossi and Anderson, 1976; Rossi *et al.*, 1976; Oshida, 1977; Carr and Reish, 1977; Pesch and Pesch, 1980). Some sublethal studies include the effect of reduced dissolved oxygen concentrations on hemoglobin concentrations (Raps and Reish, 1971), growth and production of oocytes (Davis and Reish, 1975), and activity levels of malate and lactate dehydrogenase (Cripps and Reish, 1973). Undoubtedly, *Neanthes* will continue to be used as an experimental animal in the future. It is easy to transport to other localities, and it is a convenient size for use as a bioassay organism, such as in body burden analyses. Hemoglobin can be extracted from the dorsal blood vessel to permit blood chemistry studies, the large ova would facilitate studies not possible with small sized ova, and its short life cycle would permit research of various types to be carried out on large numbers of individuals. A laboratory population of this species was established in 1964, and since that time it has undergone 70–100 generations, making this population an isogenetic strain.

CAPITELLA CAPITATA

OCCURRENCE

Capitella capitata is a cosmopolitan species that is generally found in estuarine waters. It has been described as a noncompetitive or opportunistic species because it flourishes in the absence of other polychaete species, especially within polluted waters. It is the most widely used indicator organism of marine pollution. Specimens generally measure less than 2 cm in length, although larger individuals occur, especially in the vicinity of domestic outfall sewers. Specimens burrow into the sediment, where they construct tubes lined with mucous. They are pale pink to blood red in color. Field collections can be made by several methods: (1) washing the sediment through a 0.5-mm sieve; (2) by bringing sediment back to the laboratory, where the material is placed in a pan with seawater and the worms allowed to crawl out; (3) collecting fouling organisms from quiet or polluted waters and allowing the worms to free themselves as in the case with sediment; or (4) suspending sediment traps (Reish, 1961) in estuarine waters for a month, after which the contained sediment is examined for specimens of *C. capitata* (as well as for other species).

SYSTEMATIC NOTES

The classical concept of *Capitella capitata* as a species has been questioned recently from various points of view (Grassle and Grassle, 1975; Reish, 1977b; Warren, 1976b). *C. capitata* has been separated into several species based primarily on the initial appearance of hooded hooks (Hartman, 1969; Warren, 1976b), but Reish (1977a) presented experimental evidence that all these species are not valid. Grassle and Grassle

(1974) introduced the concept of sibling species in which the various forms are separated on the bases of fine detail of the hooded hooks, the first appearance of hooded hooks, presence or absence of genital hooks, and the size and number of eggs laid. Perhaps the best solution to the present dilemma is calling the worm *Capitella capitata* and noting, if possible, which stage it belongs to according to the scheme of Grassle and Grassle (1975).

LIFE HISTORY

Eisig (1898) was first to describe the life history of *C. capitata.* Others have contributed additional data, including Thorson (1946), who summarized all known data on the species. More recently, Reish (1974) established a laboratory colony of *C. capitata* and described the life history for sibling species Type 1 of Grassle and Grassle (1975). This species is sexually dimorphic with the male possessing copulatory setae on segments 8 and 9 (Figure 13–4a). The female lacks such specialized setae (Figure 13–4b). Copulation occurs but is rarely observed. Sperm are transferred to the female with fertilization occurring either internally or at the time of discharge. Eggs are laid within the female's loosely constructed mucoid tube (Figure 13–4c). The fertilized egg remains fixed presumably by a mucoid secretion until the trochophore stage (Figure 13–5a). There is a considerable amount of variation in the number of eggs laid, but under laboratory conditions it usually ranges from 150 to 250.

The female incubates the fertilized eggs during the early developmental stages. Incubation consists of periodic body undulations by the female that circulate water through the tube. The fertilized eggs are initially white in color and measure about 250 μ in diameter. As development continues over the next 4 days, the embryos become darker in color and at the trochophore stage they appear gray-green in color.

The trochophore stage is reached about 4–6 days after egg-laying (Figure 13–5a). The trochophore is capable of moving freely within the tube either by ciliary movement or by contraction of longitudinal muscles. It may either swim free of the tube and become planktonic or proceed directly into the metatrochophore stage (Figure 13–5b) and begin to form its own tube as a side branch from its parent's tube. The planktonic trochophore, if it occurs, is of short duration, and the individual soon settles to the substrate and develops into the metatrochophore stage. The metatrochophore stage lasts 1–2 days before resembling a juvenile adult (Figure 13–5c). Growth is rapid and temperature dependent. Eggs begin to develop in the coelom of the maturing female in about 20 days with fertilized eggs being laid at 25–40 days at 20°C.

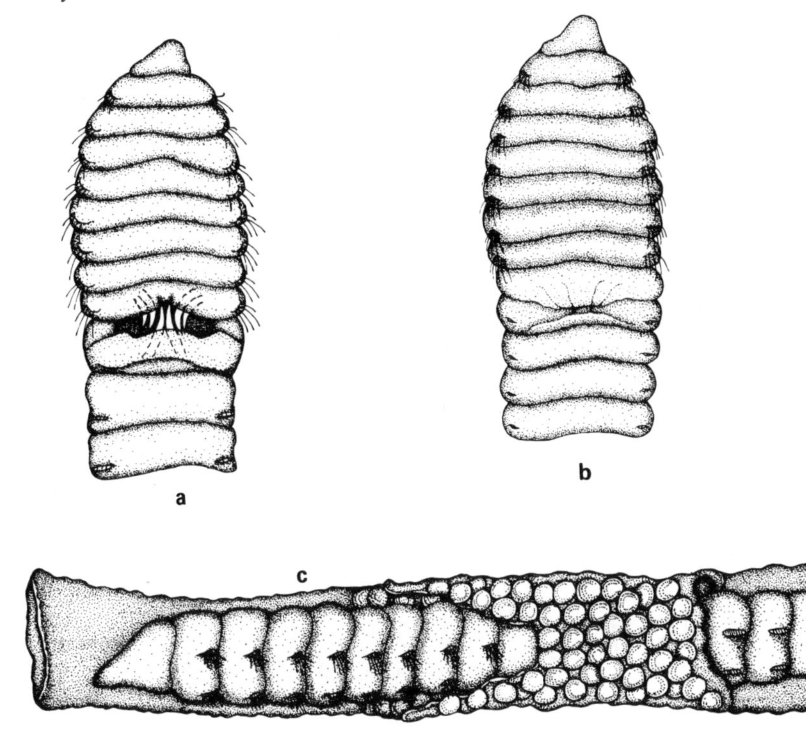

FIGURE 13–4 *Capitella capitata*. a. male showing copulatory setae of segments 8 and 9; b. female; c. female incubating eggs within her mucoid tube.

Both sexes are capable of reproducing more than once. Females have been observed to have had three successful egg layings under laboratory conditions. The second egg-laying occurs about 5–10 days after the first broods have left the tube. Under laboratory conditions both sexes appear to lose their vitality with increasing age as indicated by the change in color of the blood. The characteristic blood red color of young adults becomes dull red in color with increasing age.

TECHNIQUES OF HANDLING *CAPITELLA CAPITATA* ADULTS

Stock colonies of adults can be maintained at room temperature with a minimum of time by utilization of 1-gal jars as aquaria. Each aquarium should have about 2,500 ml of filtered seawater and be provided with aeration. Many adults (10–30) can be placed in a single aquarium. About 0.15 g dried tetramin flour is added to each aquarium per week.

If specimens are not being removed from a particular aquarium, it

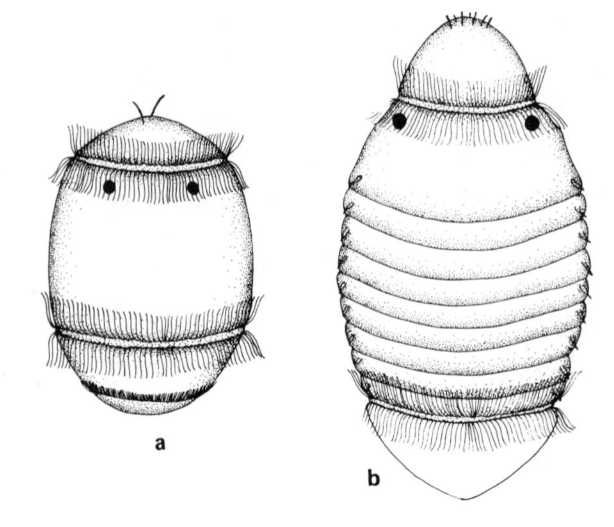

FIGURE 13–5 *Capitella capitata*. a. trochophore larval stage; b. metatrochophore larval stage; c. juvenile.

will be necessary to divide the population into two or more aquaria every 3 wk. Water should be changed at this time. If specimens are being removed periodically for experimentation, it will not be necessary to divide the population, but the water should be changed about every 3 wk.

TROCHOPHORE LARVAE

The trochophore larvae of the lecithotrophic eggs of this population do not need any special handling nor special techniques; the female provides the essential care during the early stages of development. Trochophores can be removed from the female's tube and pipetted into separate gallon aquaria. This procedure is especially useful if a large population of this species is needed. Approximately 0.1 g of tetramin flour should be added as food. Additional food should be added a week later. These larvae will soon become juveniles and adults and can be treated in the same manner as indicated above.

USE OF *CAPITELLA* IN EXPERIMENTAL RESEARCH

Capitella capitata has been used in both laboratory and field environmental studies. Reproduction (Reish and Barnard, 1960) of *Capitella* was studied under field conditions in Los Angeles Harbor. Grassle and Grassle (1974, 1975) studied the settlement of *C. capitata* following an oil spill; later they were able to distinguish six sibling species on the bases of the electrophoretic patterns of eight enzymes. Laboratory investigations have included the effects on survival of reduced dissolved oxygen, reduced salinity, increased concentrations of nutrients, and heavy metals (Reish *et al.*, 1974b; Warren, 1976a; Reish, 1977b), detergents (Foret-Montardo, 1970; Bellan *et al.*, 1972), and petrochemicals Rossi *et al.*, 1976; Carr and Reish, 1977). Some of these studies have examined the effect of the pollutant on reproductive success as measured by the number of eggs laid (Foret-Montardo, 1970; Bellan *et al.*, 1972; Reish, 1977a), including induction of abnormal bifurcated larvae (Foret-Montardo, 1970; Reish *et al.*, 1974a; Reish, 1977a).

While *Capitella* is a pollution tolerant opportunistic species, undoubtedly, it will continue to be used in marine bioassay studies in the future. The short life cycle of 30–45 days is convenient for studies involving a complete life history. Trochophore larvae are easily obtained and could be utilized in certain types of studies. The discovery of six sibling species with different methods of reproduction presents interesting avenues to pursue. The open blood system with hemoglobin as the oxygen carrier may offer distinct possibilities in the study of blood chemistry (Wells and Warren, 1975).

CTENODRILUS SERRATUS

OCCURRENCE

Ctenodrilus serratus is a minute polychaete that is widespread throughout the temperate regions of the world. Undoubtedly, it is more prevalent but is overlooked because of its small size. It is difficult, if not impossible, to observe this species in the field because of its small size. The most convenient way to collect *Ctenodrilus* is to gather clumps of fouling organisms, especially mussels and associated fauna, from boat floats or pilings and bring this material to the laboratory. Examine a mussel under the dissecting microscope and look for a small reddish-black worm 5 mm or less in length (Figure 13–6a). Because of the dark color of the worm and mussel, it is often more expeditious to look for empty mussel shells and look for *Ctenodrilus* on the inside; the worms show up against

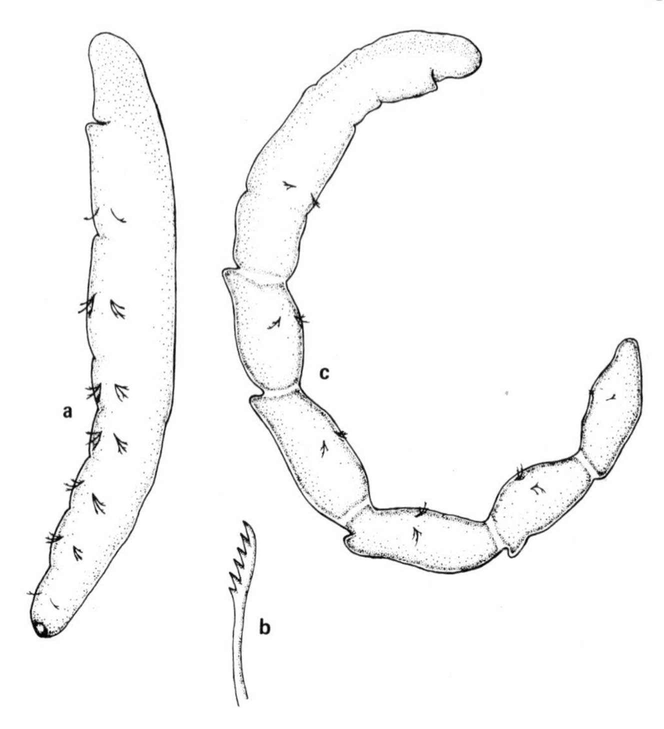

FIGURE 13–6 *Ctenodrilus serratus*. a. adult; b. simple seta; c. asexual reproduction with five buds forming by transverse fission.

the white background of the nacreous layer of the shell. This species lacks parapodia and moves over the surface in an earthworm-like fashion. If this technique is unsuccessful, suspend a wood block and sediment bottle trap (see Reish, 1961, Plate 2, Figures 1–4) from a boat dock in protected waters. Remove 1 mo later. Either examine the wood block under the dissecting microscope, or, if the accumulation of settled organisms is larger, scrape this material from the block and examine this material for *Ctenodrilus* under the dissecting microscope. If unsuccessful in locating *Ctenodrilus,* examine the sediment within the jar.

To verify the species, examine the specimen under a compound microscope. The most conspicuous features of the worm is the small number of setigerous segments and its characteristic seta. The seta (Figure 13–6b) is simple, and each is distally expanded and provided with four–six serrations along one side. Setae are of one type and generally number three–four per side per segment.

SYSTEMATIC NOTES

Family Ctenodrilidae is known for only three genera with the genus *Ctenodrilus* containing the cosmopolitan *C. serratus* and a second species. The unique, simple, coarsely denticulated seta make it an easily identifiable species.

LIFE HISTORY

Ctenodrilus serratus has been reported to reproduce in a variety of ways, by protantric hermaphroditism, internal gestation, and by transverse fission (Korschelt, 1931). Asexual reproduction by transverse fission (Figure 13–6c) is the common method. The life cycle is generally completed within about 20 days at 20°C.

TECHNIQUES OF HANDLING *CTENODRILUS SERRATUS*

Since this species has such a short life cycle, it is unnecessary to use special techniques with the different stages. Stock colonies can be maintained at room temperature (20°C) with a minimum of effort. Worms can be cultured in a variety of containers. Petri dishes are particularly useful for establishing small cultures. Place 10–25 specimens in a petri dish together with 50 ml of Millipore-filtered seawater and add one pasteur pipette drop per worm of an *Enteromorpha* flour–seawater suspension of a concentration of 350 mg flour to 25 ml seawater to each dish. Each dish will provide 200–500 specimens within a month. If these specimens are not going to be used in an experiment, subculture these worms. Larger cultures can be established in gallon jars with a similar feeding schedule. Millipore-filtered seawater should be used and aeration provided. Seawater should be changed on a monthly basis and the colony fed at this time. Cultures of *Ctenodrilus* in gallon jars should be reestablished about every 3 mo.

USE OF *CTENODRILUS* IN EXPERIMENTAL RESEARCH

Ctenodrilus serratus has not been used to any great extent as an experimental organism in the past. Earlier papers were primarily concerned with the reproductive aspects of the worm (Korschelt, 1931). Recently, *Ctenodrilus* has been used to study the long-term effects of heavy metals (Reish and Carr, 1978; Petrich and Reish, 1979) and petrochemicals (Carr and Reish, 1977), which has involved the entire life cycle. While the usefulness of this species has limitations because of its small size, the rapid life cycle and small size permit bioassay studies involving a complete life cycle. Its small size may make it a convenient organism in which to study metabolic pathways or nutritional requirements. Its asexual method of reproduction also offers several unique possibilities for experimental studies.

14

Laboratory Culture of *Aplysia*

MARILYN SWITZER-DUNLAP *and* MICHAEL G. HADFIELD

INTRODUCTION

Opisthobranch molluscs of the family Aplysiidae, commonly known as sea hares, inhabit shallow water marine environments throughout the world. Aplysiids are herbivorous, and sizes of adults of different species vary from less than a gram to greater than 10 kg. Although more than two-thirds of the species in the genus *Aplysia* inhabit tropical or subtropical waters (Eales, 1960), temperate species have been better studied.

Aplysiids are being used extensively as research models in a variety of biomedical disciplines including neurobiology, behavior, hormonal control of reproduction, circadian rhythms, developmental biology, and others. Growing research use of these animals has intensified field collection and spurred efforts to rear them through their entire life cycles in the laboratory. The aims have been conservation, a large, reliable supply of research animals, and provision of specimens of known parentage, age, and background.

The aplysiid life cycle can be conveniently divided into five phases: (1) embryonic, (2) larval, (3) metamorphic, (4) juvenile, and (5) adult. Capsules of internally fertilized ova are laid in gelatinous masses by adult aplysiids. Embryonic development occurs within these masses. In most species, planktonic veliger larvae hatch from the masses and spend approximately 1 mo (under laboratory conditions) feeding and growing. Subsequently, the planktonic larvae settle on specific algae where they metamorphose into benthic juveniles. When metamorphosis is complete, juveniles commence feeding on seaweed and grow to sexual maturity. Adult aplysiids usually have an extended reproductive phase characterized by repeated spawning and during which they continue to feed and grow until they reach a maximum size often associated with peak spawn production. Thereafter, the animals' food intake decreases, weight and spawning decline, and death follows. Within this basic pattern, there is tremendous variability in the rate and size at which specific events occur.

The rearing of the veliger larvae of aplysiids is the most difficult aspect of their laboratory culture. However, in the past 5 yr, several successful methods for rearing larval aplysiids have been devised. The following species, all with planktonic larvae, have been cultured through all phases of their life cycles: *Aplysia californica* (Kriegstein *et al.*, 1974); *A. juliana, A. dactylomela, Dolabella auricularia,* and *Stylocheilus longicauda* (Switzer-Dunlap and Hadfield, 1977); and *A. brasiliana* (Strenth and Blankenship, 1978). Bridges (1975) raised *Phyllaplysia taylori,* a species with direct development, from eggs to subadults.

Franz (1975) provided a general discussion for culture of opisthobranchs. Additional papers on culture of other opisthobranchs with planktotrophic larvae that contain information pertinent to rearing aplysiids are the following: Franz (1971); Harris (1975); Kempf and Willows (1977); Perron and Turner (1977); Chia and Koss (1978); and Harrigan and Alkon (1978a,b).

Unless otherwise indicated, the following discussion concentrates on techniques that we have developed in our laboratory for the culture of a variety of aplysiids available in Hawaii: We have recently used these same techniques to successfully culture *Aplysia californica.* Information on procedures used in other laboratories is included with appropriate references. With minor modifications, our techniques can and have been used to rear a number of other opisthobranch species in Hawaii (cephalaspids, sacoglossans, and nudibranchs) and elsewhere.

CARE OF ADULT APLYSIIDS

AQUARIUM SELECTION AND MAINTENANCE

Adult aplysiids are maintained in a variety of seawater trays and aquaria supplied with either a continuous flow of fresh seawater or recirculating seawater. Except at coastal facilities, most research animals are maintained in recirculating systems and most of our comments will be so directed.

Those inexperienced in managing closed-system marine aquaria should read the first sections of this manual, and for additional information refer to Spotte (1973, 1979a,b) and/or King and Spotte (1974). Commercially available aquaria of at least 40-liter capacity are adequate for maintaining aplysiids. Our aquaria are supplied with a subgravel–airlift system of the type that uses airstones for the airlift. The airlifts are powered by high-capacity air pumps. Filtrant of dolomite or crushed coral rock of 2–5 mm diameter size is placed over the subgravel filter. When buildup of fecal matter becomes a problem for animals fed *ad libitum,* we add to the system an external power filter (Supreme Aqua-king) containing activated charcoal and polyester fiber.

Water quality is critical in maintaining any marine animal. If good-quality natural seawater is available, it should be used. Some coastal water may be heavily polluted with oil, pesticides, and/or sewage and should be avoided. We have had good success rearing juvenile and adult aplysiids (but not larval stages) in artificial seawater. Our experience has been with Instant Ocean Sea Salts® made up according to the manufacturer's instructions except that distilled water is substituted for tap water to avoid possible contamination from copper in our tap water supply. Other synthetic seawaters may do as well. In all cases, approximately 25 percent of the water in the aquarium should be replaced monthly and more frequently if buildup of nitrogenous waste products and particulate matter is a problem.

TEMPERATURE

Aquaria in our laboratory are kept at room temperature (24–26°C), which is within the normal temperature range of Hawaiian marine waters. To rear temperate species, it may be necessary to cool the water although we were able to maintain juvenile and adult *A. californica* at 24°C. Temperatures used to culture various species are shown in Table 14–1. Aquaria may be put in temperature-controlled rooms or aquaria with their own refrigeration systems may be built or purchased. We have had good experience with Instant Ocean Culture Systems (CS–30) for maintaining subambient temperatures. Refrigerated aquaria from other sources may do as well.

ANIMAL DENSITY

The density at which animals can be maintained in recirculating systems varies with size of animals, type of food, and frequency of feeding. Aquaria with animals being fed *ad libitum* can hold fewer animals than those with animals on a limited food regime. As a general rule, under optimal feeding, animals weighing less than 10 g can be kept at one per

TABLE 14–1 Minimal Duration in Days of the Embryonic, Larval, and Juvenile Phases and the Resultant Shortest Generation Times for Seven Aplysiid Species

Species	Embryonic Phase +	Larval Phase +	Juvenile[a] Phase =	Generation Time	Temperature (°C)	Source
Aplysia brasiliana	8	30	65	103	21–25	Strenth and Blankenship, 1978
Aplysia californica	9	34	86	129	22	Kriegstein *et al.*, 1974
Aplysia californica	11[b]	38	61	110	24–26	Switzer-Dunlap and Hadfield, unpublished
Aplysia dactylomela	8	30	66	104	24–26	Switzer-Dunlap and Hadfield, 1977, 1979
Aplysia juliana	7	28	33	68	24–26	Switzer-Dunlap and Hadfield, 1977, 1979
Dolabella auricularia	9	31	180	220	24–26	Switzer-Dunlap and Hadfield, 1977, 1979
Phyllaplysia taylori	30	none	>60	not known	14.5	Bridges, 1975
Stylocheilus longicauda	6	30	27	63	24–26	Switzer-Dunlap and Hadfield, 1977, 1979

[a] Includes the 2–5 day metamorphic phase; see text.
[b] Eggs collected at 16°C and warmed to 24°C over a 24-h period.

4 liters, animals from 10 to 100 g at one per 20 liters, and animals larger than 100 g at one per 40 liters.

FOODS

Each aplysiid species has its own algal food preferences; some feed on only one or a few algae, while others eat a greater variety. Both growth and spawning rates are functions of the algal diet (Carefoot, 1967, 1970). In our laboratory, animals are fed fresh seaweeds because they are readily available. Table 14–2 presents available data on known preferred foods of various aplysiid species. Although animals grow best on diets of fresh algae, when they are not available there are alternatives. *Aplysia juliana* will feed on oven- or air-dried *Ulva fasciata*. Many inland laboratories maintain adult *A. californica* and *A. brasiliana* on diets of commercially available dried red algae (Table 14–2). Although Winkler (1959) reported success in maintaining *A. californica* on diets of celery tops and parsley leaves, our attempts to feed garden produce (lettuce, spinach, and celery tops) to *A. juliana* were unsuccessful. If fresh algae are not available, commercially dried seaweeds seem to be the best substitutes.

SPAWNING

In young aplysias, mature sperm may be present before oocytes are fully grown (Smith and Carefoot, 1967), but for most of their reproductive lives, aplysiids are simultaneous hermaphrodites. Fertilization is internal and as animals are apparently unable to self-fertilize (Thompson and Bebbington, 1969), adults should be kept in groups of two or more to ensure mating and production of fertilized eggs. Solitary animals that have never mated will spawn unfertilized eggs.

Well-fed adult aplysiids can be expected to spawn readily, although the frequency of spawning varies among species. For instance, *Aplysia juliana* and *Stylocheilus longicauda* spawn daily or every other day during peak reproductive life, whereas *Dolabella auricularia* spawns only every 10 to 20 days (Switzer-Dunlap and Hadfield, 1979). Animals on restricted diets may spawn less often than well-fed animals.

Egg-laying in sexually mature aplysias can be induced by injecting them with extracts of the parieto–visceral ganglion (=abdominal ganglion) or of the "bag cells" associated with the ganglion (Kupfermann, 1967, 1970, 1972; Strumwasser *et al.*, 1969). Animals may be refractory to extracts if they have recently spawned. Temperature may be a factor in the ability to spawn. Whereas Strumwasser *et al.* (1969) found that

TABLE 14–2 Fresh and Dried Algal Foods for Adult Aplysiids

Aplysiid Species	Algal Foods[a]		Reference
	Fresh	Dried	
Aplysia brasiliana		"Laver," dried *Porphyra* (R)	Strength and Blankenship, 1978
Aplysia californica	*Ulva* sp. (C)		Kriegstein *et al.*, 1974
Aplysia californica	*Plocamium coccineum* (R) *Laurencia pacifica* (R)		Audesirk, 1975
Aplysia californica	*Laurencia* spp. (R) *Acanthophora spicifera* (R)	"Dulse," dried *Rhodymenia palmata* (R)	Jahan-Parwar, 1972 Switzer-Dunlap and Hadfield, unpublished
Aplysia dactylomela	*Laurencia* sp. (R) *Spyridea filamentosa* (R) *Acanthophora spicifera* (R) *Ulva fasciata* (C) *Ulva reticulata* (C)		Switzer-Dunlap and Hadfield, 1977, 1979
Aplysia juliana	*Ulva fasciata* (C) *Ulva reticulata* (C)		Switzer-Dunlap and Hadfield, 1977, 1979
Aplysia juliana	*Enteromorpha compressa* (C)		Usuki, 1970
Aplysia juliana	*Undaria pinnatifida* (P)		Saito and Nakamura, 1961
Aplysia juliana	*Ulva fasciata* (C)		Switzer-Dunlap and Hadfield, unpublished
Dolabella auricularia	*Spyridea filamentosa* (R) *Acanthophora spicifera* (R)		Switzer-Dunlap and Hadfield, 1977, 1979
Phyllaplysia taylori	diatom films on *Zostera*		Bridges, 1975
Stylocheilus longicauda	*Lyngbya majuscula* (Cy)		Switzer-Dunlap and Hadfield, 1977, 1979

[a] The algal class is given in parentheses: C, Chlorophyta; Cy, Cyanophyta; P, Phaeophyta; R, Rhodophyta.

egg-laying in response to ganglionic extracts was seasonal in *A. californica* maintained at 14°C, Kriegstein *et al.* (1974) reported the same species spawned spontaneously throughout the year when held at 18°C. Temperature is also an important factor in determining the age at sexual maturation of juveniles (see below).

SPAWN MAINTENANCE

In our laboratory, spawn is collected daily from tanks that house adults. A thin plastic ruler facilitates removal of intact egg masses from tank walls. Egg masses are cleaned of clinging algae and debris and maintained in 600- or 1,000-ml glass beakers filled with natural seawater that has been filtered through a Cuno-Aqua Pure (No. AP110, 5 μm pore size) filter. Embryos in egg masses kept in artificial seawater appear to develop normally but are smaller at hatching than veligers from parts of the same mass kept in natural seawater. Veligers hatched from spawn kept in artificial seawater and transferred to natural seawater for the larval phase will survive through metamorphosis. Spawn is usually maintained at a density of about 1 g (wet weight) per 100 ml of seawater. Spawn of Hawaiian aplysiids is routinely maintained at room temperature (24–26°C) with gentle aeration supplied through a disposable pasteur pipette; airstones may also be used. Beakers are covered with Parafilm to reduce evaporation, and the water is changed daily.

Within certain limits, lowering the temperature will slow, and increasing the temperature will accelerate, embryonic development of aplysiids. Aeration speeds embryonic development and also provides for more synchronous development within an egg mass resulting in complete hatching of an entire mass over a short time span. Table 14–1 shows the duration of the embryonic phases of species that have been cultured.

Glassware used for maintaining spawn and culturing larvae is used only for that purpose. Such glassware is washed with hot tap water (no detergents or chemicals), rinsed with distilled water, and allowed to dry before it is used again.

FEEDING AND CARE OF LARVAE

SEAWATER

Natural coastal seawater (salinity 35‰) filtered through Millipore prefilters (Cat. No. AP2504700) is used in all of our larval cultures. If seawater is suspected to have high bacterial contamination, it may require filtration through 0.22 or 0.45 μm Millipore filters. If the latter is necessary, filtered water should be aerated before introduction of larvae. All our attempts to rear larvae to metamorphosis in artificial seawater (Instant Ocean Sea Salts) have been unsuccessful. Although larvae do grow in artificial seawater, the rate of growth is slow and survival is extremely poor. Neither aging artificial seawater nor inoculating it with small volumes of natural seawater followed by aging has improved its suitability for larval rearing.

LIGHT AND TEMPERATURE

Larval cultures are supplied with continuous light from 30-W fluorescent bulbs set at distances of 10–20 cm from culture vessels. Larval cultures of Hawaiian aplysiids are maintained at room temperature (24–26°C), which is within the range of temperatures larvae would encounter under natural conditions. Larvae of temperate species may require lower rearing temperatures (see Table 14–1) although those of *A. californica* showed good growth and survival at 24–26°C in our laboratory.

CONTAINERS

After testing containers of varying sizes and shapes, 600- and 1,000-ml beakers were chosen as our standard larval rearing vessels. Stirring by rotating or reciprocating paddles either provides no advantage or is detrimental to larval survival, so cultures are maintained without stirring or agitation. Beakers are covered with plastic wrap to prevent evaporation and to keep out dust and insects.

LARVAL DENSITY

We culture aplysiid larvae at initial densities of 0.5 to 1.0 per ml. Strenth and Blankenship (1978) cultured *A. brasiliana* at 0.167 to 0.5 larvae per ml, and Kriegstein *et al.* (1974) cultured *A. californica* at a much lower density—10 per liter. Newly hatched larvae are either counted individually into culture vessels or small aliquots of a dense culture are counted and the volume of culture necessary to produce a given density is determined.

ANTIBIOTICS

Use of antibiotics in our larval cultures significantly lowers larval mortality. Survival to metamorphosis without antibiotics is usually 10–20 percent, but is 25–50 percent with them. A stock mixture of 0.6 g Penicillin G (Sigma) and 0.5 g Streptomycin Sulfate (Sigma) in 100 ml of Millipore (0.22 μm) filtered seawater is prepared and frozen in small vials in convenient quantities. The stock mixtures are thawed just prior to use and are added to larval cultures at a concentration of 1.0 ml per 100 ml of culture water to yield a final concentration of 60 μg/ml Penicillin G and 50 μg/ml Streptomycin Sulfate. Antibiotics are added at each water change. Once larvae have reached maximal size (21–24 days in our cultures), antibiotics are omitted for the remainder of the culture period. No detrimental effects from the antibiotics have been noted.

PREVENTING ENTRAPMENT OF VELIGERS AT AIR–WATER INTERFACE

A major problem encountered in culturing opisthobranch veligers is the entrapment of their shells at the water surface. Young veligers tend to swim upward at hatching and are particularly prone to this problem. Flakes of cetyl alcohol (1-hexadecanol) sprinkled sparingly on the surface of the culture effectively decrease the surface film and have no apparent toxic effect (Hurst, 1967). Kriegstein *et al.* (1974) described complicated sealed culture vessels to deal with the same problem, but we prefer the use of cetyl alcohol flakes.

TRANSFER OF LARVAE

We use a simple procedure to rinse and concentrate freshly hatched veligers and to transfer them at each water change without exposure to an air–water interface. The apparatus for larval transfer shown in Figure 14–1 consists of a 400-ml polypropylene (Tri-pour) beaker, the bottom of which has been replaced with 41 μm or 73 μm mesh (Nitex, Cat. No. HD 3–73 and HC 3–41), and a 600-ml glass beaker. The mesh is attached to the bottomless polypropylene beaker with Hot Glue (Sears). To effect transfer, the mesh-bottom beaker is placed inside the glass beaker and water is added to a level of 1–2 cm above the mesh of the inner beaker. Then a culture containing larvae is poured slowly into the inner beaker. While the rim of the inner beaker is lifted several centimeters, water can be repeatedly added to the inner beaker and removed from the outer one until the small volume of water above the mesh containing the larvae is clean (Figure 14–1). Flakes of cetyl alcohol from the old culture continue to float on the water surface during the changing process and prevent the majority of larvae from being trapped at the surface; larvae that do get caught are knocked down with gentle streams of seawater from a pipette. The beaker apparatus is next placed on a small

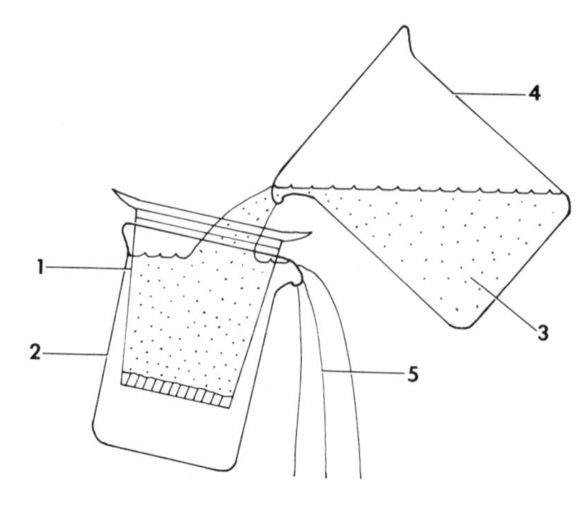

FIGURE 14–1 Apparatus and method for transferring or concentrating larvae. The transfer apparatus consists of a (1) filter basket (a 400-ml polypropylene (Tri-pour) beaker, the bottom of which has been replaced with 41 or 73 μm Nitex® mesh) placed inside a (2) 600-ml glass beaker. Water is added to a level 1–2 cm above the mesh bottom of the filter basket (1) and then water containing veligers (3) from a larval culture (4) is poured slowly into the basket. Larvae are retained within the basket and by tipping the glass beaker water to be discarded is allowed to overflow (5). Larvae are rinsed with fresh filtered seawater and concentrated in a small amount of water above the mesh. The transfer apparatus is next placed on a small light table and veligers are pipetted into fresh seawater.

light table that facilitates detection of the larvae against the lighted background. Veligers are then pipetted into small dishes for microscopic examination and counting, or into new culture vessels.

Strenth and Blankenship (1978) have developed a technique for rearing *Aplysia brasiliana* veligers that requires no transfer of larvae after the cultures are started. They "condition" lidded preparation dishes (Pyrex #3250, 100 × 80 mm) for several weeks prior to use by adding 300 ml of paper-filtered natural seawater to each dish, covering it, and leaving it in normal room light. During the conditioning period, diatoms and/or algae grow on the inner surface of the dish. Before addition of veligers, the dish is emptied of old water, rinsed with filtered water taking care not to remove the biological film, and refilled with 300 ml of freshly aerated 0.2 μm filtered seawater. Fifty to 150 veligers are added to each dish, the lids are replaced, and the dishes are placed 5 to 10 cm from a fluorescent lamp. Larvae are fed *Isochrysis galbana* that has been centrifuged to remove the algal culture media, and then resuspended in filtered seawater and allowed to stand for 1–4 days before use. Larval cultures receive 1 ml of the resuspended *I. galbana* culture and are vigorously mixed once daily. Strenth and Blankenship (1978) reported survival rates of 50 percent with minimal care using this technique.

LARVAL FOODS AND THEIR CULTURE

The unicellular algae *Pavlova lutheri* (Droop) Green (formerly *Monochrysis lutheri*) and *Isochrysis galbana* Parke are the best foods known for aplysiid larvae. In our laboratory, *A. juliana* grow best on *P. lutheri,* while other species do about equally well on either species. *Aplysia californica* (Kriegstein *et al.,* 1974) and *A. brasiliana* (Strenth and Blankenship, 1978) have been reared on *Isochrysis galbana.*

Starter cultures of algae may be obtained from a number of sources listed by Stein (1973); our initial cultures were supplied by the Food Chain Research Group, Scripps Institute of Oceanography, La Jolla, California. The microscopic algae are cultured in Provasoli's (1968) ES medium in continuous light supplied by 20-W cool white fluorescent bulbs in an incubator at 18°C. Directions for preparing the ES enrichment media are given in Table 14–3. Unaerated stock cultures of unicellular algae are started weekly in 125 ml screw-cap erlenmeyer flasks containing 50 ml of enriched seawater. Working cultures are started every 3–4 days in 500-ml-wide-mouth erlenmeyer flasks containing 300 ml of enriched seawater. These are plugged with two-holed rubber stoppers each fitted with an inlet and outlet made from glass tubing for aeration (Figure 14–2). Cotton plugs are put into the inlet and outlet tubes prior to sterilization. All flasks and the filtering apparatus are steam sterilized followed by filter sterilization of the enriched seawater. Guillard (1975) provides an excellent account of techniques for and problems of culturing microscopic algal species used to feed invertebrate larvae.

Larval cultures are fed with *Pavlova lutheri* and/or *Isochrysis galbana* at a density of 10^4 cells/ml. Algae from working cultures are used for feeding only, while the algal populations are still actively growing; that is, before maximal cell density is attained. The volume of algal culture needed to feed the larvae is determined from measurement of the algal population in the working cultures by counting cells in a hemacytometer

TABLE 14–3 Method for Preparing Provasoli's ES Medium for Culture of Phytoplankton

1. Enrichment Medium

To 730 ml of distilled water add:

$NaNO_3$	3.5 g	
Na_2 glycerophosphate	0.5 g	
Thiamine	0.005 g	
Tris buffer	5.0 g	
Fe solution	250 ml	see below
Vitamin B_{12} solution	10 ml	see below
Biotin solution	10 ml	see below
PII metal mix	250 ml	see below

Adjust pH to 7.8 with NaOH or HCl. Filter-sterilize by filtering through a 0.22 μm Millipore filter. Refrigerate in sterile dark bottles. Note that final volume is 1,250 ml.

Twenty ml of the above medium is added to 1 liter of 0.22 μm Millipore-filtered seawater to make the culture medium.

2. Fe Solution (EDTA 1:1 *M*)

To 500 ml of distilled water add:

Fe $(NH_4)_2(SO_4)_2 \cdot 6H_2O$	0.351 g
Na_2EDTA	0.300 g

Solution = 0.1 mg Fe/ml

3. Vitamin B_{12} Solution

To 1,000 ml of distilled water add:

Vitamin B_{12}	0.01 g

Solution = 10 μg/ml

4. Biotin Solution

To 1,000 ml of distilled water add:

Biotin	0.005 g

Solution = 5 μg/ml

5. PII Metal Mix

To 1,000 ml of distilled water add:

H_3BO_3	1.14 g
$FeCl_3 \cdot 6\ H_2O$	0.049 g
$MnSO_4 \cdot 4\ H_2O$	0.164 g
$ZnSO_4 \cdot 7\ H_2O$	0.022 g
$CoSO_4 \cdot 7\ H_2O$	0.0048 g
Na_2EDTA	1.0 g

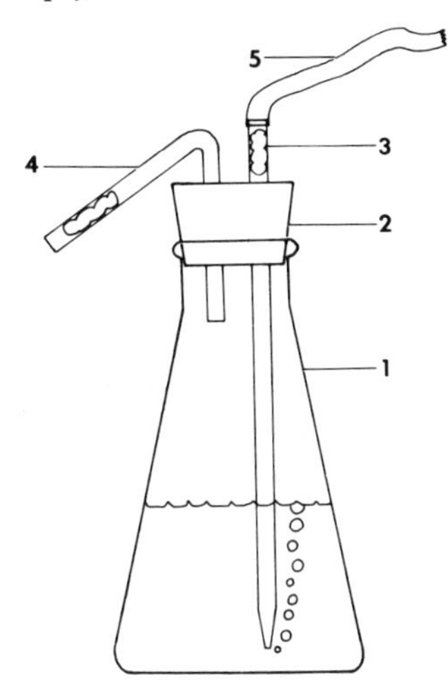

FIGURE 14–2 Algae culture container. (1) 500-ml-wide mouth erlenmeyer flask; (2) rubber stopper; (3) inlet tube with cotton plug; (4) outlet tube with cotton plug; (5) airline tubing.

or other counting device. Algae are added to the larval cultures without removal of the culture media. Larval cultures are routinely fed every 3 days, after water changes.

Growth and survival rates of veligers are the best indicators of the nutritive value of the various unicellular algal species. When testing new foods, we normally count and measure the larvae at every other water change. Larval size, defined as maximal shell dimension (= shell length), is measured with a calibrated ocular micrometer on a compound microscope.

LARVAL GROWTH AND DEVELOPMENT

Extensive descriptions of morphological and behavioral changes associated with larval growth and development of aplysiids are available in other sources (Kriegstein *et al.*, 1974; Kriegstein, 1977a,b; Switzer-Dunlap and Hadfield, 1977; Strenth and Blankenship, 1978; and Switzer-Dunlap, 1978). In each species, the larval phase includes a period of rapid shell growth to a species-specific size followed by a period of arrested shell growth during which other morphological developments occur. The sizes of veligers at hatching and metamorphosis for eight aplysiid species are given in Table 14–4. During the growth period, the shell and velar lobes enlarge, the eyes appear, and the larval heart develops and commences beating. Once maximal shell size is attained, the following occur: The mantle fold (responsible for shell growth) becomes retracted from the shell aperture (Figure 14–3); the propodium develops as a swelling in the middle of the foot; the eyes and left digestive

TABLE 14–4 Shell Lengths at Hatching and Metamorphosis of Aplysiids Cultured under Laboratory Conditions

Species	Average Shell Length at Hatching (μm)	Shell Length at Metamorphosis (μm)	Reference
Aplysia brasiliana	140	375–400	Strenth and Blankenship, 1978
Aplysia californica	125	400	Kriegstein *et al.*, 1974
Aplysia dactylomela	144	310–315	Switzer-Dunlap and Hadfield, 1977
Aplysia juliana	125	315–330	Switzer-Dunlap and Hadfield, 1977
Aplysia parvula	105	500[a]	Switzer-Dunlap, 1978
Aplysia pulmonica	128	330–340	Switzer-Dunlap, 1978
Dolabella auricularia	148	290–300	Switzer-Dunlap and Hadfield, 1977
Stylocheilus longicauda	103	325–340	Switzer-Dunlap and Hadfield, 1977

[a] Determined from larvae that were taken from plankton hauls and metamorphosed in the laboratory.

diverticulum enlarge and become darkly pigmented; and clear droplets accumulate in the left digestive diverticulum and in the region surrounding the stomach. When these developments are complete, the larva is competent, thus, capable of metamorphosis in response to an appropriate stimulus. A competent veliger of *A. juliana* is shown in Figure 14–3. Competent larvae of *Aplysia californica* are characterized by six irregularly shaped red spots located on the right side of the outer perivisceral membrane and a red band overlying the mantle margin (Kriegstein, 1977b; personal observations); similar pigmented areas are not evident in competent larvae of Hawaiian aplysiids or *A. brasiliana* (Strenth and Blankenship, 1978).

The minimal length of the larval phases varies from 28 to 34 days depending on the species under consideration and the culture technique. Lengths of larval phases for the various species are given in Table 14–1.

METAMORPHOSIS

When larvae appear competent to metamorphose based on the criteria discussed above, a small group (usually 20) is tested. Veligers are placed

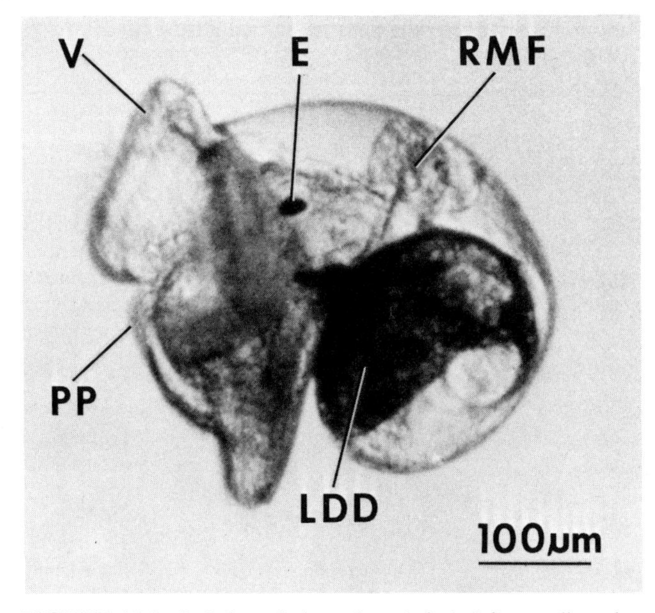

FIGURE 14–3 Left lateral view of an *Aplysia juliana* veliger that is competent to metamorphose. E, eye; LDD, left digestive diverticulum; PP, propodium; RMF, retracted mantle fold; V, velum (after Switzer-Dunlap and Hadfield, 1977).

in small dishes (stender or petri) of seawater (with cetyl alcohol sprinkled on the surface) containing small fresh pieces of an appropriate seaweed, usually the species of algae preferred as adult food. Dried algae have not been successful in triggering metamorphosis because they rapidly foul the water. While some aplysiids will settle and commence metamorphosis on several algal species, usually only one or a few algae will promote good postlarval growth. Preferred settling substrata for species currently cultured are listed in Table 14–5.

If larvae are ready to metamorphose, when placed with seaweed they will soon settle and crawl on the alga. If the larvae are not yet competent or the seaweed is not attractive, they may crawl briefly and then resume swimming. Larvae preparing to metamorphose will crawl around for up to an hour, attach to the seaweed by the metapodium, and retract almost completely into the shell where they will stay for a day or more. The metamorphic phase, from settling to beginning of radular feeding, takes from 2 to 5 days depending on the species. During the first few hours after settling, the ciliated preoral and postoral cells of the velum are shed. Soon contractions of the two-chambered adult heart will begin,

TABLE 14–5 Preferred Algal Settling Substrata for Planktotrophic Species of Aplysiids

Species	Preferred Algal Substratum for Metamorphosis (Algal Class[a])	Reference
Aplysia brasiliana	*Callithamnion halliae* (R)	Strenth and Blankenship, 1978
Aplysia californica	*Laurencia pacifica* (R)	Kriegstein *et al.*, 1974
Aplysia californica	*Laurencia majuscula* (R)	Switzer-Dunlap and Hadfield, unpublished
Aplysia dactylomela	*Laurencia* sp. (R)	Switzer-Dunlap and Hadfield, 1977
Aplysia juliana	*Ulva fasciata* and *U. reticulata* (C)	Switzer-Dunlap and Hadfield, 1977
Aplysia parvula	*Chondrococcus hornemanni* (R)	Switzer-Dunlap, 1978
Dolabella auricularia	Unidentified blue-green (Cy)	Switzer-Dunlap and Hadfield, 1977
Stylocheilus longicauda	*Lyngbya majuscula* (Cy)	Switzer-Dunlap and Hadfield, 1977

[a] C, Chlorophyta; Cy, Cyanophyta; R, Rhodophyta.

followed within a day by cessation of beating of the larval heart. Radular feeding signals the end of the metamorphic process.

During metamorphosis water is changed daily in the small dishes; care is taken not to dislodge the minute aplysiids from the seaweed. If the algal pieces deteriorate, larvae are removed and replaced with fresh fronds.

JUVENILE CARE AND DEVELOPMENT

Juveniles are fed fresh pieces of the algal species that triggered their metamorphosis until they are from 1 to 2 mo of age, at which time they can be fed dried algae. As juveniles grow, they are moved to successively larger bowls of seawater. Ten to fourteen days after settling, juveniles are put into baskets suspended in aquaria. These baskets are similar to those used for transferring larvae but are made with 1-liter propylene beakers and larger mesh (425 μm) (HC 3–425 Nitex). Alternatively, baskets are made of Plexiglass pipe with a mesh bottom. The propylene beakers are floated in aquaria with collars of styrofoam. To suspend

Plexiglass baskets, holes are first drilled on opposite sides near the top of the basket and a rod or dowel is placed through the holes to support the basket on a styrofoam collar.

Minimal durations of juvenile phases for six aplysiid species raised in the laboratory are shown in Table 14–1.

The rate of growth and maximal size attained by aplysiids are functions of a variety of factors among which are genetically determined size limits, type and amount of food, and temperature. Aplysiid species range from small ones like *Stylocheilus longicauda,* which usually weigh less than 5 g in Hawaiian waters, to *Aplysia vaccaria,* which reaches sizes up to 10 kg in California waters.

The effect of diet on growth and maximum size is well illustrated by studies we have made of *Aplysia juliana.* In the field in Hawaii, individuals of this species rarely weigh more than 50 g, whereas in the laboratory, animals fed *ad libitum* with *Ulva* spp. often grow to 200 g or more. In the laboratory, *A. juliana* fed *Ulva fasciata* grow larger than animals fed *U. reticulata.* Individuals reared on dried algal diets or on restricted intake of fresh algae grow slower, mature later, and are smaller than animals fed large quantities of fresh algae.

Both laboratory and field studies indicate that temperature is an important factor affecting the rate of growth, age at sexual maturity, maximal size, and life span of aplysiids (Smith and Carefoot, 1967; Beeman, 1970). In our laboratory the effects of temperature on various life history parameters have been investigated with Hawaiian populations of *Aplysia juliana, A. parvula,* and *Stylocheilus longicauda.* Ambient temperature of nearshore Hawaiian waters ranges from 22–28°C during the year. In all three species, animals reared at 20°C grew slower, matured later, reached larger sizes (usually at least twice as large), and lived longer than their siblings reared under similar conditions at 28°C (Sarver, personal communication; Switzer-Dunlap and Hadfield, unpublished). Thus temperature appears to be a useful tool to either speed development for rapid generation time or slow development to produce larger animals.

CONCLUSIONS

Choice of an aplysiid species for culture will depend on a variety of factors including: location of the laboratory; availability of "seed" animals; and, most importantly, the nature of the proposed research. Each species has its advantages. For instance, *A. californica* is available commercially, reaches large sizes, and is the subject of a vast literature. For life history or aging studies, smaller species with more rapid generation times like *A. juliana* and *S. longicauda* (Table 14–1) are good candidates. These two species are also found worldwide in warm waters making them available to laboratories in a number of countries. *Aplysia brasiliana* is a moderate-sized species available in Florida and along the Gulf Coast and exhibits both swimming and burrowing among its behaviors. At present *Dolabella auricularia* and *Aplysia dactylomela* do not seem to be among the best candidates for culture. The slow growth, late maturity, and large size of *D. auricularia* (Table 14–1) make this species less practical for laboratory culture. Our current low survival rate with postlarval *A. dactylomela* recommend against it for cultivation. Other aplysiid species that have not yet been cultured may also be good candidates for rearing using the techniques described herein. Finding the appropriate alga to stimulate settling will be a major factor in rearing any species.

The amount of work it will take to maintain a culture facility will depend greatly on the quantity and size of animals needed for research. In our laboratory we usually keep 5–15 adults of each of four aplysiid species. To maintain these animals, raise their replacements, and provide large numbers of competent larvae for tests on settling we estimate the following times are devoted to various chores: culture of larval foods, 5 h/wk; transfer of larval cultures, 1 h/wk per culture and we usually have 8–10 cultures of 640 larvae each at any one time; collection of algae including travel time, 6 h/wk; cleaning tanks, weighing animals, etc., of adults, 5–10 h/wk; collecting and caring for spawn, 1 h/wk; caring for juveniles, 2 h/wk; preparing antibiotics and algal culture media, washing glassware, and miscellaneous tasks, 5 h/wk. These tasks represent 35–40 h of work per week. At least a full-time technician or several part-time employees are needed to do all the tasks.

ACKNOWLEDGMENTS

We wish to thank Dr. William F. Van Heukelem and Mr. Stephen Kempf for their assistance in development of our laboratory rearing techniques. Dr. Dale Sarver provided us with his unpublished data on temperature effects on growth, fecundity, and longevity of *Stylocheilus longicauda.* Dr. Sarver's dissertation studies greatly enhanced our understanding of *Aplysia juliana* in the field; to him we are grateful. We thank Dr. Isabella Abbott of the Botany Department, University of Hawaii, for her assistance in identification of algal species. This project was supported by Contract No. RR–4–2168 and Grant No. RR–0–1057 from the Division of Research Resources, National Institutes of Health.

15

Methods of Obtaining and Handling Eggs and Embryos of the Marine Mud Snail *Ilyanassa obsoleta*

J. R. COLLIER

The egg of the marine prosobranch gastropod *Ilyanassa obsoleta* (Say, 1822) was the subject of one of the earliest experimental studies of development by American embryologists. At Woods Hole, H. E. Crampton (1896) demonstrated that the polar lobe, an anuclear vegetal region of ooplasm that is nearly separated from the egg just before first cleavage, was required for the normal development. This early study, followed by the detailed analysis of Wilson (1904) with the scaphopod *Dentalium* and Clement (1952, 1956) with *Ilyanassa*, contributed significantly to the establishment of the essential role of the egg cytoplasm in differentiation.

The purpose of this chapter is to describe the methods of obtaining and rearing *Ilyanassa* embryos. The experimental and biochemical embryology of *Ilyanassa* has been reviewed a number of times (Collier, 1965, 1966, 1976; Clement, 1971, 1976; Cather, 1971; see also recent biochemical studies of Raff *et al.*, 1976; Collier, 1977; Kidder *et al.*, 1977).

TAXONOMY AND GEOGRAPHICAL DISTRIBUTION

The prosobranch neogastropod family Nassariidae (Iredale, 1916) consists of two genera, *Nassarius* (Dumeril, 1806) and *Ilyanassa* (Stimpson, 1865); in the genus *Nassarius* there are, in North America, three subgenera and at least 31 species, while in the genus *Ilyanassa* there is only the type species *obsoleta* (Say, 1822). *Ilyanassa* may be readily distinguished from *Nassarius* by the absence of cirri on the posterior end of the *Ilyanassa* foot in contrast to the presence of two short cirri on the foot of *Nassarius* (see Abbott, 1974).

Ilyanassa inhabits marine and estuarine intertidal mud flats along the eastern coast of North America from the Gulf of St. Lawrence to Cape Kennedy in northern Florida; on the West Coast it is found from Vancouver, British Columbia, to central California (Abbott, 1974). In the estuarine environment *Ilyanassa* inhabits seaward areas where the salinity is greater than 15 percent (Scheltema, 1964); in both the marine and estuarine environment *Ilyanassa* is found in great abundance on mud flats.

LIFE CYCLE OF *ILYANASSA*

Cytochemical and ultrastructural studies of *Ilyanassa* oogenesis have been reported by Appelgate (1968), Taylor and Anderson (1969), Collier (1975), McCann-Collier (1977), and Gerin (1976a,b), and the cleavage, embryogenesis, and veliger morphology of *Ilyanassa* has been previously described as cited below. Clement (1952, 1962) has shown that the cell lineage of *Ilyanassa* is similar, except for the much larger polar lobe of *Ilyanassa,* to that of *Crepidula,* as is the structure of the veliger larva (Atkinson, 1971). The RNA content, DNA content, and cell number of all stages of embryogenesis have been reported (Collier, 1975), as have the nucleic acid, protein and yolk contents of the egg, and the molar distribution of RNA throughout development (Collier, 1976; see this reference for histological sections of the *Ilyanassa* embryo and Clement [1976] for photographs of early cleavage). Werner's (1955) paper on the anatomy and development of *Crepidula fornicata* is valuable for the morphology of a prosobranch veliger.

The time required for oogenesis has not been established. However, when the ovaries from snails collected in mid-September are examined, only small previtellogenic oocytes in early meiosis (stages 1 and 2 as described by M. M. Collier [1975]) are present; egg laying begins 5 to 6 wk later. Thus, stage 1 and stage 2 oocytes require 5 to 6 wk to develop into mature oocytes.

Ilyanassa sexes are separate. During copulation the long reflexed spatulate penis is inserted into the mantle cavity of the female, where the sperm are received by a sperm receptacle. As eggs pass along the oviduct, they are fertilized by sperm from the sperm receptacle and encased in an egg capsule containing from 50 to 250 eggs. Within this capsule the eggs are suspended in a viscous capsular fluid. Although the capsular fluid probably contributes to the maintenance of a suitable environment for embryogenesis, it is not essential for postcleavage stages of embryogenesis. Normal development occurs in seawater after removal of uncleaved eggs from their capsule. The capsular fluid is also not required for fertilization, as shown by the successful fertilization of oocytes from dissected ovaries (Mirkes, 1972).

At 19°C fertilized eggs differentiate into veliger larvae within 6 to 7 days; at this temperature escape of the veligers from the capsule occurs a few days later. Hatching is promoted by a secretion from the veliger of a labile hatching substance that loosens the egg capsule plug (Pechenik, 1975).

The veliger larvae, in the presence of a satisfactory substratum (Scheltema, 1961) and, when provided with the diatom *Phaeodactylum tricornutum* or *Nitzchia closterium* as a food source (Scheltema, 1962), undergo rapid growth and metamorphose within 3 to 5 wk into adult snails. Sexual maturity is reached when the snails have grown to 12 to 14 mm, which occurs after a third summer's growth among natural populations of snails on Cape Cod (Scheltema, 1962). The life span of the adult snail is from 5 to 10 yr, and they are reproductively active as long as they live (C. E. Jenner, personal communication).

The breeding season of *Ilyanassa* is correlated with the seasonal rise in seawater temperature. Egg laying in the natural habitat has been observed from June to August in Maine (Sindermann, 1960); from late April to mid-July on south Cape Cod, Massachusetts (Great Pond); on north Cape Cod (Barnstable Harbor) from April to late August (Jenner, 1956a); and in North Carolina from late December through May (Jenner, 1956a; Sastry, 1971).

Pseudohermaphroditism, i.e., hermaphroditism only in secondary sexual characteristics, has been reported in *Ilyanassa* by Jenner and Jenner (1977). They report that the most frequent situation in pseudohermaphroditic populations of snails is for the female to have a penis, variable in size but often as large as the male penis; males with a small foot gland (the organ of the female used to mold the egg cases) occur much less frequently. The frequency of pseudohermaphroditic females ranges, among different populations, from 0 to nearly 100 percent (Jenner and Jenner, 1977).

PARASITISM

Ilyanassa is an intermediate host to species of trematodes whose definitive hosts are birds and fishes. Vernberg *et al.* (1969) have recorded eight species of larval trematodes from snails collected at Beaufort,

North Carolina; seven of these species appeared as double infections. *Ilyanassa* are infected by the miracidium, which produces sporocysts in the liver and gonad of the snail. Cercariae develop in the sporocysts and emerge as free-swimming larvae. Sindermann (1960) reported a periodicity of 3 to 4 days in the emergence of cercariae and that feeding starved snails produced extensive cercarial emergence within 48 h.

Although cercariae are not able to infect snails, the sporocysts continue to produce cercariae and their release from infected snails continues over a very long period of time, probably for the life of the snail (C. E. Jenner, personal communication); accordingly, it should not be anticipated that the cercarial population will disappear, although it may diminish, from snails kept in the laboratory.

These larval trematode parasites are of interest because they (1) produce a cercarial dermatitis ("swimmer's itch") in people, (2) affect egg laying and generally debilitate the snail, and (3) sometimes occur in sufficient numbers to contaminate biochemical preparations extracted from the liver and gonad.

The cercariae of the avian schistosome, *Austrobilharzia variglandis,* to which *Ilyanassa* is an intermediate host, are responsible for a form of human dermatitis (Stunkard and Hinchliffe, 1952; Sindermann and Gibbs, 1953; Sindermann, 1960; and Grodhaus and Keh, 1958). These cercariae do not mature in man, but their penetration of the skin causes intense itching and inflammation of the skin. Aside from the nuisance of "swimmer's itch" on the hands and arms, one should be aware that susceptibility to these cercariae is variable and that more extreme reactions could occur in some individuals. The wearing of long gloves while collecting egg capsules in the laboratory is a useful preventive measure if good gloves of sufficient length are available. Mounting the razor blade used for scraping egg capsules from the aquaria on a long piece of wood is useful for obtaining small numbers of eggs. An immediate brisk and thorough drying of hands and arms reduces infection, because the cercariae tend to penetrate the skin from small droplets of seawater. An effective means of reducing the level of cercariae in snail tanks is to filter the seawater daily for 1 to 2 h with a Diatom Filter as discussed below.

When snails are brought into the laboratory heavily infected with larval trematodes, egg production may be delayed and fewer eggs are produced. The liver and gonads may both be so heavily infected with trematodes that these parasites may contribute significantly to nucleic acids and proteins extracted from these organs. Further, the protein content of the liver (hepatopancreas) of parasitized snails is reduced by about 50 percent (Schilansky *et al.*, 1977). Accordingly, the dissected snail should be inspected for the presence of trematodes before these organs are used for biochemical preparations.

It is, depending on one's quantitative requirements for eggs, worthwhile to sample several populations of snails and to select for the lowest frequency of infection by trematodes. Grodhaus and Keh (1958), reporting on a population of *Ilyanassa* in San Francisco Bay, California, found that among snails from 5 to 19 mm long only 2.4 percent were infected in comparison to 53.6 percent for snails 20 to 29 mm long; the median length of this population was 20 to 24 mm. Sindermann (1960) observed that snails remaining in the high-tide zone during September and October have a very high frequency (up to 32 percent) of larval schistosome infection; snails near and below mean low tide had a low frequency of infection throughout the year. These ecological observations should be of value in collecting snails with lower levels of infection. Caution should be exercised in selecting smaller snails, as this may lead to a skewed sex ratio since the male tends to be somewhat smaller than the female.

COLLECTION AND TRANSPORTATION

Ilyanassa often occurs in great numbers on mud flats and may be easily collected at low tide from early spring until late fall. Collecting equipment includes boots, a bucket, and paper towels. Precautions should be taken to protect feet and ankles from "swimmer's itch." On a mud flat, recognizable by its cover of black oozy mud, hundreds of snails can be picked up in a few minutes. The snails may be transported semidry if they are covered with seawater-soaked towels and protected from excessive heat.

During the winter months the snails move out into deeper water and are rarely available on the intertidal flats; however, a persistent collector, equipped with high boots, a long-handled scoop, and chill-resistant hands, may be rewarded by dredging in deeper water along mud flats where *Ilyanassa* is known to occur.

Ilyanassa may be safely air-freighted if packaged in groups of a hundred or so in a 6 × 8 in. polyethylene bag containing wet toweling and an air pocket. If shipments are made during warm weather and freighting delays are anticipated, the polyethylene bags of snails should be shipped in a thermos container; otherwise, the packaged snails may be shipped in any moisture-proof container. Shipment of snails abroad usually requires prior arrangement with an airline and a letter from the sender stating that live specimens of no commercial value are being shipped.

MAINTENANCE OF SNAILS IN THE LABORATORY

Snails may be maintained in 40- to 80-liter aquaria of running or recirculating seawater. Sixty to one hundred snails in a 80-liter aquaria will provide an abundance of egg capsules. All-glass aquaria are the most desirable because of the ease with which egg capsules can be scraped from the surface. Any aquarium cement should be removed from the inside corners of the aquaria so that egg capsules can be easily removed from these areas.

Snails will survive and lay eggs at temperatures as high as 30°C (Sastry, 1971), but a temperature of 18 to 20°C is optimum. It is advantageous to keep the snails at the same temperature used for rearing embryos.

The aquaria should be kept covered to reduce evaporation; the salinity should be checked every few days by refractive index or with a hydrometer. The refractive index and specific gravity of seawater from Woods Hole, Massachusetts (a source of seawater that I have used for rearing embryos and as a standard for maintaining optimal conditions in recirculating seawater systems) is at 20°C, 1.3374 and 1.025, respectively. For optimal egg laying, it is important to maintain this salinity by the periodic addition of distilled water to the tanks.

The pH of my Woods Hole standard seawater is 8.1. A pH range of 7.6 to 8.1 in recirculating seawater systems has proved acceptable in my laboratory. Aeration and the presence of crushed oyster shells in the filtration system is essential in maintaining a satisfactory pH.

It is expensive to have large quantities of seawater shipped from any great distance. I have used seawater from a variety of sources for maintenance of snails, but have always kept a separate supply of Woods Hole seawater for rearing embryos. Artificial seawater mixtures, e.g., Instant Ocean, are satisfactory for use in recirculating seawater systems, but are, in my experience, less satisfactory than natural seawater for rearing embryos. Seawater should be collected, if possible, from beaches exposed to the open ocean or from more seaward areas if collection is to be made from a harbor. Obviously contaminated seawater, which is often encountered in navigation channels and marinas, should be avoided. The best criterion for selecting suitable seawater is the presence of an abundant fauna in the area of collection.

Regardless of the type of recirculating seawater system used, an excellent supplementary filter is the Diatom Filter (Vortex Innerspace Products, 3317 E. Bristol Road, P. O. Box 7048, Flint, MI 48507). This filter is portable and can be easily used for emergency filtration. It removes trematode cercariae, copepods, protozoa, and, if charged with both diatomaceous earth and activated charcoal, the yellow-green algal pigments that accumulate in recirculating seawater systems.

EGG PRODUCTION

If large numbers of egg capsules are required daily, it is essential that the snail tanks be kept clean. Algae growing on the sides of the tanks and in the tubing used for recirculation of seawater should be removed. In two 80-liter aquaria connected to one 40-liter filter tank, I have 200 snails that provide 100 to 200 capsules of uncleaved to four-cell embryos daily. Generally, I make a complete change in seawater only once a year and usually replace the filtering material (sand and crushed oyster shells) twice during this time interval. If the tanks and snails are properly attended, one collection of snails will continue to produce an abundance of egg capsules throughout the year (from late October until July or August). Occasionally, I have kept the same collection of snails for more than 1 yr, and they would begin egg production again in the fall.

Snails brought into the laboratory between late October and the middle of July usually begin egg production within a week to 10 days if they are placed in a well-filtered seawater system and fed as outlined below. Egg capsules are not deposited below 10°C nor above 30°C; between these extremes the time required for snails to begin reproductive activity is linearly correlated with temperature (Sastry, 1971). Snails collected after the end of their natural breeding season generally require 3 to 4 mo before they will resume reproductive activity. This refractory period probably results from the time required for the maturation of previtellogenic oocytes as discussed above.

Because the sexes of *Ilyanassa* are not readily distinguishable by external criteria the copulatory position of the snails, i.e., female with the foot extended on the glass plate of the aquarium and the male with its foot along the side of the female, is a convenient means of selecting males and females. The correctness of a selection is evidenced by observing the withdrawal of the penis from the mantle cavity of the female as the male is gently pulled away from the female.

Frequently, when snails are first placed in a sea tank they will exhibit a "schooling behavior" characterized by a continuous movement around the tank. This pattern of behavior, described in natural populations by

Jenner (1956b, 1957, 1958, 1959), usually continues for a few days, during which time they do not feed.

Light is not a determining factor in egg production but illumination for 10 to 12 h a day does enhance egg laying (C. E. Jenner, personal communication).

FEEDING

Although *Ilyanassa* is structurally adapted for a carnivorous diet, i.e., it has a protrusible proboscis, radula, and other typical carnivore structures, in its natural environment it is a deposit-feeder of ingested sand and mud. Adaptations for a herbivorous diet are the presence of a crystalline style and the presence of the enzymes required for the metabolism of algae (Jenner, 1956c; Brown, 1969). For laboratory cultures of *Ilyanassa,* it is convenient to take advantage of their carnivorous adaptations and to feed them marine clams, which are available at most fish markets. Clams may be kept alive in the filter tank where they aid in filtration. I have also used freshwater mussels, and Clement (1971) has used oysters and slices of frozen shrimps as snail food. The most convenient and consistently satisfactory food source has been the marine clam *Mercenaria mercenaria.*

For optimal egg production snails should be fed daily. Feeding on alternate days is also acceptable but there will usually be a decline in egg production the day after feeding. The amount of food should be adjusted so that it is totally consumed, e.g., 1 clam on the "half shell" per 100 to 200 snails. The uneaten portion should be removed after 2 or 3 h.

EGG COLLECTION

In the laboratory the *Ilyanassa* female cleans a small area of the aquarium with her radula and deposits an egg capsule onto this freshly cleaned surface; egg capsules are deposited singularly or in a row of several. In a row of egg capsules deposited by a single female there is sequential development, with the eggs of the first capsule being slightly older than its neighbor. When a capsule is first deposited the eggs usually have an intact germinal vesicle. In some cases, the eggs may have progressed to the early maturation divisions and contain one or two polar bodies when they are laid. Occasionally eggs will be much further along in development; this indicates that the snails have been disturbed and in extreme cases that the condition of the seawater is far from optimal.

The egg capsules are tightly cemented to the glass plates of the aquarium and are covered by mucus that is gradually dissolved by the seawater. Capsules may be conveniently scraped from the sides of the aquarium with a single-edged razor blade at an angle 30 to 45 degrees from the glass surface, without any damage to their contents. Egg capsules may be removed from the corners of the tank, in which snails deposit most of their capsules, by scraping first along the surface of one side of the tank and then making a second pass perpendicular to the first. The recovery of egg capsules from the corners requires, as mentioned above, that all aquarium cement be removed from the inside corners of the tank.

The detached egg capsules fall to the bottom of the tank, where they may be collected by siphoning with a piece of tubing into a tea strainer (either metal or nylon mesh) suspended over a bucket or carboy. Egg capsules should be washed by spraying with seawater while they are retained in the strainer.

REARING OF EMBRYOS

A separate supply of seawater should be set aside for rearing embryos, and it should be filtered through a Millipore membrane (0.45 μm pore size) before use. Penicillin and streptomycin (100 to 500 μg per ml of each) may be added to the seawater to prevent bacterial growth; these antibiotics should be prepared fresh daily, as penicillin is unstable in solution. The lower concentration of antibiotics is suitable for routine cultures; the higher concentration is desirable when embryos are to be pulsed with radioactive precursors. If cultures are kept clean and freshly filtered or pasteurized seawater is replaced daily, embryos may be reared without the use of antibiotics.

A set of glass dishes with covers, e.g., 37 × 25 mm and 48 × 33 mm Wheaton dishes, should be available for rearing embryos. Dishes should be washed in detergent, scrubbed with a piece of cheese cloth, rinsed in running tap water, and air dried by inverting over a clean towel. Small plastic dishes are convenient for rearing embryos in egg capsules, but are not suitable for culturing embryos removed from the egg capsule as the eggs and embryos stick to some plastics. However, organ tissue culture dishes (dish number 3010 from Falcon, Oxnard, California) are excellent for rearing embryos in small volumes. These dishes have an absorbent ring of filter paper that provides a moist chamber when soaked with water, and the embryos do not stick to the surface. A pair of watchmaker's forceps, e.g., Dumont # 5 as supplied by Ladd Research Industries, Burlington, Vermont, and a pair of fine scissors (iridectomy scissors as supplied by Clay Adams, Parsippany, New Jersey) are required for handling and cutting egg capsules. Pipets suitable for the transfer of eggs and embryos are made from 3-mm glass tubing, one end of which has been heated and pulled to a capillary of about 0.5 mm in diameter. Egg pipets are operated by applying suction or pressure through a mouth hose of small rubber tubing attached to the pipet.

Washed egg capsules should be examined with a dissecting microscope, and the capsules containing eggs or embryos of the desired stage transferred to a separate dish of filtered seawater. The embryos may then be reared in the capsules at 19°C in a constant-temperature incubator. I collect egg capsules each day and select 50 or more capsules containing either eggs or two- or four-cell embryos and keep them at 19°C. This regime, if followed daily, results in a "farm" of embryos at all stages of development. I stage embryos numerically from 0 to 7, in which 0 is the uncleaved egg and 1 to 7 is the age in days of embryos reared at 19°C. These stages have been previously described and illustrated (Collier, 1965, 1975, 1976). The schedule of the meiotic and early mitotic divisions has been reported by Morgan (1933) and Cather (1963); Costello and Henley (1971) also list the times for early cleavages.

Eggs develop normally when reared outside of the egg capsule in seawater prepared as described above. Removal of eggs from the capsule is best accomplished by holding the egg capsule in a pair of forceps and cutting off one end of the capsule with a pair of scissors. After cutting the capsule it is desirable, but not essential, to leave the opened capsule in seawater for several minutes before attempting to remove the eggs. During this time the capsular fluid, which is moderately viscous, will be dissolved in the seawater. Eggs are then removed by holding the capsule in a forceps and *gently* directing a stream of seawater from an egg pipet into the opened end of the egg capsule. This operation, which may need to be repeated a few times, will flush the eggs out of the capsule. The removal of eggs from the capsule should not be attempted when eggs are in the trefoil stage (the stage when the third polar lobe is nearly severed from the egg), as it will result in the detachment of the third polar lobe and damage to the lobeless egg. Removing eggs or early cleavage stage embryos from the egg capsule is perhaps one of the more difficult operations in working with *Ilyanassa* embryos. Success in this maneuver requires that the cutting of the egg capsule produce a fairly large hole, that the flushing of the eggs out of the capsule and all subsequent transfers of the eggs be done very gently. To provide a cushion when eggs are transferred into a dish of seawater, the dish should be three-quarters full, the egg pipet should be introduced at an angle just beneath the seawater, and the eggs should be allowed to flow by gravity out of the pipet. *Do not blow out the pipet! Ilyanassa* eggs are fragile during these early stages, and rough handling invariably injures them. Damage to the egg is often not immediately evident but will cause aberrant cleavage and/or delayed cytolysis of the macromeres, most frequently the D macromere.

Once eggs have been successfully removed from the egg capsule, they will develop into normal veliger larvae within 7 days at 19°C. Embryos reared outside of the capsule should be changed to fresh seawater daily; for embryos left in the capsule, a daily change of seawater is desirable but less critical.

The *Ilyanassa* egg capsule is resistant to proteolytic enzymes (trypsin and pronase) and hyaluronidase (Collier, unpublished observations). The egg capsule of *Buccinum undatum* is a glycoprotein (78 percent protein, 8 percent sugars, and 2 percent hexosame; Hunt, 1966). Flower *et al.* (1969) have demonstrated that this egg capsule protein has a tightly banded structure, and this periodic banding of the capsular protein along with the presence of four layers of protein fibers in the capsule probably accounts for the remarkable resistance of egg capsules to enzymatic and other chemical treatments (see review by Brown, 1975). These observations suggest that removal of the egg capsule by enzymatic means is unlikely to be successful; any efforts in this direction should be focused on digestion of the capsular plug as reported by Pechenik (1975).

POLAR LOBE FORMATION AND EARLY CLEAVAGE

The formation of polar lobes during early cleavage is a striking and significant feature of *Ilyanassa* embryogenesis. The polar lobe, variously designated as antipolar or yolk lobe in the earlier literature (Morgan, 1933), is an anucleate protrusion of vegetal cytoplasm that appears and disappears several times during the early cleavages of *Ilyanassa* and several other proterostomes, such as *Chaetopterus, Sabellaria, Dentalium, Bithynia,* and *Mytilus.* In all of these, the removal of the polar lobe has a marked effect on differentiation; the role of the polar lobe in differentiation and the biochemistry of the *Ilyanassa* egg have been studied extensively. This work is covered in the reviews mentioned at the beginning of this chapter. The widespread presence of the polar lobe indicates that it is not an embryological oddity, rather it is an extreme example of a widespread *modus operandi* among animal embryos.

Ilyanassa eggs are about 160 μm in diameter and, when viewed by transmitted light, are opaque, except for a small clear area in the animal hemisphere that marks the position of the germinal vesicle in the freshly laid eggs. When the egg is viewed by reflected light, the highly refractile yolk platelets are seen to occupy the vegetal two-thirds of the egg, and the animal third of the egg contains less refractive material that, in section (Crowell, 1965; Pucci-Minafra *et al.,* 1969), is seen to consist of mitochondria and lipid droplets. In eggs that have been fertilized, the germinal vesicle breaks down within 15 to 50 min after egg laying (Morgan, 1933; Costello and Henley, 1971), and the metaphase spindle of the first meiotic division forms and moves to the surface of the animal pole as the first polar lobe is formed. After the first meiotic division produces the first polar body, the first polar lobe recedes (Cather, 1963). A second polar lobe appears just before the metaphase of the second meiotic division, within 20 min the second polar body is formed, and shortly thereafter the second polar lobe recedes. A third polar lobe appears during the anaphase of the first cleavage. As the first cleavage furrow appears at the animal pole, there is a subequatorial constriction of the third polar lobe. The advancement of the first cleavage furrow and the constriction of the third polar lobe continues to divide the egg into two nucleated blastomeres in the animal hemisphere and the polar lobe at the vegetal pole of the egg. This is the *trefoil* stage, and it is followed by the completion of the cleavage furrow and the fusion of the polar lobe to the CD blastomere. The latter event accounts for the unequal division of the *Ilyanassa* egg at the first cleavage. Subsequently, a fourth polar lobe is formed by the CD blastomere prior to the second cleavage and the asymetrical distribution of this polar lobe into the D macromere produces the characteristically larger D quadrant of the four-cell stage. Finally, a fifth polar lobe appears in the vegetal region of the D macromere just before the third cleavage. The fourth and fifth polar lobes are less conspicuous than earlier ones as the constriction separating them from the CD and D macromere is less pronounced.

Table 15–1 is a time schedule for the events outlined above; time is measured from the appearance of the polar lobes. Staging from the trefoil stage is often convenient, but for some purposes it is desirable to stage from earlier events.

For further details of early cleavage, see Clement (1952, 1976) for photographs of early cleavage. For the role of microfilaments in polar lobe formation, see Conrad *et al.* (1973) and Raff (1972); also see Conrad *et al.* (1977) for changes in membrane potential during polar lobe formation and cleavage and Conrad and Williams (1974) for the role of calcium and potassium ions in polar lobe formation.

ISOLATION OF POLAR LOBES AND BLASTOMERES

Polar lobes and the early blastomeres may be isolated by severing the connecting cytoplasm or cleavage remnant with a fine glass fiber or by

TABLE 15–1 Schedule of Meiotic Divisions, Polar Lobe Formation, and Cleavage in the *Ilyanassa* Egg [a]

Stage	Minutes From:		
	First Polar Lobe	Second Polar Lobe	Trefoil
First polar lobe	0		
First polar body	10		
Disappearance of first polar lobe			
Second polar lobe	30	0	
Second polar body	50	20	
Third polar lobe	135	105	
Construction of third polar lobe	160	130	
Beginning of first cleavage furrow	175	145	
Trefoil	190	160	0
Completion of cleavage and fusion of polar lobe with the CD blastomere	210	180	20
Fourth polar lobe	240	210	50
Second cleavage	275	245	85
Fifth polar lobe	290	260	100
Third cleavage	375	345	185

[a] This schedule was compiled from the data of Cather (1963) for eggs at 23°C.

agitation in calcium–magnesium low seawater (Clement, 1952). Macromeres, up to the 25-cell stage, and the first quartet of micromeres can be deleted by puncturing the cells with a glass needle (Clement, 1962, 1967).

While polar lobes can be isolated in larger numbers by agitation in Ca–Mg low seawater, when there are reasons for avoiding exposure to this ionic environment, isolation by a glass fiber is useful. This method is best described by Clement (1952): "The polar lobe can be successfully removed with a glass hair at the extreme trefoil stage when it is attached by a slender thread of protoplasm. Fewer eggs cytolyze if the lobe is not completely cut off at one stroke; it is better to press across the connection two or three times at intervals of a quarter of a minute or more, and then cut the lobe off. A number of eggs may be lined up in an operation dish and carried through the delobing process simultaneously." Similarly, the freehand deletion of micromeres is described by Clement (1967) as follows: "By using the large 1 D macromere as a landmark, the individual micromeres may be identified by their relative positions with respect to the macromeres and to one another. In the present experiments, they were removed, freehand, with a glass needle. For operation, an egg was isolated in a small glass dish and placed on the stage of a dissecting microscope with substage illumination, at a magnification of 80 diameters. At the appropriate stage, either by agitating the dish or by the use of a gentle current of water from a fine pipette, the egg was rolled into a position that brought the desired micromere into profile view. The micromere was then torn with a rapid downward flick of the needle tip. A spurt of released granules usually heralded a fatal injury to the cell. The injured cell swelled and, after some minutes, detached from the embryo. It is advisable to wait until 15 or 20 min after a micromere has been formed before attempting to remove it; otherwise injury may spread to the macromere and result in its death also."

I isolate polar lobes in Ca–Mg low seawater; my Ca–Mg low seawater contains 25.17 g/liter of NaCl, 0.7 g/liter KCl, and 3.07 g/liter of Tris-HCl, and the pH is adjusted to 8.2. To ensure the recovery of lobeless eggs without subsequent separation of the lobeless blastomeres it is desirable to add 10 to 15 percent (v/v) of normal seawater to this solution. Similarly, all isolates should be removed from this mixture as soon as possible and carefully placed in a dish of normal seawater.

My routine for isolating polar lobes is to select 25 to 30 capsules of uncleaved eggs and then remove the eggs from each capsule in a small dish of Ca–Mg low seawater. I follow the development of each dish of eggs by repeatedly checking their progress toward the trefoil stage under a dissecting microscope. When a group of eggs reach the trefoil stage, their polar lobes are isolated by picking up, in an egg pipet, those eggs in which the maximal separation of the third polar lobe has occurred and gently "bouncing" them off the bottom of the dish. Generally 10 or so eggs from a capsule of 80 to 100 eggs will have reached the proper stage at any given moment, and polar lobes will be detached from 3 or 4 eggs during the "first bounce." The isolated polar lobes and lobeless eggs are quickly transferred to a dish of normal seawater, and those eggs that retained the polar lobe are quickly recovered and repeatedly pipeted onto the bottom of the dish until the polar lobes are detached. This operation requires a minute or so, and by this time other eggs in this group are sufficiently advanced to be selected for delobing. When the delobing of the eggs in one dish is completed, a quick check is made of the remaining dishes of eggs, and another dish containing eggs at the proper stage is selected for delobing.

The successful processing of 25 to 30 capsules of eggs can be done in 4 to 5 h, and one may expect to obtain a thousand or more lobeless eggs and polar lobes. Two points should be kept in mind when isolating polar lobes: (1) rarely will polar lobes be successfully isolated from all of the eggs in a capsule, and (2) eggs from some capsules will develop beyond the trefoil stage before there is an opportunity to remove the polar lobe. An excess of egg capsules and a relaxed and patient mode of working will generally provide a good yield of isolates. There is considerable variability among egg capsules in the ease with which polar lobes may be isolated. An average of 30 to 40 lobeless isolates per egg capsule is a reasonable yield.

The use of a New Brunswick rotary shaker is a useful aid in isolating polar lobes. Dishes of eggs that arrive at the trefoil stage when one is engaged in operating on other eggs may be agitated at 150 rpm on the shaker, as described by Newrock and Raff (1975). This agitation is often effective in isolating polar lobes and permits a moderate increase in the yield of isolates. While I frequently keep a shaker at hand for this purpose, I have not found its exclusive use to be more effective than the regime described above.

If close synchrony of the developing isolates is required they may be kept at 10 to 12°C after isolation and then transferred to 19°C after all operations have been completed.

MISCELLANEOUS METHODS

Cather (1967) described methods for carbon marking of blastomeres and for transplanting polar lobes onto ectoblast isolates. Styron (1967) has reported a method for compressing the *Ilyanassa* egg so as to repress the formation of the polar lobe, and Parrish and Parrish (1962) have designed an ingenious system for the differential exposure of selected portions of the egg surface to various solutions.

Ebstein *et al.* (1965) have devised a culture medium that has permitted the culture of epithelial and fibroblast-like cells obtained from *Ilyanassa* embryos.

REARING AND METAMORPHOSIS OF VELIGER LARVAE

The rearing and metamorphosis of the *Ilyanassa* veliger has been carried out in the laboratory by Scheltema (1961, 1962, 1964, 1967), and the following account of metamorphosis is based on his observations.

After emergence from the egg capsules, larvae are collected onto a

100-mesh (140-μm) stainless steel screen (bolting cloth or Nitex cloth is also satisfactory), placed into suitable vessels containing seawater, and aerated. The larvae are reared at room temperature (20 to 25°C) and are fed either *Phaeodactylum tricornutum* or *Nitzchia closterium* (the latter is available from Carolina Biological Supply Company, Burlington, North Carolina) at a concentration of 200,000 cells per ml. The seawater is changed every 2 or 3 days by removing the veligers on screens as described above, and the fresh seawater is charged with a food organism. The diatoms used for food may be reared as described by Ketchum and Redfield (1938) or as described in Chapter 6.

When larvae are reared for 2 to 3 wk after hatching from the egg capsule and then transferred to a container whose bottom is just covered with a "natural substratum," an average of 70 percent of the larvae will metamorphose into juvenile snails within a week. Natural substratum is a bottom sediment upon which juvenile snails are found in nature. An appropriate substratum is essential for the metamorphosis of a large proportion of the larvae (Scheltema, 1961).

Metamorphosis of the *Ilyanassa* veliger is accompanied by the *in toto* detachment of the velum and followed by a darkening and spiralization of the shell to produce a juvenile. The young snail grows rapidly during the first few months after metamorphosis and attains sexual maturity upon growing to a length of 12 to 14 mm, in about 2½ yr (Scheltema, 1964).

ACKNOWLEDGMENTS

I thank Dr. Charles E. Jenner and Martha G. Jenner who have, for many years, instructed me in the biology and natural history of *Ilyanassa*. I thank Dr. Norman Levin for his assistance with the parasitological literature on *Ilyanassa*.

16

Laboratory Culture of Marine Bivalve Molluscs

MATOIRA H. CHANLEY

INTRODUCTION

Bivalve culture techniques have become better defined since the early work of Loosanoff and Davis (1963), and with this there is an increased interest in the cultivation of this group of molluscs, both for experimental and commercial purposes.

All bivalves are filter feeders, and their culture almost always requires a copious supply of seawater with suspended food. Consequently, except for very small species, a running seawater system, while not absolutely essential, is certainly desirable. In some areas there may be sufficient algal food in the natural seawater to support captive bivalve populations (Hidu, 1975), but most often additional food, usually in the form of cultured algae, must be provided on a continuous basis.

LIFE CYCLE

Though there are several modes of reproduction, most familiar marine bivalve molluscs are dioecious animals that do not usually display outward sexual dimorphism. With appropriate environmental stimuli both sexes spawn freely into the water where fertilization, embryonic, and larval development occur. The typical marine bivalve mollusc passes through two larval stages, the trochophore and the veliger, before setting down to a benthic juvenile and adult existence. Both the embryo and trochophore are nonfeeding, or more precisely yolk-feeding (lecithotrophic) stages, and for this reason are favored by many investigators for short-term toxicity studies. The veliger is the shelled, swimming, planktotrophic stage. The velum serves the dual function of locomotion and food-gathering for the larval shellfish. Near the end of the larval stage, the typical molluscan foot develops. At this stage the little, still microscopic, pediveliger is capable of both swimming and crawling. Soon the velum is lost. Then the animal has become a juvenile, or is said to have "set."

There are several variations to the typical reproductive mode. Some, like the bay scallop, *Argopecten irradians,* used in oil and heavy metals toxicity studies (APHA, 1976), are hermaphrodites. The same animal will generally spawn first as a male and then as a female, and often will release eggs and sperm simultaneously, during the same spawning period. Some hermaphrodites, like the European oyster, *Ostrea edulis,* spawn first as a male and then in the next spawning season as a female. This second, or alternate, type of hermaphrodite is commonly encountered in cold-water environments and appears to be an energy-conserving measure. Egg production is much more consumptive of energy than is sperm production, and the continuous alternation of sexes allows the animal to grow during the male season in environments of limited resources.

Some marine bivalve molluscs are larviparous, a greater or lesser degree of larval development taking place in a brood chamber, the gill chamber, or mantle cavity of the female. Only the male spawns freely into the water. The female takes in the sperm with the seawater, and the eggs are fertilized internally. Some degree of such protected early development is more characteristic of deepwater and high-latitude species. All degrees of development can be found from release as fairly early veligers, as in the commensal clam, *Montacuta percompressa,* to rather complete development (Chanley, 1969). Larvae of both the New Zealand bluff oyster, *Ostrea lutaria,* and the Chilean oyster, *Ostrea chilensis,* set within a few hours of their release.

One can also find the curious situation where the male is no more than a parasitic, functional gonad in the mantle cavity of the female, as in *M. percompressa* (Chanley and Chanley, 1970).

As with other poikilothermic animals, the length of time for larval development varies not only with the genetic makeup of the species, but also with several environmental factors. Temperature and adequate food are usually the two most important such parameters. The eastern oyster, *Crassostrea virginica,* at typical East Coast (Atlantic) temperatures during its spawning season (24–27°C), will pass from fertilized egg, through

TABLE 16–1 Bivalve Molluscan Species Recommended for Bioassay Studies (APHA, 1976)

Scientific Name	Common Name
Argopecten irradians	Bay scallop
Crassostrea gigas	Pacific oyster
Crassostrea virginica	Eastern oyster
Macoma balthica	Macoma
Mercenaria mercenaria	Quahog
Mulinia lateralis	Coot clam
Mytilus edulis	Blue mussel
Ostrea lurida	Olympia oyster
Rangia cuneata	Road clam
Spisula solidissima	Surf clam

the trochophore stage, to become a veliger in about 18 h. At colder temperatures up to 48 h may be required for many species to develop to the veliger stage. If temperatures are too cold the eggs will not develop normally. Eggs of *C. virginica* do not develop properly below 18°C. Growth-inhibiting properties of some pollutants can be discerned at sublethal levels in short- and long-term bioassays using embryonic and/or larval growth (Walne, 1970; Waldichuk, 1974; Calabrese *et al.*, 1977; APHA, 1976).

The general procedures described here are suitable for culturing embryos, larvae, juveniles, and adults of the species recommended by APHA (1976) for bioassay studies (Table 16–1) and many others. Some adaptation of these general procedures may be necessary for cultivation of species from highly specialized environments.

WATER QUALITY

Whether for purposes of bioassay or general life-cycle studies, the seawater used for the laboratory culture of marine bivalve molluscs must be from a high-quality source and handled in such a manner that its quality is not impaired before use. Thus far no artificial seawater has been developed that will support the growth and development of marine bivalves through complete life cycles regularly, although short-term maintenance has been accomplished with several simple-formula artificial seawaters.

Seawater can be transported, if necessary, from a site of good-quality seawater for an inland marine research project. Because endemic bacterial populations will flourish in recently containerized seawater (Zobell and Anderson, 1936), one should plan on using it immediately after collection (within 8 h). If this is not practical, water may be used satisfactorily if filtered and given long-term (generally a month or more) storage in lidded inert containers of the largest volumes available.

In some areas saltwater wells are a very practical source of good-quality culture medium, but not always. Since virtually all metals are toxic to bivalve larvae, it is best to use only plastic throughout the system wherever it contacts the seawater. Pumps should have plastic or nylon impellers.

For the brood stock and larger juveniles, the natural phytoplankton in the incoming water will provide some food. However, the most abundant phytoplankton are usually too large for the larvae. For embryonic, larval, and early juvenile culture the seawater should be filtered. For large cultures the water is usually filtered through a 5–20-μ felt bag or candle filters. Sometimes a coarser prefilter or settling tank is desirable when water is extremely turbid. For experimental work this filtered water can be passed through a 1-μ or submicron filter.

Usually the cause of most culture problems are bacteria. Bacteria are present in small numbers (10^3–10^4 per ml) in natural seawater and usually cause no trouble. However, when brought into the laboratory and provided with culture vessels and larvae as suitable substrate, the numbers rise dramatically in all but the most nutrient-poor waters. Experience has shown that when the bacterial count (as determined on PGY media) reaches 10^7 per ml, a pathogenic condition exists regardless of bacterial species (Leibovitz, 1978). To avoid prolonged larval contact with such levels, the culture water should be changed daily. The culture vessels should be cleaned well with a chlorine solution (10 ppm using Chlorox or a similar 5 percent hypochlorite solution) and thoroughly rinsed before returning the larvae to them. To provide the least-possible vessel surface in proportion to seawater volume, use the largest practical vessels, preferably cylindrical with a domed or conical bottom.

MAINTENANCE OF BROOD STOCK AND INDUCED SPAWNING

All studies with embryos and most with larvae are initiated by spawning the parent animals in the laboratory. Most bivalves become sexually mature while still very small. Often, for experimental purposes, sufficient gametes can be obtained from animals less than one-third full size. Spawning is most easily induced by collecting the animals at the beginning of their natural spawning season, transporting them to the laboratory quickly, and placing them in seawater at a temperature just a few degrees warmer than that at collection. Sometimes this is all that is required to initiate a spawning. If spawning does not occur within a half-hour or so after the animals start pumping, a suspension of stripped sperm should be added. This is obtained by opening the animal carefully and raising the mantle. The mature male gonad usually will be full-appearing and whitish, basically covering the central dorsal portion of the soft body. Sperm can be extracted easily by piercing the gonad with a Pasteur pipette. Diluted in a beaker of seawater, active stripped sperm can be a powerful spawning stimulus.

Except for *Macoma balthica* and possibly also *Rangia cuneata,* most of the species listed in Table 16–1 are usually easy to spawn if ripe (Chanley, 1975). For those that are more difficult, additional techniques may have to be applied. If rise in temperature accompanied by sperm suspension does not initiate a spawning, drop the temperature for a short while and then raise it again. Sometimes simply changing the water works. Some bivalves spawn when moved to slightly cooler water after first an hour or two at the elevated temperature.

In the summer when the ambient temperature is near the maximum tolerable temperature it may not be possible to raise the temperature. In these instances, particularly with north temperate species, refrigeration for ½–1 h first may bring about success. However, southern species often will not tolerate this treatment, but gentle cooling with a cool water or ice bath followed by the elevation of temperature and addition of stripped sperm may be successful. Castagna's (1975) technique for spawning conditioned Virginia populations of the bay scallop *Argopecten irradians* includes the following. Place 1–2 animals in each of several-liter Pyrex baking dishes of filtered seawater in a warm-water bath at 24–26°C to induce maximum pumping. The temperature is then raised to 30°C, for only a few minutes, and dropped back to 24°C. The scallops will usually spawn as the temperature is dropping, between 26–28°C. As these animals are functional hermaphrodites, temperature manipulation is the only stimulus used.

Other techniques that have on occasion initiated spawning include electric shock, injection with KCl, NaOH, or NH_3OH, pricking or mildly injuring an adductor muscle, addition of Kraft Mill effluent (Chanley, 1975), and vigorously shaking the animals.

If the animals are not ripe at collection, or for genetic and other studies, the culturist may want to keep them in the laboratory until they are in spawning condition. For long-term storage, particularly of larger specimens, a suitable outside area with abundant natural phytoplankton is very desirable, possibly even essential. Access to a dock or raft would be most useful. Clams that burrow in a substrate can be held in trays of suitable sand hung from a raft or dock. Oysters can be held on strings of cultch (the substrate oyster spat attach to; usually oyster shell) or cultchless in trays without sand. Mussels can be kept on fibrous ropes. Scallops can be kept in cages if water circulation is good enough and they aren't too crowded. Some knowledge of the adult natural habitat should enable the culturist to house a brood stock population.

Adults of small species can often be maintained in the laboratory with ease and success. Rhodes *et al.* (1975) have maintained many generations of *Mulinia lateralis* in the laboratory in flowing seawater providing them with a 2″ layer of beach sand substrate.

Some populations of some species will develop gonads almost immediately after the spawning season and overwinter in this condition. These animals are generally fairly easy to condition to spawn before the next regular spawning season. A general procedure is to bring the animals into the laboratory and gradually raise the temperature to that encountered in their natural environment a few weeks before spawning. This would be the conditioning temperature. For example, Castagna (1975) conditions *A. irradians* at 18–22°C for 3 to 6 wk before spawning them at 26–28°C. During the conditioning period the animals must be well fed or they will resorb the gonadal material, metabolizing it as a source of energy.

Animals that overwinter without gonads cannot be induced to develop gonads quickly. For example, *M. lateralis* are generally gametogenically inactive for about 3 mo after spawning (Rhodes *et al.*, 1975). Following this rest period they can be conditioned easily by gradually increasing the temperature, from the cold ambient, several degrees a day until 18–20°C is reached. Well-fed, they will condition in 2–5 wk at that temperature. The ripe gonads of *M. lateralis* can be seen through the animal's thin shell. The full male gonad will be white; that of the female pink, red, or orange.

Spawning of animals collected just prior to the normal spawning season can often be delayed considerably by transporting the animals to more northern waters where the ambient temperature does not reach spawning temperatures. However, more states are restricting such shipments, and such stalling tactics may have to be undertaken in the laboratory. For this a system for chilling the seawater is necessary. Some caution must be exercised so that the animals are not crowded, that the water is changed at fairly frequent intervals, and that the new water is at the same temperature as the old water. The temperature should be raised gradually over several days before spawning is induced. During this time sufficient food must be supplied so that the animals do not resorb their gonads (Chanley, 1975). Unfortunately, some active, and especially some southern species, often will not tolerate this treatment.

To maintain an animal in spawning condition more than a very few weeks requires an abundance of algae. Some species, notably scallops, are difficult to keep in satisfactory spawning condition for more than a few days without extremely large volumes of seawater and algae. To condition an animal to spawn, in the laboratory, requires even more food and water.

For those animals that brood their larvae, it is necessary to maintain the adults until the larvae are released. Walne (1966) working with small *Ostrea edulis* keeps the brood-stock at mid-depth in tanks of seawater, running at 10–15 liters per hour, with the continuous addition of algae, at about 2 liters per day per 10–15 oysters. Spawning of the males is induced by heating to 20°C. Hidu (1975) working with the same species in Maine reports spawning from February to October by maintaining adults at 6 per 60-liter tub in standing aerated seawater of 20–22°C with algae added. Larvae are then released in 30–40 days. In Maine the natural spawning season is August and September. This technique has also been used with *Ostrea lurida* (Loosanoff and Davis, 1963).

CULTURE OF FERTILIZED EGGS

For normal development spawned eggs should be screened (to eliminate debris) and diluted as soon as possible. Excessive storage at high densities leads to increased percentages of abnormal development. Culture density varies with egg size. Small eggs (60 μ or less), such as those of the American oyster, *Crassostrea virginica,* can safely be cultured at densities of 50 per ml, or a little higher. Larger eggs, those over 60 μ, such as those of the bay scallop, *Argopecten irradians,* and the quahog, *Mercenaria mercenaria,* should not be cultured more densely than 30 per ml.

The screen size should be large enough to allow the eggs to pass, yet retain most of the considerable tissue and debris that accompany spawning. Eggs with an extensive membrane system require a very coarse screen.

Castagna (1975) recommends screening and counting the eggs and then adding the sperm to the diluted egg culture at a rate of 6 ml sperm suspension per liter of egg culture. Although self-fertilized eggs of *A.*

irradians develop normally, if the purpose of cultivation is to raise subsequent generations, self-fertilization should be avoided. For *Mulinia lateralis* excessive sperm leads to polyspermy and arrested development (Rhodes *et al.*, 1975). For many species, however, polyspermy is not a problem (Culliney *et al.*, 1975).

For culture of marine bivalve eggs and embryos, the two most significant environmental parameters generally are temperature and salinity. In the absence of other more precise data, the environmental temperatures and salinities where and when the population reproduces, or the temperature and salinity of gametogenesis, is adequate. This will be a little lower than the spawning temperature, but not greatly different. For a cold-water species such as *Placopecten magellanicus,* it will be in the 10–15°C range, while for a typical north temperate U.S. East Coast species such as *M. mercenaria,* it would be 24–28°C. Either too warm or too cold will result in a high percentage of abnormal development or failure to develop (Culliney *et al.*, 1975).

Many adult estuarine bivalve species can tolerate wide fluctuations in salinity. Eggs and larvae are apparently as tolerant as the adults, but cannot survive unfavorable conditions as long. Generally, salinity through embryological development should be maintained as near to spawning salinity as practical. Thus if gametogenesis occurred at 30 ppt, eggs might develop at 20–35 ppt, but if gametogenesis occurred at 20 ppt the eggs might develop at 15–30 ppt. According to Castagna (1975) working with Virginia bay scallops, eggs must be cultured above 20°C with an optimum at 26–28°C. Possibly Massachusetts scallop eggs will have lower optima and Florida eggs higher. Their minimum tolerable salinity is 22.5 ppt with the optimum at 28–30 ppt. Oceanic and Pacific coast species generally are less tolerant of salinity variation than Atlantic inshore and estuarine species. Embryos of the sea scallop, *P. magellanicus,* develop normally between 25–35 ppt.

Eggs must not be overcrowded. A quick and easy way to estimate culture density is to thoroughly mix the eggs in a reasonable quantity of filtered seawater and count all the eggs in a 1-ml random sample. If there are 300 or less, count another 1-ml sample. If they agree fairly well, use the average. If the two counts differ significantly, take a third sample and average the two close ones. Thus, an average of 200 eggs (75 μ diameter) per ml in a 10-liter bucket means 2,000,000 total eggs and will require 67 liters (or more) of filtered seawater for their culture. To discourage bacterial growth it is desirable, in addition to water filtration and screening the eggs, to use as large a culture container as possible. This will present the least surface for bacterial growth (Zobell and Anderson, 1936). For example, if there are 150,000 fertilized eggs of 75 μ diameter it will be better to culture them in one 5-liter bucket than in five 1-liter beakers. If the initial count is over 300, counting accuracy is virtually impossible and the 1-ml sample should be diluted and a subsample counted.

Seawater is not changed during embryonic development. This is generally 15–48 h depending on species and temperature.

Eggs of bivalve molluscs may be cultured for acquiring larvae and older stages, or for the study of or experimentation with the eggs themselves. Oyster eggs have been used in many short-term pollution studies (Davis, 1961; APHA, 1976) because cell division usually begins about 1 h after fertilization.

LARVAL CULTURE

Once the larva has become straight-hinged, or acquired the typical D shape of the early veliger, larval culture has begun. At this point daily water changes and feeding are initiated. Although larvae have often been cultured successfully by changing water at less frequent intervals, the chances of failure increase.

The water should be removed from the culture vessel through a screen of sufficiently fine mesh to retain these small larvae. Generally a 25- or 35-μ screen will accomplish this. As the larvae grow, larger mesh screens may be used. The larvae should be washed from the screen into a convenient volume of clean-filtered seawater immediately.

Screening should be done in two aliquots, separating the suspension larvae from those on the bottom. Take the suspension larvae first. This aliquot will contain those larvae swimming actively in the water column. The second, bottom deposits will contain dead and moribund larvae as well as some healthy ones. The two aliquots should be screened and washed into separate containers. Examine the suspension larvae first. Measure a representative sample to estimate rate of growth and look at the gut color to see if they have been feeding well. It should reflect the general color of the algae used as food. Look at their general apparent health and condition. Return these larvae to their cleaned containers of freshly filtered seawater and feed as soon as possible to ensure most rapid growth.

Examine the bottom deposits next. Under healthy conditions, this is normally a small percentage (5–20 percent or so) of the total larval population. Only at setting would one expect to find a large percentage of healthy larvae on the bottom. Thus, if there are significant numbers in the bottom deposits, that is immediate information that some culture condition is probably adversely affecting the larvae. It may be inappropriate temperature or salinity, excessive bacteria, or some other problem. Microscopic examination of the bottom deposit larvae will sometimes reveal the cause. Mucus at the shell margin indicates irritation (physical, chemical, or bacterial); bacteria, if populations are high, can sometimes be seen as a shimmering mass; protozoa are often large enough to be seen; fungal hyphae can be recognized with practice. If the *Sirolpidium zoophthorum* fungus shows up, throw the culture out. Davis and Chanley (1956) attributed certain seasonal culture failures to dissolved substances associated with dinoflagellate blooms. Millar and Scott (1968) also experienced repeated summer failures using techniques successful during the rest of the year and theorized that possibly breakdown products or external metabolites from plankton blooms were the cause. They found that using Fuller's Earth or magnesium trisilicate, adsorbing agents, larvae could then be cultured successfully.

The culture vessels should be sanitized with a 10-ppm chlorine solution and well rinsed before the larvae are returned to them. While the larvae are concentrated, they should be counted following the same procedures used for counting the eggs. However, to facilitate counting these active swimmers they must be killed or slowed. Reduction of salinity by diluting the 1-ml sample on the slide with fresh water slows the larvae without killing them, or the larvae in the sample may be killed with a drop of formalin.

The bottom-deposit larvae may be preserved to provide a culture record or discarded. (See the Appendix to this chapter for preservative formulas.) If there are few total larvae, they may be cultured, but should be cultured separately from the suspension larvae so as not to spread disease. A portion of the healthy ones will be up in suspension the next day and can be returned to the main suspension group then. Examine both bottom deposit groups separately.

A number of antibiotics have been used to control bacteria in larval cultures. Probably the most satisfactory is chloramphenicol at 30 mg/liter. Dihydrostreptomycin at 50–100 mg/liter is also useful and stimulates larval growth. Unfortunately, it does not have as broad a spectrum as chloramphenicol, and bacteria rapidly develop resistant strains (D'Agostino, 1975). Although some investigators regularly add antibiotics to their culture water (Walne, 1966), with conditions of good hygiene this should not be necessary.

Larvae cannot be cultured as densely as the nonfeeding eggs and embryos. As with egg culture, larval culture density generally varies with the larval size. A useful guide for temperate species is provided in

TABLE 16–2 Larval Culture Densities (Chanley, 1975)

Larvel Size	Number per Milliliter
50–100 μ	15
100–200 μ	8
200–300 μ	5
Over–300 μ	1

Table 16–2. This initial larval count is also necessary to determine the percentage of eggs resulting in normal development.

It should not be necessary to count the larvae at every water change. The culling that occurs during the changing process will automatically reduce the density if the remaining larvae are redistributed to the same volume of water. Larvae should, however, be counted periodically and at that time culture volumes redetermined.

Veliger larvae are shelled and much more tolerant of fluctuations in temperature and salinity than are the eggs, embryos, and trochophores. Although they will tolerate a wide temperature range, larvae will grow satisfactorily over a much narrower range. The optimum varies not only with the species, but also with the population of a species, particularly for widely distributed species such as the American oyster, *Crassostrea virginica,* which is found from Canada to Florida.

Culture is most successful if the animals are brought through their larval development as rapidly as possible. Thus, culture at near-optimum temperature is best. There are exceptions. Frequently, particularly in nutrient-rich seawater, bacterial growth is also most rapid at larval culture temperature optima. Most successful larval culture, then, takes place at a few degrees below the temperature of most rapid growth.

One technique developed by Culliney *et al.* (1975) to determine the temperature preferred by cultured larvae is as follows. Put the larvae in a tall cylinder of seawater set in a cold-water bath. A gradient to room temperature will form, and the larvae will tend to swim in a band at the preferred temperature. Thus it was discovered that, although *Xylophaga atlantica* spawned at 8–9°C, the early veligers preferred a 11–13°C temperature range. Older larvae were found randomly throughout the available temperature range (9–16°C).

Estuarine species of bivalve molluscs are particularly well-adapted to withstand great variability in both temperature and salinity. For removal of external growths of epiphytes and epifauna on their shells, a few minutes dip or rinse with fresh water is often successful. Older larvae can be dipped in a weak (3 ppm) chlorine solution to remove such growths and kill bacteria on their shell surfaces. Walne (1966, 1974b) recommends a 5-min dip.

As with the eggs, to discourage bacterial growth the largest practical culture container should be used. Healthy larvae will swim in an upward spiral to the surface. Thus, a tall rather than wide vessel is preferred. In the absence of turbulence, they will often form rafts on the surface of the water. A round shape, such as provided by a bucket or cylinder, is preferable to a square or rectangular one, because larvae then pile up in the corners (Culliney *et al.*, 1975). A conical or domed bottom is most frequently and successfully used particularly with the large culture vessels.

Aeration is not necessary to keep healthy larvae in suspension. Aeration, however, appears to be useful occasionally. The reasons for this are obscure. Oxygen demands of the larvae are very low, and aeration should not be needed ordinarily to maintain adequate levels of oxygen. Recent experiments indicate that aeration may actually cause an increase in bacterial populations. Possibly the air bubbles provide additional surface for bacterial growth.

Rate of larval growth varies considerably with species and culture conditions. Most species set (end larval growth) in 1 to 2 wk. However, species with lecithotrophic development throughout their larval period may set in 2–3 days, and some incubated species set within hours of their release. Species with small eggs, such as *C. virginica* and the pholads, tend to have a longer larval period, and it is not unusual for their larvae to remain pelagic for 3–6 wk or longer. Regardless of species, if larvae are measured to the nearest 5 μ length they will show increased growth daily if culture conditions are adequate. Often a length increase of 10 μ per day is possible, and occasionally larvae have been known to increase in length from 60 to 300 μ in only 8 days. Large larvae tend to grow more rapidly than smaller ones. If larvae do not grow at least 5 μ in length every 2 days, something is wrong.

SETTING AND JUVENILE CULTURE

During larval growth and development, many anatomical changes take place. Development of an eye spot is one indication that metamorphosis, or setting, is imminent. The eye spots appear near the middle of the animal near both left and right valves as a dark black irregular spot 5–15

TABLE 16–3 Cultch and Other Substrata for Setting and Juvenile Culture

Species	Setting: Cultch Required/Desired	Substrate for Juvenile Culture
Argopecten irradians	Vertical surfaces such as fiberglassed slats, sheets of mylar None needed	Trays fitted with vertical slats
Crassostrea gigas	Strings of shell Piles of tiles A variety of artificial cultches	On cultch, or stripped from cultch in trays
Crassostrea virginica	Same as for *C. gigas*	Same as for *C. gigas*
Macoma balthica	?	In false bottom trays or in trays of sand after juveniles are larger than sand
Mercenaria mercenaria	None needed	Same as *M. balthica*
Mulinia lateralis	None needed	Cultchless in trays, or same as *M. balthica*
Mytilus edulis	None needed Fibrous polypropylene rope or nylon netting often used	Cultchless in trays or suspended on fibrous ropes
Ostrea edulis	Same as *C. gigas*	Same as *C. gigas*
Ostrea lurida	Same as *C. gigas*	Same as *C. gigas*
Rangia cuneata	None needed	Same as *M. balthica*
Spisula solidissima	None needed	Same as *M. balthica*

μ in diameter. The eye spot is a patch of light-sensitive pigmented cells. Larvae of many species, however, either do not produce an eye spot or it remains inconspicuous. At about this time the foot develops and the larva is referred to as a pediveliger. It can both swim and crawl. Sometimes setting can be enhanced with a temperature rise of 3–4°C (Chanley, 1975). With some species, for example oysters, it is useful to provide a cultch material if continued culture is desired (Table 16–3). Some species can delay setting for considerable periods in the absence of satisfactory cultch. Fortunately, most species do not appear to be so selective. Oysters, at setting, cement themselves to a solid surface, and if suitable cultch is not provided they will cement themselves to the culture vessel. Although scallops do not form a permanent attachment at setting as oysters do, they do prefer a substrate (Castagna, 1975).

At setting bivalve molluscs give up their planktonic existence and become bottom dwellers. For this reason their culture vessel and substrate requirements change. A more or less flat tray is often most useful. However, for collection of the setting European oysters, Walne (1966) hangs strings of mussel shell in his larval tubs. He also catches the set on black PVC "tiles."

Setting, or metamorphosis, is presumably a very critical period because of the considerable stress from the rapid and extensive physiological changes taking place. Regardless of the reason, animals are especially prone to unfavorable conditions at this time, and generally disease problems are most severe with setting larvae.

Once set the food requirements of the now young juveniles increase greatly and they become much hardier. Daily water changes, however, should be continued at least until the juveniles are large enough to be seen with the unaided eye. Once they attain a length of 400–500 μ flowing raw, (unfiltered or very coarsely filtered) seawater can supplement cultured algae for these voracious eaters.

Handling of juvenile molluscs will vary greatly with the size, the generation time, and the purpose of the study, as well as the nature of the animal. If the goal is full-grown adults of a commercial species, such as oysters or quahogs, an outdoor growing area is really necessary unless a special facility is constructed to provide the tremendous amounts of food and water required. In the culture of the European oyster, for example, Walne (1966) treats his new spat just like larvae for 3–6 wk, then switches them to a system of recirculated seawater to which algae is constantly added. Water is exchanged twice a week. When the spat are 2–3 mm, they are moved to outdoor tanks of running seawater; and when 8–10 mm, they are removed from the collectors, kept another 2–5 wk in the running seawater, then planted at the oysterage.

For small species, with sufficient laboratory space available, the juveniles can be cultured inside. Rhodes *et al.* (1975), working with *Mulinia lateralis,* kept the set in static cultures until they were 0.5 mm maximum dimension, then moved them to running seawater trays with algae added. *M. lateralis* goes through an egg-to-egg cycle in 39–135 days with an average generation time of 60 days. A fully grown animal is 15–20 mm long, but both male and female spawn at 2.7 mm (Rhodes *et al.*, 1975). *Petricola pholadiformis* has a similar short generation time and small size at maturity.

See Table 16–3 for some useful setting and juvenile culture substrates.

Gemma gemma, which does not have pelagic larvae, can be reared throughout its life cycle in individual 50–100 ml containers with regular water changes and the addition of 10–30 ml cultured algae daily. Probably most bivalves that are not much larger than 5 mm can be reared in similar manner. Several leptonacid and/or eurycinacean species have been cultured in this manner. It is quite possible that some of these species may have shorter generation times than *M. lateralis, P. pholadiformis,* or *G. gemma.*

CULTURING MICROALGAE AS FOOD FOR BIVALVE MOLLUSKS

Culturing microalgae is a necessary adjunct to culturing bivalve larvae and recent set. Instructions can be found in many references (Guillard, 1975; Stein, 1973) and Part I, Chapter 6. What follows is a brief set of procedures adapted for the production of algae specifically as food for bivalve larvae.

Initial stock cultures may be acquired from a number of sources. One of the most reliable is Dr. Robert R. L. Guillard, Biology Department, Woods Hold Oceanographic Institution, Woods Hole, Massachusetts. Although he maintains an extensive collection, the beginning larval culturist would be advised to concentrate on *Isochrysis galbana* or *Pavlova* (*Monochrysis*) *lutheri,* species widely considered to be the best larval foods. Other algal species widely and successfully used include *Thalassiosira pseudonana, Chaetoceros calcitrans, Skeletonema costatum,* and *Tetraselmis suecica.* All can be cultured in the same manner, except that the diatoms require silicate in the media. Diatoms are especially desirable for larger larvae and juveniles.

The general algal culture procedure is to start with the initial (test tube) culture of each species and use it as an inoculum for several more, using aseptic techniques. Culture each species separately. The number maintained will depend on food demands. A 2-ml inoculum into 10 ml of media in another test tube will give good rapid algal growth. For sake of illustration, let us assume 10 stock cultures (test tubes) of each species have thus been established. Start a new set of 10 stock cultures by inoculating each of 10 new tubes with 2 ml from each of the present stock tubes. Thus, the old tubes now have 10 ml each and the new ones 12 ml. After the new tubes show algal growth, use the old ones to start larger volume cultures. Use each 10-ml starter culture to inoculate a 50-ml culture. This can be in Erlenmeyer flasks, medicine bottles, or other convenient sterilizable containers. Later, each 50-ml culture will become the inoculum for a liter culture, and so on until algal production matches larval culture needs. A very small-scale study probably would not need culture volumes greater than 2 liters each. A very large project might require 1,000 to 50,000 liter cultures, or even larger.

Small cultures are handled using general aseptic techniques. The culture medium and vessels are sterilized together and inoculation is done aseptically. For larger cultures this becomes a bit more difficult. Test tube and 50-ml cultures are most easily sterilized by autoclaving for 15 min at 15 psi. For larger cultures, sterilization of the vessel and seawater by chlorination, with dechlorination just before inoculation, is most practical. The nutrient is added later, aseptically, from a stock concentrate. Sufficient Clorox (or other similar 5 percent hypochlorite bleach solution) to produce 10-ppm residual chlorine is adequate. Chlorine contact time varies with the volume of seawater; 2 h is adequate for volumes up to 2–2½ liters, 4 h for volumes to 15–20 liters, and overnight for larger volumes. Dechlorination is accomplished with sodium thiosulfate. One-fourth the volume of Clorox used, of a 1 *N* solution, is sufficient. Thus, if it required 10 ml of Clorox to produce 10 ppm of residual chlorine in 20 liters of seawater, it would take 2.5 ml of 1 *N* NaS_2O_3 to dechlorinate. Algae are very tolerant of excess thiosulfate, but not of excess chlorine.

Algae have four basic requirements for growth: light, suitable temperature, carbon dioxide, and a suitable nutrient mixture. Most species cultured as food for bivalve larvae, including those listed above, grow well under continuous illumination. Although algae tolerate a wide range of light intensities, something on the order of 3,500 lux is required to provide good, rapid growth (Guillard, 1975). Temperature should be maintained at 20 ± 2–4°C. Ordinarily carbon dioxide will not become limiting in small cultures. In the larger cultures sufficient CO_2 is provided by aeration. Aeration also keeps the cells in suspension and helps maintain a stable pH. Sometimes 2 percent CO_2 is added to the incoming air to increase the density of algal cultures.

An algal nutrient medium, such as Guillard's (1975) f/2 is useful for maintenance of the small, autoclaved cultures. For the larger cultures, those sterilized by chlorination, use of a nutrient concentrate is easier. See Part I, Chapter 6, for formulas for algal culture media.

Since bivalve larvae apparently grow equally well on a wide range of algal concentrations of the species mentioned above, and tolerate well the algal culture medium itself, feeding is quite simple. If feeding axenic algae from sterile cultures, simply add the desired volume of algal culture to the larval culture after the water change. Generally this will be 1/20 to 1/40 the larval culture volume. If the feeding cultures are exposed to the air, they will not be axenic. In this case add chloramphenicol to the algal culture 48 h before feeding it. At 30 mg/liter bacterial counts drop to near zero after 48 h and stay so for about a week before they begin to rise slowly.

Larvae frequently grow better on a mixture of algae than a single species. Algae may be mixed just before feeding, or nearly the same effect can be achieved by feeding different species on successive days. Generally larger larvae utilize a greater variety of foods than smaller larvae. Either *I. galbana* or *P. lutheri* can be used to raise any species through its larval stages, and most of the algae mentioned can be used for most species. Nonetheless, for optimum growth feed *I. galbana* and/or *P. lutheri* for the first 2–3 days, then begin to include the diatoms. Use the *T. suecica* only for older larvae, juveniles, and adults. The algal species cannot be grown at the same time in the same culture vessels.

APPENDIX: PRESERVATIVES FOR BIVALVE LARVAE

For fixation and preservation of marine zooplankton, recommended by UNESCO:

5 ml formalin buffered to pH 8 with sodium glycerophosphate
1 ml propylene phenoxytol
10 ml propylene glycol distilled water to make 100 ml

Carriker's Solution:

5 ml commercial formalin
0.25 g sodium tetraborate
50 g sucrose
filtered seawater to make 500 ml

17

Methods for Rearing Crabs in the Laboratory

C. G. BOOKHOUT *and* J. D. COSTLOW, JR.

INTRODUCTION

Larvae of Brachyura, or true crabs, are found in most estuarine and marine environments and form an important component of the plankton. Before the studies of Lebour in 1927 and 1928, practically all of the published life histories of crabs were based on reconstructions from the plankton. The procedure usually involved obtaining freshly hatched larvae from an ovigerous crab of known identity and rearing the larvae as far as possible. When unsuccessful in rearing larvae to a later stage, which was usually the case, larvae were sought in the plankton that were slightly more advanced but matched the general characteristics of laboratory-hatched zoeae. This procedure resulted in incomplete life histories, and at times descriptions of reconstructured life histories included the larval stages of several species (Hopkins, 1944). Lebour (1927) was not able to rear freshly hatched crab larvae on algae from the plankton or on a culture of *Nitzschia*, but found that oyster larvae measuring 0.17–0.18 mm across, *Teredo* larvae measuring 0.05–0.06 mm across, and *Pomatoceros* larvae measuring 0.08 mm across constituted suitable food on which to rear *Inachus dorsettensis* Pennant and *Macropodia longirostris* (Fabricius) from the time of hatching to the megalopa. Crab stages were reared in the laboratory from late zoeae, and megalopa from the plankton. Lebour (1927) also recognized that crab larvae would not eat dead copepods that sank to the bottom.

Some years after Rollefsen (1939) discovered that *Artemia salina* (L.)

nauplii were satisfactory food for postlarval plaice, flounder, and cod, they were used to rear larvae of other fish, shrimp, crabs, and hermit crabs. From the late 1950's to the present, representatives of most all groups of decapod Crustacea have been reared in the laboratory on *Artemia* nauplii alone or, in case of small decapod larvae, in combination with supplementary food sources, such as sea urchin embryos. With the techniques now available, it is possible to obtain hundreds of larvae of known age and stage from a parent of known identity. From a thorough description of these larvae, it should be possible to identify planktonic species with a greater degree of assurance than previously. With a large number of larvae available, it is now possible to use crab larvae for embryological, ecological, physiological, and biochemical investigations.

The account that follows is intended for investigators who may wish to study some phase of the larval biology of decapods. It gives a brief outline of the life history of Brachyura and methods of rearing crabs in the laboratory, together with pertinent information concerning equipment which might be used, as well as possible sources where the equipment might be purchased. With minor modifications other groups of decapods, such as shrimp and hermit crabs, may be reared using the same techniques.

LIFE HISTORY OF BRACHYURA

LARVAL STAGES AND CHARACTERS

True crabs (Brachyura) pass through two to eight zoeal stages and one megalopa stage before developing into a first-stage juvenile crab.

For the benefit of readers who are not familiar with larval anatomy of crabs, a brief description of the general characteristics of crab zoeae, characters that are most frequently used to stage zoeae and characters of megalopa will be given.

A zoea is laterally compressed and characterized by a cephalothorax and a segmented abdomen plus a telson. The functional appendages throughout zoeal development are antennules, antennae, mandibles, maxillules, maxilla, and first and second maxillipeds. Rostral, dorsal, and lateral carapace spines are usually present. The rostral spine is directed ventrally, and the dorsal spine curves gently backwards. Lateral spines may be present, but they are absent in zoeae of some species.

Zoeae may be staged by many characters, but the following are most frequently used.

The *first zoea* has sessile eyes joined to the carapace, a five-segmented abdomen, plus a telson and four natatory, plumose setae extending from the exopodites of the first and second maxillipeds.

The *second zoea* has stalked eyes, a five-segmented abdomen, or a six-segmented abdomen if the second stage is the terminal one. Six or seven plumose setae are on the terminal ends of the first and second maxillipeds. The second zoea is the last zoeal stage in some species, such as the spider crabs, *Libinia dubia* Milne Edwards (Sandifer and Van Engle, 1971), *Hyas araneus* (Linnaeus) and *H. coarctatus* Leach (Christiansen, 1973), and the commensal crab in *Chaetopterus variopedatus* tubes, *Polyonyx gibbesi* Haig (Gore, 1968). In most families of true crabs there are more than two zoeal stages.

The *third zoea* of Brachyura has a sixth abdominal segment that in earlier stages was fused with the telson. The plumose setae on the terminal end of the first and second maxillipeds have increased, usually by two, over the number present in the second zoeal stage. The third zoeal stage is the terminal one in the grapsid crab, *Sesarma reticulatum* (Say) (Costlow and Bookhout, 1960b).

Zoeae in the *next to the last zoeal stage* of true crabs, which have three or more zoeal stages, have small rudimentary pleopod buds on abdominal segments two to six, a third pair of rudimentary maxillipeds, and five pairs of rudimentary pereiopods that extend slightly below the carapace.

Zoeae in the *last stage*, whether it be the second or seventh, have well-developed leg and pleopod buds.

Other characters used to distinguish zoeae of families or species of true crabs are discussed by Bookhout and Costlow (1974b) in a paper on the larval development of *Portunus spinicarpus* (Stimpson).

The *megalopa* is the stage of development between the last zoeal and the first crab stage. It is dorso-ventrally depressed, and all of the appendages of the cephalothorax and abdomen are present and functional. The dorsal and lateral carapace spines are generally absent, and the rostral spines are greatly reduced or absent. The abdomen is extended rather than folded under the cephalothorax as in all crab stages.

Megalopa of different species may be differentiated from one another by characters of spines on the cephalothorax and legs, by setal patterns on the appendages, and other characters (Bookhout and Costlow, 1974a,b).

BREEDING SEASONS

Most species of crabs show distinct annual breeding cycles in temperate climates and are ovigerous (carrying eggs) from late spring to the end of summer, or to early fall. In semitropical and tropical climates the breeding season is extended, and in some cases it is possible to obtain ovigerous crabs throughout the year.

Crabs belonging to Families Xanthidae and Grapsidae are common in estuaries and in shallow offshore waters. Their larvae are easy to rear, and they are often used in experimental studies. Periods when a few crabs of each family are ovigerous are given below.

Rhithropanopeus harrisii Gould, a xanthid, is ovigerous in North and South Carolina from April to October, and as early as January and February in Florida. This species is an excellent bioassay organism, for it furnishes larvae that can withstand salinities from 2.5 to 40.0‰ at a temperature of 25°C (Costlow *et al.*, 1966). Furthermore, survival to the first crab stage under optimum salinity and temperature conditions ranges between 90 and 100 percent (Bookhout *et al.*, 1976). The original range of *R. harrisii* was from New Brunswick, Canada, to Brazil, but this species has been introduced to the West Coast of the United States and to parts of Europe (Williams, 1965). The xanthid, *Panopeus herbstii* Milne Edwards, the common mud crab, is ovigerous from late spring to September in the Carolinas and throughout the year in Florida. Its range is from Massachusetts to Brazil (Williams, 1965). Hairy crabs, *Pilumnus sayi* Rathbun and *P. dasypodus* Kingsley, are ovigerous in the Carolinas from late April to August, but as early as March in Florida (Williams, 1965). *Pilumnus sayi* has a range from North Carolina through the Gulf of Mexico and West Indies, and *P. dasypodus* may be found from North Carolina to Brazil.

Crabs belonging to the Family Grapsidae, such as *Sesarma cinereum* (Bosc) and *S. reticulatum* (Say), have been reared in the laboratory and their larvae have been raised under experimental conditions of temperatures, salinity, and light (Costlow and Bookhout, 1960a,b, 1962; Costlow *et al.*, 1960). *Sesarma cinereum*, the wharf crab, is ovigerous from May to November in North Carolina. Its known range is from the Chesapeake to Venezuela (Williams, 1965). *Sesarma reticulatum* may be found in muddy salt marshes in the ovigerous condition during the summer months. Its known range is from Cape Cod to Texas (Williams, 1965). On the Monterey Peninsula of California *Pachygrapsus crassipes* Randall, the striped shore crab, is in berry (carrying eggs) between March and August with the peak in July, whereas *Hemigrapsus nudus* (Dana), another member of the Family Grapsidae in the same area, is ovigerous from October to May with the peak in January (Boolootian *et al.*, 1959). The range of *P. crassipes* is from Oregon to the Gulf of California; Galapagos Islands and Chile; also Japan and Korea. The range of *Hemigrapsus nudus* is from Sitka, Alaska, to the Gulf of California.

COLLECTION AND TRANSPORTATION

All crabs mentioned in the previous section, with the exception of species of *Pilumnus*, can usually be collected by wading. When small ovigerous crabs are collected near shore, each crab should be placed in separate glass or plastic jars three-fourths full of clean seawater from the site of collection. Thus crowding is avoided, and there is no danger of one crab tearing the eggs from another. When trawling for larger crabs by power boat, it is best to limit the hauling time to approximately 20 min in order not to interfere with aeration of the eggs by mother crabs, and not to damage the eggs by compression of other animals in the bag of the trawl. As the trawl load is emptied on the deck, ovigerous crabs should be immediately placed into a large tank of running seawater to avoid damage to eggs from drying. To prevent loss of smaller crabs from the trawl sample, they should be placed in plastic buckets or smaller containers, but the seawater should be changed as often as necessary to avoid exposing crabs to more than a few degrees change in temperature and to lack of oxygen.

Small ovigerous crabs, such as *Rhithropanopeus harrisii*, may be shipped safely by air freight from almost any U.S. airport to another if properly collected and packed. We have received shipments of this species from California and also from Florida. Our shipments from Florida were often collected, packed, and sent to Beaufort, North Carolina, in the course of a day. Each ovigerous crab was placed in a double plastic bag, ½ to ¾ filled with seawater from the site of collection. Twenty or thirty bags were then placed in a large stout plastic bag and packed with newspaper in a strong styrofoam box. As soon as the crabs were unpacked, each one was placed in a large glass finger bowl (19.4 cm diam.) of filtered seawater of the same salinity as the shipping water and aerated. The survival of larvae hatched from these crabs ranged from 96 to 100 percent to the crab stage (Bookhout *et al.*, 1976). Larger species of crabs, such as *Callinectes sapidus*, are more difficult to ship in a viable condition. The mother crab and its eggs probably suffer from lack of oxygen and accumulation of metabolic contaminants.

STAGING DEVELOPING EGGS

After ovigerous crabs are brought to the laboratory samples of their eggs should be staged to determine the approximate time hatching might take place. Freshly laid eggs can be distinguished from developed eggs by gross changes in color. For example, the "sponge" (the entire egg mass) of the blue crab, *Callinectes sapidus* (Rathbun), is light yellow when laid and very dark brown to black when about ready to hatch. The eggs of *Rhithropanopeus harrisii* are greenish when laid and change to yellow white with black pigment when ready to hatch. *Pachygrapsus crassipes* are maroon to dark maroon when laid and brown to dark brown when ready to hatch, whereas those of *Emerita analoga* (Stimpson) are bright yellow orange when laid and dull yellow when ready to hatch. Initially the eggs show no segmentation, but cleavage follows. This is followed by a yolk-free, transparent area and yolk-containing areas. Eye pigment appears and is followed by pigment in the body. Pigment intensifies, but there is still much yolk. Yolk is reduced to two small areas and zoea larva may be seen (Boolootian *et al.*, 1959). Completely developed embryos show a heart beat several days before hatching . The shape of the egg changes from spherical early in development to ovoidal as the time of hatching approaches. Movement of the abdomen indicates hatching is ready to take place.

HATCHING

Hatching begins by the bursting of the outer membrane caused by a swelling of the inner one. This may involve osmotic swelling of the inner membrane or the swelling of the larva itself. As the outer membrane is sloughed off the stalk of the egg is carried posteriorly. The bursting of the inner membrane and the emergence of the larva are apparently due to the movements of the larva itself or to the currents produced by the movement of the mother's pleopods, which wash the zoea out of the inner membrane (Davis, 1965).

REARING BRACHYURA IN THE LABORATORY

NATURAL HATCHING IN THE LABORATORY

To obtain newly hatched larvae from small ovigerous crabs, each crab should be placed in a large, glass finger bowl (19.4 cm diam.) containing filtered seawater of the same salinity to be used for rearing experiments. Each day thereafter the ovigerous crab should be changed to a clean finger bowl containing fresh seawater. This procedure should be continued until hatching occurs. In the event eggs are immature, *Artemia salina* (L.) nauplii may be added to the seawater as food for the mother crab. The bowls of crabs should be covered and maintained at a temperature to be used for rearing and with a light regime of approximately 12 h light and 12 h darkness.

DEVELOPING BRACHYURAN EGGS *IN VITRO*

If one wishes to hatch eggs from a large crab, such as the blue crab, *Callinectes sapidus* (Rathbun), or the stone crab, *Menippe mercenaria* (Say), the method or a modification first described by Costlow and Bookhout (1960) should be used. It involves removing pleopods that bear the egg mass or "sponge" from a crab and placing them in a large finger bowl (19.4 cm diam.) containing filtered seawater of the salinity to be used for rearing larvae. Each egg-bearing pleopod should then be washed in several clean bowls of seawater, and strands of eggs should be removed from each pleopod with fine scissors. These may be further dissociated with glass needles, washed in seawater, and placed in an acrylic or plastic compartmented box with each compartment half filled with seawater of a salinity to be used for rearing.

The covered compartmented boxes currently in use are made of clear acrylic-based material and were obtained from Mail Order Plastic, 56 Lispenard St., New York City, NY 10013. We use their 800 Series—Overall Size: 13⅛″ × 9″ × 2⁵⁄₁₆″, with 24 compartments, each of which measures 2″ × 2″ × 2¹⁄₁₆″. Each of the compartments of the box holds 135 ml. If one wishes to hatch eggs and rear larvae in a variety of salinities and temperatures, the eggs should be gradually changed into lower or higher salinities before placing them in the final salinity of the compartmented box. Strands of eggs may also be placed in a 2,000-ml flask half filled with filtered seawater. Compartmented boxes and flasks are then tied onto a variable speed shaker. Sixty to 110 oscillations per minute is sufficient to oxygenate the water in the containers. The variable speed shaker may be installed in a constant temperature cabinet or maintained in an air conditioned room with a temperature intended for rearing. When employing this method, we select crabs with eggs that have eyes and a visible heart beat. Usually these eggs hatch within 3 days, and hence there is no necessity to change the seawater. If the eggs are immature and to be left in containers for a longer time, the water should be changed every other day or whenever the water becomes cloudy. If the compartments are not crowded with eggs, 100 percent hatching may result. There is some variability in hatching rate, and occasionally one or two compartments may contain deteriorated eggs with ciliates of fungi. Larvae are selected from compartments in which the majority of eggs have hatched and larvae are swimming towards the light.

Before this method was employed larvae were obtained from an ovigerous crab that had been isolated in a bowl or glass jar. With small crabs this method may be satisfactory but numerous difficulties can arise if one wishes to obtain larvae from any large crab. For example, if a few eggs of an ovigerous crab are infected with fungi, the whole mass of eggs will not become infected if they are segregated in different compartments of a box. When an ovigerous crab is confined in a small aquarium, or glass cylinder of seawater, she will frequently remove the entire egg mass with her pereiopods shortly before hatching or even after a few larvae have hatched. Furthermore, the newer method allows the investigator to subject the eggs of one female to a variety of salinities, or other controlled environmental factors. Thus, the eggs are acclimated to the conditions under which the larvae will be reared. Provenzano (1967) has used a modification of this method with considerable success, not only to develop isolated eggs of Brachyura, but eggs of caridean and penaeid shrimps, anomurans, and lobsters as well.

WATER

The use of clean, uncontaminated seawater is essential for rearing. We collect high-salinity seawater in 5 gal polyethylene carboys 5 mi offshore. In the laboratory seawater is stored in 50-gal polypropylene storage cylinders after being filtered through a polypropylene calibrated strainer bag with a mesh size of 5 μ. Water for rearing purposes is drawn from the large resevoir and passed through a 5-μ polypropylene strainer bag into a 5-gal polyethylene carboy with a spigot.

The GAF calibrated filter bag is manufactured by GAF Corporation, Glenville Station, Greenwich, CT 06830.

Salinities above 35‰ and low-salinity water may be obtained from a carboy of frozen seawater. The first seawater to melt has a salinity that is very high. As more seawater melts, the salinity becomes lower. If the supernatant brine is removed, the last seawater to melt will be low-salinity water. Low- and high-salinity water procured in this manner will have all the components of natural seawater, including trace substances. If small reductions in salinities are desired, however, they may be made by adding glass distilled water.

Salinities may be determined by the use of Coast and Geodetic Survey hydrometers that give the density of seawater. These may be obtained from Emil Greiner Co., 20–26 North More St., New York, N.Y., together with a glass cylinder and liquid-in-glass thermometer. There are three hydrometers graduated for the following ranges of density: 0.9960 to 1.0110, 1.0100 to 1.0210, and 1.0200 to 1.0310. The range of the three hydrometers is sufficient to include density of any sample from fresh water to any saline water that might be encountered. The density reading is corrected for temperature as listed in Seawater Temperature and

Density Reduction Tables, U.S. Department of Commerce, Coast and Geodetic Survey, Special Publication No. 298 by Zerbe and Taylor (1953). By the use of Table 5 of this pamphlet, salinities in parts per thousand (‰) can be determined from the density obtained. Before using the hydrometers, they should be checked against standard seawater (chlorinity) samples obtained from Standard Sea-Water Service, Charlottenlund Slot, DK-2920 Charlottenlund, Denmark.

Salinities may also be obtained to within 0.5‰ by the use of an American Optical Company refractometer designed to determine salinity directly from as little as 1 ml of seawater. This instrument may be obtained from a number of distributors.

In the event it is not possible to obtain natural ocean water for rearing, synthetic sea salts, such as Instant Ocean, Seven Seas Marine Mix, or other brands, may be used. Roberts (1975) reported that Instant Ocean Artificial Sea Salts (Aquarium Systems, Inc., Mentor, Ohio) prepared in distilled water is more suitable for rearing crab larvae to the crab stage than any other artificial sea salt that he tested at Aquatic Sciences. Nevertheless, he found that crab larvae will not survive to the crab stage as well in this medium as in natural seawater.

FOOD

Rollefsen (1939) discovered that *Artemia salina* nauplii were satisfactory food for postlarval plaice, flounder, and cod. In subsequent years they have been widely used throughout the world for rearing fish, shrimp, hermit crabs, and crabs. Dry brine shrimp eggs may be obtained commercially in pressure packed cans and may be kept in the laboratory for several years with only a slight reduction in the viability. Kuenen (1939) noted there was considerable variability in *Artemia* from various locations in the world. Bookhout and Costlow (1970) reared four crabs, *Rhithropanopeus harrisii, Hexapanopeus angustifrons* (Benedict and Rathbun), *Libinia emarginata* (Leach), and *Callinectes sapidus* on freshly hatched *Artemia salina* nauplii from two sources, salt pools in the vicinity of San Francisco Bay, California, and the Great Salt Lake in Utah. Survival of *R. harrisii* from the time of hatching to the first crab stage was 94 to 98 percent when fed on California *Artemia* nauplii, but when reared on Utah *Artemia* nauplii none survived to molt to the first crab stage and those that reached the megalopa stage were deformed. There was also a marked difference in survival of the other three species of crabs that were fed on the two diets. In the case of *Hexapanopeus angustifrons* 69 percent survived to first crab stage when reared on California *Artemia* nauplii compared to 0 percent on Utah

Artemia nauplii, survival of *Libinia emarginata* to first crab was 63 to 5 percent on Utah *Artemia* nauplii, and survival of *Callinectes sapidus* larvae on *Artemia* nauplii with a supplement of *Arbacia* embryos was 34 to 24 percent. There was evidence that the survival and normality of developmental stages of four species of crabs might have been due to differences in the amount of DDT in the nauplii from California and Utah. There was three times the amount of DDT in the Utah nauplii compared to the amount in the nauplii hatched from California eggs. Little (1969) also reported that *Palaemon macrodactylus* Rathbun could be reared on San Francisco Bay Brand *Artemia* nauplii, but not on *Artemia* nauplii from Great Salt Lake, Utah.

There are some crab larvae that are so small that they cannot be reared on freshly hatched *Artemia* nauplii. We have reared *Callinectes sapidus* larvae to first crab stage on a diet of one drop of *Artemia* nauplii and one drop of *Arbacia punctulata* embryos per finger bowl of 10 larvae (Costlow and Bookhout, 1959; Bookhout and Costlow, 1975; Bookhout *et al.*, 1976; Bookhout and Monroe, 1977). In 1959 survival was low, ranging from 1 to 8 percent, but in 1975, 1976, and 1977 survival to the first crab stage averaged from 52 to 55 percent. Sulkin and Epifanio (1975) reported that the first two zoeal stages of *Callinectes sapidus* could be reared on the rotifer *Brachionus plicatilis* (Muller) ranging in length from 45 to 180 μm in length with better survival and faster molting rate than when reared on *Lytechinus variegatus* (Lamarck) gastrulae with an average diameter of 110 μm or *Artemia salina* nauplii, which measured 250 μm. They implied that there might be characteristic differences in the diets used, but concluded that one of the criteria for a successful prey organism in the rearing of early stage blue crab larvae is a size of not more than 110 μm. Christiansen and Yong (1976) reared the larvae of *Uca pugilator* (Bosc) from hatching to the megalopa stage on three different diets: (1) newly hatched *Artemia salina* nauplii, (2) the rotifer *Brachionus plicatilis* and a ciliate *Euplotes* sp., and (3) a combination of the first two diets. Survival to the megalopa on diet (1) was 90.0 percent, diet (3) 93.8 percent, and diet (2) 22.5 percent. They, too, concluded that, although a mixed diet of (1) and (3) may have provided a higher nutritional level, the poor results observed in a diet of rotifers and ciliates probably indicated that the food organisms were too small to be caught by larvae of *Uca pugilator* rather than a nutritional deficiency.

Although Sulkin and Epifanio (1975) found that *Brachionus plicatilis* provided a good diet for the first two zoeal stages of *Callinectes sapidus*, later Sulkin (1975) reported that rotifers alone would not sustain development to the megalopa. Trochophores of *Hydroides dianthus* (Ver-

Species: ____________ Date Hatched: __________ #____% to Meg. __________
Number: ____________ Salinity: __________ #____% to Crab __________
Bowl Size: ____________ Temperature: __________ #____% Meg. to Crab __________
# of Bowls: ____________ Light: __________ Days H to Meg. __________
Larvae/Bowl: ____________ Diet: __________ Meg. to Crab __________
Total # Larvae: ____________ (Days) H-C __________

Comments:

	BOWL			BOWL			BOWL			BOWL			BOWL		
Day	L	M	D	L	M	D	L	M	D	L	M	D	L	M	D
1															
2															
3															
4															
5															
6															
7															
8															
9															
10															
11															
12															
13															
14															
15															
16															
17															
18															
19															
20															
21															
22															
23															
24															
25															
26															
27															
28															
29															
30															

FIGURE 17–1

rill) or a combination of rotifers for the first 14 days and freshly hatched *Artemia salina* thereafter was an adequate diet to complete development. Sulkin and Norman (1976) found that *Artemia salina* was a better diet for rearing *Rhithropanopeus harrisii* and *Neopanope* sp. than the rotifer *Brachionus plicatilis* when based on survival and molting rates. *Rhithropanopeus harrisii* may be reared to the crab stage on freshly hatched *Artemia salina* in concentrations of 2, 5, 10, 20, or 40/ml, or in excess. The minimum concentration for good survival and optimum rate of molting appears to be 10/ml for stage 1 zoea and 5/ml for zoeal stages 2–4 (Welch and Sulkin, 1974).

METHODS OF REARING CRAB LARVAE

Static Cultures Using Finger Bowls

As soon as hatching occurs, 10 zoeae are placed in each of 5 to 10 finger bowls (8.9 cm diam.) with 50 ml of filtered seawater. A drop of freshly hatched *Artemia salina* nauplii is added to each bowl if the zoeae to be reared are large, such as those of *Rhithropanopeus harrisii* or *Menippe mercenaria* (Say). If the zoeae are small, such as the zoeae of the Subfamily Portuninae to which the species *Callinectes* and *Portunus* belong, a small drop of *Artemia* nauplii is added, as well as a small drop of a supplementary food source such as one of the following: *Arbacia* or *Lytechinus* embryos; rotifers, *Branchionus*; trochophores of *Hydroides dianthus*; or barnacle nauplii. Living zoeae are transferred daily to clean bowls with fresh media and food. Larvae are transferred with a dropping pipette with a wide mouth opening so as not to injure spines, or any other part of the larva. At the end of each 24-h period and before transferring zoeae, the number of living larvae, the number and stage of each exuvium, and dead larvae are recorded on a data sheet as shown in Figure 17-1. Hence, this series of 50 to 100 larvae is called a "check series." For experimental investigations, three to four replicate series should be run with larvae for each series from a different mother crab.

Static Cultures Using Compartmented Boxes

Zoeae may also be reared as "check series" in compartmented boxes using the same procedure as outlined above for finger bowls. From 1 to 10 larvae may be maintained in each compartment of a box, depending on the size of the compartment. When we first reared *Callinectes sapidus* from hatch to first crab, we placed one zoea in each compartment, thus we could keep records of when each zoea molted by the presence of an exuvium, the number of stages by the number of successive exuvia found, the time in days from one stage to the next, and the time from hatch to megalopa. After we became familiar with the life history of the blue crab, we reared subsequent series in finger bowls with 10 larvae per bowl. It is easier to find exuvia in the bottom of a finger bowl than in the bottom of each compartment of a box when rearing as many as 10 larvae in each unit.

Rearing Megalopa in Static Cultures

When a last-stage zoea from a finger bowl or compartmented box molts to a megalopa, it is placed in 1 of the 18 units of a clean compartmented acrylic box containing 25 ml of the same medium and food as used to rear zoeae in the "check series." The medium from each compartment is removed daily by a pipette with a large bulb and replaced by fresh media and food. This procedure is repeated daily until a megalopa exuvium is seen and the first crab stage is reached. The purpose of isolating each megalopa is to avoid cannibalism. Records are kept of the day the last-stage zoea molted to a megalopa and the day it molted into a first crab stage, or the day it died in the megalopa stage (see Figure 17–2).

The clear acrylic boxes used as containers for megalopa have hinged covers. They may be obtained from Mail Order Plastic, 56 Lispernard St., New York, NY 10013, by ordering "200" Series—Overall Size: 8¼″ × 4½″ × 1⅜″. Each of the 18 compartments is 1¼″ × 1¼″ × 1³⁄₁₆″.

Static Mass Cultures of Zoeae

We rear mass cultures of 700 to 800 crab zoeae in large glass finger bowls (19.4 cm diam.) with 700 to 800 ml of the same media as the "check series." Zoeae of these cultures are also changed daily to clean bowls containing filtered seawater of the same salinity as the "check series." If the zoeae are large, a dropping pipette full of *Artemia* nauplii is added to each bowl daily. If the zoeae are as small as *Callinectes* larvae, a dropping pipette full of *Artemia* nauplii and sea urchin embryos, or other supplemental food source, as mentioned above, is added to each bowl daily. When megalopa appear in a large bowl, each one is placed in a division of a compartmented box and reared as was described for the megalopa from the "check series." Mass cultures are a source of larvae for study of zoeae in the living and fixed condition, or the source of larvae for residue analysis in pollution experiments.

The method of rearing crab larvae in small finger bowls or compart-

	Date to Megalopa	Date to 1st Crab		Date to Megalopa	Date to 1st Crab
1			31		
2			32		
3			33		
4			34		
5			35		
6			36		
7			37		
8			38		
9			39		
10			40		
11			41		
12			42		
13			43		
14			44		
15			45		
16			46		
17			47		
18			48		
19			49		
20			50		
21					
22					
23					
24					
25					
26					
27					
28					
29					
30					

FIGURE 17–2

mented boxes is particularly valuable for the investigator who wishes to study the effect of environmental factors on the development of crabs. The method also furnishes larvae for morphological, ecological, physiological, biochemical, and endocrinological studies. The technique requires much time and considerable manpower, and is, therefore, not suitable for large-scale production of crab larvae and juveniles (mariculture). Within recent years attempts have been made to develop a satisfactory recirculating seawater system and a flowing water apparatus for large-scale culture of crab larvae.

Recirculating System for Rearing Crab Larvae

Sastry (1970) has developed a recirculating seawater system for rearing crab larvae. The main features of the system will be given here, but for details the reader should consult Sastry's paper, which is listed in the bibliography.

The flow of filtered water from a seawater reservoir to a larval culture tank was a siphoning action with slight positive pressure due to aeration of the seawater reservoir. Water from the culture tank was elevated through a charcoal filter by a variable speed peristaltic Masterflex tubing pump (Cole-Parmer, New York) and returned through 15 μ and 0.5 μ Orlon-would cartridge filters and an Ultraviolet Sterilization Unit to the resevoir. The optimum seawater temperature for best survival of larvae in the culture tank was controlled by placing the seawater resevoir in a constant temperature water bath. A timer controlled fluorescent lamp provided 14-h light and 10-h darkness. Initially, recently hatched crab larvae were added to the seawater of the culture tank and supplied with *Artemia* nauplii daily.

Sastry (1970) made a comparison of percent survival and time from hatch to first crab stage of *Cancer irroratus* (Say) and *Panopeus herbstii* when reared in a recirculating system and in compartmented boxes. Survival of 4,800 first zoeae of *C. irroratus* to the first crab stage in the recirculating seawater system was 0.06 percent in 43–58 days, whereas survival of 54 first zoeae to first crab stage was 33 percent in 37–58 days in compartmented boxes. During the first week zoeae were dispersed in the water column of the culture tank of the recirculating system, but in the second week and later there was a large number of crab larvae and some algae on the bottom. An increasing number of larvae appeared on the bottom with time and high mortality of zoeae and megalopa followed. Survival of 4,000 first-stage zoeae of *P. herbstii* to the first crab stage in the recirculating seawater system was 3.47 percent in 28–37 days, whereas in compartmented boxes survival of 36 first-stage zoeae of *P. herbstii* to the first crab stage was 52.7 percent in 19–34 days.

Sastry is of the opinion that larval survival rates in the recirculating seawater system may be expected to improve with further refinement of the system.

Flowing Water Apparatus for Rearing Crab Larvae

Buchanan *et al.* (1975) developed a flowing water apparatus for rearing *Cancer magister*. For details of the system, the reader should refer to the paper of Buchanan *et al.* (1975). The system was designed for chronic toxicant exposure studies, but it does not have to be limited to such investigations. It employs a flow reversal principle and includes culture containers in the form of 250-ml beakers with a 15-mm hole near the bottom with a Nitex screen on the inside and a second Nitex screen on the outside separated by 3–4 mm. Of course, seawater filters and sterilizers, pumps, and temperature control apparatus are also incorporated into the system.

Ten first-stage *C. magister* larvae were placed into each of eight beakers and fed 700 *Artemia* nuaplii three times a week during the period when they were in the first to the third zoeal stages; in later stages they were provided with 1,000 *Artemia* nauplii three times per week. Zoeae were transferred to clean sterilized beakers three times per week, but mortalities were recorded daily. Seawater flowed through the system on the average of 185 ml min^{-1}.

Survival of a group of *C. magister* zoeae reared in seawater with 1.0 ppt acetone in the flowing water apparatus for 50 days was 88 percent as compared to a survival of 85 percent for larvae reared in static culture. Between 50 and 60 days there was an increase in mortality in both systems, which seemed to be associated with molting to the fifth or last zoeal stage. Nevertheless, the survival at 60 days was essentially the same in both systems, 73 and 74 percent, respectively.

As far as we are aware, the apparatus described is the first successful flow system for rearing crab larvae.

FEEDING AND CARE OF CRAB OR JUVENILE STAGES

Most carnivorous decapods may be reared in the laboratory from the first crab stage to the late juvenile, or even to sexual maturity, if they are maintained in larger containers as they grow, supplied with filtered seawater of optimum salinity, and given a diet of shrimp or clam meat after the fourth instar.

We have reared the first four crab stages of *Rhithropanopeus harrisii* and *Callinectes sapidus* in a plastic box with 18 compartments, each containing 35 ml of seawater. As the juveniles became larger they were transferred to a 24-compartmented plastic box, each box containing 135 ml of seawater, later to small finger bowls (0.9 cm diam.), then to large finger bowls (19.4 cm diam.), and finally to tanks of running water.

During the first four instars, crabs will survive and molt if fed *Artemia* nauplii alone. After the fourth instar, they will continue to grow and molt if fed *Artemia* nauplii plus pieces of shrimp, which have previously been frozen, or pieces of fresh clam meat. If any pieces of shrimp or clam meat are not eaten after 3 or 4 h, they should be removed together with seawater and replaced by clean filtered seawater to avoid fouling of the medium.

CLEANING

If one wishes to rear crustacean larvae successfully, meticulous attention must be given to cleaning all containers and pipettes that come in contact with larvae. Glassware and transfer pipettes should be washed daily. It is our custom to rinse bowls and compartmented boxes and transfer pipettes in cold water. Finger bowls and compartmented boxes are then scrubbed inside and out with a good-quality brush and cake "Bon Ami" or "Alconox" powder, both of which are nontoxic. This should be followed by sufficient rinses in hot water until there is no trace of the cleaning agent. Finally, bowls and compartmented boxes should be rinsed thoroughly with distilled water and allowed to air dry. Under no circumstances should bowls or compartmented boxes be dried with a towel that has been washed with laundry soap. Pipettes may be washed with Bon Ami or Aloconox solution by drawing the solution in and out of the pipette. Rinsing with hot and distilled water should be done in the same fashion and pipettes allowed to dry in a rack.

If experiments are being done on the effect of heavy metals or pesticides on the development of decapods, it is essential that test containers, volumetric and transfer pipettes, etc., are cleaned before any use and daily after use. Procedures differ depending on the chemicals and bioassay organisms used. It is recommended, therefore, that the procedures given by the U.S. Environmental Protection Agency, Environmental Research Laboratory, Office of Research and Development, Gulf Breeze, FL 32601, entitled "Bioassay Procedures for the Ocean Disposal Permit Program," (EPA–600/9–76–010, May 1976) be followed.

CULTURE CABINETS

Culture cabinets may be fabricated to particular specifications, or purchased in different sizes from manufacturers of growth chambers. We are using six culture cabinets that were constructed for us by the Duke University Maintenance Department and Six Sherer Model Cel 255–6 growth chambers manufactured by Sherer-Gillet Company, Environmental Division, Marshall, MI 79068.

Our fabricated culture cabinets are 7′ L × 3′4″ D × 4′4″ H outside and 6′ L × 2′11″ D × 3′3″ H inside. The insulated box has two doors on each side that allow us to obtain culture bowls from either side of the cabinet. Each door measures 2′5½″ L × 3′5⅜″ H. Macbeth Examolites are suspended from the top of the cabinet by chains. They have a spectral distribution similar to daylight and can be regulated to provide 12 h of light and 12 h of darkness or any combination thereof. Each box has a fan and a heating and cooling element at each end of the cabinet. The difference in temperature at any one part of the cabinet is not more than 0.7°C. The cabinets may be set for a uniform temperature or regulated to provide fluctuating temperatures during a 24-h period. By maintaining some cultures in a constant-temperature cabinet and others in a cabinet with temperatures that fluctuate, we can compare survival rates of crab larvae when reared under the two conditions.

Each Controlled Environmental Chamber purchased from Sherer-Gillett Company has exterior dimensions of 5′ L × 2′4¾″ D and 2½′ H and interior dimensions of 3′ L × 2′ D × 2′7″ H. The chamber has one door that measures 2′½″ L × 2′6″ H. Temperature and lights are individually programmed. The temperature range is 7 to 43°C with full lights. Accessories and optional equipment, such as exterior mounted recording thermometer, temperature recorder, etc., may be purchased from the company.

Growth chambers that are more expensive and sophisticated than those we obtained from Sherer-Gillett Company may be purchased from different companies. If an investigator has a limited budget, he can use a BOD refrigerator type of constant-temperature cabinet. These may be obtained from many scientific suppliers. If a uniform temperature is not required, cultures may be maintained in an air conditioned room.

NARCOTIZING AND FIXATION

In our opinion, the best means of narcotizing a living zoea or megalopa is to add a small drop of 2-phenoxyethanol ($C_6H_5OCH_2OH$) (Eastman Kodak Co.) on the end of a glass needle to a watch glass containing a minimum of seawater and the larva to be studied. Whole larvae and molts may be fixed in 7 percent formalin buffered with hexamethylene–tetramine, briefly rinsed in distilled water and preserved in 70 percent ethanol or ethylene glycol. Media that have been used for dissection and wet mounts include ethanol and glycerine, 85 percent lactic acid, and Hoyer's medium.

18

Procedures for Maintaining Adults, Collecting Gametes, and Culturing Embryos and Juveniles of the Horseshoe Crab, *Limulus polyphemus* L.

GEORGE GORDON BROWN *and* DAVID L. CLAPPER

INTRODUCTION

Procedures for maintaining adult *Limulus*, and collecting and fertilizing gametes *in vitro,* have been developed over the last few years (Schrank *et al.*, 1967; Brown and Knouse, 1973; Clapper and Brown, 1980). As a result, large quantities of gametes can be easily obtained from both sexes throughout the year, thus making *Limulus* highly suitable for examination of developmental processes.

Horseshoe crabs belong to the arthropod class Merostomata. Early literature reflects disagreements over the number of extant species and their relationships. Currently, four species (representing three genera) are recognized (Waterman, 1958), and identifing morphological characteristics are clearly presented by Sekiguchi and Nakamura (1979). *Limulus polyphemus* L. is the North American representative (found along the Atlantic and Gulf coasts); the closely related asiatic species, *Tachypleus tridentatus* Leach, *T. gigas* Müller, and *Carcinoscorpius rotundicanda* Latreille are found in the western Pacific along the Asiatic coast and in the northern Indian Ocean (Waterman, 1958; Sekiguchi and Nakamura, 1979). All these species have probably existed since the late Mesozoic.

Interest in the developmental biology of horseshoe crabs extends over a century. Until recently, developmental events after blastulation had received the most attention, and three species had been examined to various degrees: *Limulus polyphemus* (e.g., Lockwood, 1870; Packard,

1885; Kingsley, 1892, 1893; Patten, 1896; Scholl, 1977); *Tachypleus gigas* (Iwanoff, 1933; Roonwal, 1944); and *T. tridentatus* (Kishinouye, 1892; Sekiguchi, 1960, 1973). Now, gamete anatomy and early developmental events are receiving great emphasis (e.g., André, 1963; Shoger and Brown, 1970; Mowbray and Brown, 1974; Tilney, 1975; Brown, 1976; Tilney *et al.*, 1979; Bannon and Brown, 1980; Brown and Clapper, 1980; Clapper and Brown, 1980).

Limulus embryo and larvae are also being utilized for a variety of other studies. These include: (1) a bioassay for ecdysteroids in which injected larvae show accelerated molting (Jegla and Costlow, 1979), (2) examining the response of larvae (and especially their blood amoebocytes) to gram-negative bacterial endotoxin (French, 1979; Bang, 1979), and (3) examining water and ion fluxes in larvae (Laughlin, 1979).

This paper addresses methods for procuring and maintaining adult *Limulus*, collecting and fertilizing gametes, and culturing and staging embryos and juveniles.

DISTRIBUTION AND NATURAL SPAWNING

Limulus polyphemus is distributed along the East Coast of North America from Maine to the Yucatan peninsula (Shuster, 1979). Spawning occurs at high tide and on sandy estuarian beaches (Shuster, 1979). Animals are in amplexus during spawning; the smaller male clasps the posterior part of the female's opisthosoma with a pair of specialized pincers. The female deposits several hundred eggs in a nest and the male is then stimulated to release semen. The spawning season varies considerably between the northern and southern populations of *Limulus*. In Massachusetts, spawning occurs from May through August, with a peak in June (Shuster, 1950); at Panama City, Florida, coupled animals in breeding condition are found in shallow water (apparently spawning) from April through October (Glendle Noble, Florida Marine Biol. Specimen Co., Inc., personal communication); and at Cape Canaveral, Florida, coupled animals are found throughout the year, with peak spawning occuring in April (Roy Laughlin, personal communication).

SOURCE OF ADULTS

Limulus can be collected during the seasons indicated or obtained from several commercial suppliers. Even during spawning season, not all animals will produce gametes, especially early and late in the season or if animals are obtained from deep water. Therefore, animals should be individually checked for gamete production before shipment. We have obtained *Limulus* from only two suppliers that will verify animals for gamete production (Florida Marine Biological Specimen Co., Inc., Panama City, Florida, and Marine Biological Laboratory, Woods Hole, Massachusetts). The animals ship very well and can survive several days in wood or styrofoam containers if they are packed in a moist material such as Spanish moss or wood shavings. Some marine algae are unsuitable because they die rapidly, deteriorate, and produce an unsatisfactory environment.

Limulus is dioecious, females being much larger (Figures 18–1,2). Males (Figure 18–3) are easily identified by their modified first pair of walking legs, which are used for clasping the female during amplexus (Figure 18–4). To obtain large quantities of gametes, we use females that are at least 19 cm (and preferably greater than 22 cm) across the prosoma (cephalothorax) and males that are 14 cm (preferably 16 cm) or larger.

MAINTENANCE OF ADULTS

We have maintained adults in two types of systems. At Iowa State University, several temperature-controlled Instant Ocean aquaria are used, each containing 570 liters of artificial seawater (ASW) (Figures 18–4,5). Aquaria and ASW were obtained from Aquarium Systems, Inc., Mentor, Ohio. A maximum of 15–20 animals are placed in each aquarium and maintained at 15°C and a light-dark cycle of about 14 and 10 h. Approximately 90 liters of ASW is changed monthly. At Hopkins, flow-through seawater is utilized in an outside concrete tank filled to a depth of about 0.3 m and containing 3,600 liters. The tank is shaded at one end to allow animals to avoid direct sunlight. At present, 57 animals are being maintained in this tank. The temperature has fluctuated daily from 15–22°C in September to 11–15°C in January. In both locations, the animals are fed all they will eat three times a week. In the flow-through tank, they consume about 25 g of squid per animal per feeding. In the case of the ASW aquaria, various types of nonoily fish (with flesh that does not float in seawater) are used, and the animals eat somewhat less.

The longevity of *Limulus* is dramatically different in the two systems. During 12 yr of use, mortality in the ASW aquaria shows a consistent trend: Few animals are lost in the first 3 mo after which mortality rate increases and few animals live beyond 6 mo. Gamete production by each animal seems little changed from arrival until death if the animals are fed well. If they are not fed, gamete production ceases by about 3 mo, although longevity is similar. The cause of death is unknown, although there is sometimes excessive bleeding from genital openings. For the last year, a different ASW was utilized (from Jungle Laboratories Corp., Sanford, Florida), and longevity seems somewhat improved.

Our experience with the flow-through system is limited to the last 18 mo. Six animals were maintained (and fed sporadically) with two deaths during the first year; an additional 32 animals were obtained 6 mo ago and 27 are still alive. Therefore, although the ASW system has proved very workable for 12 yr, the flow-through system is probably superior.

In both systems, we see no seasonality in gamete production. If animals are producing gametes upon arrival, they usually continue until death. If they are not producing upon arrival, fewer than half have been brought into breeding condition with either system.

DISEASES

Little is known about diseases afflicting *Limulus*. Bang (1979) discusses some circulatory system and amoebocyte responses to gram-negative bacteria and briefly mentions some diseases of the gills observed at Woods Hole. The three major difficulties we have encountered are: (1) formation of surface lesions in the genital region, (2) hemorrhaging from gonopores (as previously mentioned), and (3) the formation of large blood-filled blisters on the genital operculum, similar to the pulmonary edema described by Bang (1979). The surface lesions become serious only on animals that are used excessively by students, who (while collecting gametes) apparently use voltages that are too high or durations of electrostimulation that are too long. The hemorrhaging and formation of blisters occurs mainly on animals from which gametes have been collected twice a week for 2 mo or longer. These animals usually die in the ASW system, but in the flow-through system they seem to recover if allowed to rest a few weeks. However, this system has not been utilized long enough to verify whether full recovery occurs.

TAGGING OF ANIMALS

All animals upon receipt are routinely tagged, thus allowing notation of the donor of gametes used for each culture or for experimental procedures. As a result, documentation has been obtained for: (1) the productivity period of each animal, (2) fertilizability and quantities of gametes produced by each animal, (3) variations in sperm motility initiation in sperm collected from different males (Clapper and Brown, 1980), and (4) variations in the egg cortical reaction in eggs from different females (Brown and Clapper, 1980). Several methods for marking animals were tried, but plastic tags are the most satisfactory. This requires drilling a small hole through the margin of the opisthosoma midway between the prosoma and telson and attaching a commercially prepared plastic tag with a nylon wire fastener. Marking animals by painting numbers directly on the prosoma proved unsatisfactory, since no paint or other marking material would last longer than 2–3 wk.

GAMETE MORPHOLOGY

The sperm is classified morphologically as a typical primitive sperm (Åfzelius, 1972) and is composed of an apical acrosome, a nucleus, and a flagellum (Figure 18–6). The unique feature is a preformed acrosomal rod that is coiled posterior to the nucleus (André, 1963; Shoger and Brown, 1970; Tilney, 1975; Tilney *et al.*, 1979). The acrosomal reaction is readily observed with phase-contrast optics at a magnification of 400×. The spherical egg (Figure 18–7) is large, approximately 1.9 mm in diameter and is surrounded by an extracellular egg envelope (Brown and Humphreys, 1971; Brown and Clapper, 1980).

GAMETE PROCURING AND PREPARATION

Although adult *Limulus* possess several pairs of formidable-appearing pincers, even the largest individuals are quite harmless. When first learning how to collect gametes, assistance in holding the animals is helpful. Alternatively, a round, flat stone such as shown in Figure 18–8 can be wedged between the prosoma (cephalothorax) and opisthosoma (abdomen) as shown in Figure 18–13 to prevent the animal from folding the opisthosoma forward and preventing access to the gonopore region. The genital operculum is then lifted to expose the gonopores (Figure 18–9 through Figure 18–16) and an electrical stimulus (3–4 V, 0.2–1.0 ma, a.c.) is applied below and slightly lateral to the gonopores (Figures 18–10,14). Great care must be taken in this procedure because excessive stimulation will cause lesions. The best method is to apply the stimulation in short intervals. An alternative procedure reported by French (1979) is to apply slight pressure anterior to the genital operculum.

After electrical stimulation, semen is collected by moving a 15 ml centrifuge tube gently over the gonopores. Semen, particularly when only small amounts are needed, can also be collected by using a Pasteur pipette. After the first few usages, the male generally responds quite easily to gentle pressure of a centrifuge tube beneath the genital opening (Figure 18–9), thus obviating the electrical stimulus. In general, for routine developmental studies 0.1 ml of semen is sufficient. Normally, this semen (10^{10} spermatozoa/ml) is diluted with seawater to a 1–10 percent sperm suspension. Since this dilution does not induce sperm motility, the sperm suspension can either be used immediately, allowed to sit at room temperature for several hours, or stored at 5°C for as long as 5 days and still be viable. It is extremely important to use well-rinsed glassware, since small amounts of detergents can induce motility. The sperm suspension can be checked for viability by mixing a drop with an egg or an egg section and determining whether sperm are motile (Mowbray and Brown, 1974; Clapper and Brown, 1980).

In procuring eggs, special care and handling must be observed in order to obtain satisfactory results. Since eggs are extremely adhesive to glass, collecting equipment should be plastic, wooden, or covered with parafilm. With a little dexterity, two wooden applicators can be used to transfer a mass of spawned eggs to a Petri dish. With these precautions, eggs are collected as follows: Several short electrical stimuli (1 s each) approximately 2 cm beneath a gonopore promotes contraction of the oviduct and spawning of 10–50 eggs from the opening (Figures 18–14,15). Since the number of spawned eggs is controlled by the duration of stimulus, 25–400 eggs can be collected from each oviduct. Eggs should be fertilized within 5–10 min after collecting or the percent of development may drop. Although many different ways of collecting and storing have been attempted, successful storage of eggs is exceedingly rare. For some reason, a form of parthenogenesis comes into play shortly after spawning, and, although development is initiated, it aborts at or before the blastula stage. Thus, for developmental studies, appropriate culturing containers should be prepared, the semen collected, and *then* the female should be spawned.

CULTURING

Culturing of embryos through to hatching is readily accomplished in 9-cm plastic Petri dishes at room temperature. Seawater (35–40 ml) is added to each dish; then eggs (10–50) are collected and placed in the seawater (Figure 18–15 through Figure 18–17). Numbers higher than

75–100 cause overcrowding and lead to variable development. Each dish is gently swirled to distribute eggs evenly in the dish. Then 1–2 drops of the previously prepared sperm suspension is added (Figure 18–18) and the dish again gently swirled. Sperm motility and attachment can be verified at this time by observing at 50× with a good dissecting microscope (using transmitted light). Most eggs will adhere slightly to the plastic after making contact and remain thus for several days. This can be an advantage, since the location of each egg can be noted. If this adherence is not desired, the fertilized eggs can later be gently nudged loose.

Fertilization is probably completed in 5 min. However, as a precautionary measure we generally do not wash away excess spermatozoa and debris until 30–45 min has elapsed. Daily changes of seawater for 3–4 days or the use of antibiotics (actinomycin D and penicillin or streptomycin) will generally keep a clean culture. After this period, changing seawater once a week will suffice. The cultures can be maintained within a temperature range of 15–32°C, depending on the rate of development desired. If delay of development is needed, refrigeration at 5–10°C will inhibit developmental processes without causing detrimental effects. Salinity varying from 50 to 125 percent of normal seawater has little effect on development. Diagnostic features of each developmental stage can be examined by viewing embyros on a dark background with one or more sources of light placed at a 90° angle to the vertical axis of the microscope.

Culturing of *Limulus* embryos and early juveniles has been highly successful. In culturing several thousand embryos, fertilization averages approximately 90 percent, with a mortality rate of approximately 5 percent during development to the first posthatched juvenile. In culturing juveniles, data at hand are limited to one trial using 20 embryos (10 eggs fertilized from each of two females). At the present time, 18 juveniles have survived to the fourth posthatched juvenile stage, and two of these have reached the fifth stage.

After the first posthatch moult, the juvenile must be fed if development is to continue. Ten juveniles should be placed in a 9-cm Petri dish containing a layer of fine sand sufficiently deep to allow burrowing. Sand prevents the growth of epiphytes on the surface of the juveniles and has been reported to reduce mortality (French, 1979). Although various foods are probably adequate (French, 1979), freshly hatched brine shrimp (*Artemia*) work satisfactorily for the first and second posthatched juveniles. *Artemia* "eggs" are sprinkled directly into the *Limulus* culture dishes. The eggs float, but the hatched nauplii (1–2 days) sink to the bottom and are consumed by the *Limulus*. Sufficient *Artemia* eggs should be added every other day to maintain a low concentration of nauplii in the culture. At least once a week, the seawater should be changed to remove the egg hulls and waste materials. The third and fourth posthatched juveniles should be placed in larger containers and the *Artemia* eggs supplemented with finely chopped squid, frozen brine shrimp, or other marine organisms. Kropach (1979) developed an aquaculture system that will probably be adequate for the entire life cycle, but the advantages are limited, since it takes several years for sexual maturity to be reached (Shuster, 1950).

NORMAL STAGES OF DEVELOPMENT

Although previous studies (Kingsley, 1892, 1893; Iwanoff, 1933; Scholl, 1977) have described many of the developmental stages in the horseshoe crab, a continuous system for designating various stages (beginning with the fertilized egg) has not been developed for *Limulus polyphemus*. Sekiguchi (1973) has published a plate showing normal development for the Japanese horseshoe crab, *Tachypleus tridentatus*, and although this species develops very similarly to *Limulus*, the rate is much slower. Furthermore, the specimens used for illustration had been fixed and stained. In our study, micrographs of living *Limulus* embryos cultured at 22–23°C were used to illustrate developmental stages. The stages, as described by Sekiguchi, have been followed as closely as possible. However, discrepancies do exist and are discussed below.

All *Limulus* developmental stages shown in Figures 18–19 to 18–47 were photographed through a dissecting microscope and are illustrated at the same magnification (22×). The diagnostic features are described in Table 18–1. Although the characterstics of most stages were illustrated, this was difficult to do for Stages 9–11 (Figure 18–31 through Figure 18–33) because of subtle differences in surface features. By comparing the description for each stage with careful examination of the living embryos, identification of these stages is possible. Fortunately, embryos in a single culture generally develop at a similar rate. However, since some variability does exist, timing of stages was designated at the point when 50 percent of the embryos reached a particular developmental stage. Overcrowding and microbial growth can greatly aggravate the variability.

Although most stages of *Limulus* are comparable to the stages of *Tachypleus* as described in Sekiguchi (1973), events in some stages differ. Stage 1 includes all events from fertilization to the appearance of nuclei near the surface. These "events" have recently been investigated (Ban-

TABLE 18–1 Description of Developmental Stages of *Limulus*[a,b]

Stage	Figure	Time from Fertilization	Brief Description
1	19	0	Unfertilized egg. The surface is smooth. Indentations are caused by oviducal packing and will eventually disappear.
1	20	30 min	The appearance, coalescence and disappearance of pits represent the egg cortical reaction. Pit appearance can usually be observed 10 min after insemination and the pits appear uniformly over the surface.
1	21	45 min	After disappearance of the pits, the surface becomes smooth. The duration of the cortical reaction is variable but generally is completed by this time.
1	22	3½ h	The first granulation of the surface begins to appear about 2½ h after fertilization and ends 1½ h later when surface becomes smooth. A second granulation cycle will occur between 5 h and 11 h after fertilization.
1	23	4 h	Smooth surface embryo between the first two granulation cycles.
2	24	21–28 h	On the surface, 8–16 large nuclei are equally distributed. They frequently appear in pairs as if in mitosis. Often, but not always, each nucleus appears to be enclosed in a cell. A third granulation cycle often makes viewing of the nuclei difficult.
3	25	28–45 h	Segmentation or yolk block formation, possibly representing the beginning of blastulation. The stage begins with deep grooves and the appearance of 4–8 large blocks and continues to about 150 blastomeres on surface. Another granulation cycle occurs on the surface of each blastomere near end of stage. Nuclei are rarely obvious, although some are observed between the large blocks.
4	26	45–58 h	Immediately after the previously mentioned granulation cycle is completed, the nucleus in each blastomere is obvious. Blastomeres vary greatly in shape and size and the surface of the embryo is quite irregular. Patches of smaller blastomeres are frequently observed in the latter half of this stage.
5	27	58–73 h	Due to rapid cell division, blastomeres are now smaller in size and more uniform in shape. They are loosely attached to one another and give a jumbled appearance to surface of embryo. However, as this stage progresses, surface becomes smoother.

6	28	3½ days	Unlike previous stage, difficult to see margins of blastomeres and nuclei appear equally distributed. Embryo appears very smooth with a few cells sticking above the surface.
7	29	4 days	Individual blastomeres can only be observed at high magnifications. The appearance of the irregularly shaped indentation represents the area which will become the germ disc.
8	30	5 days	Area around previous indentation is now uplifting and is approximately ⅙ the diameter of the embryo. This area is the developing germ disc.
9	31	6 days	Germ disc increases in size to approximate ¼ diameter of embryo. The disc is still an uplifted area and has equal margins on all sides. Unfortunately, Stages 9, 10, and 11 are very difficult to reproduce photographically, as the germ disc is difficult to observe. Margin is not obvious in light micrograph.
10	32	7 days	Germ disc is now approximately ½ diameter of embryo and can be observed as an uplifted area if embryo is viewed laterally. This stage is unfortunately variable since germ disc does not always appear as an uplifted area but sometimes quite smooth.
11	33	8 days	The germ disc definitely has a marginal groove which is much more pronounced along one edge. If viewed laterally, this particular edge appears as a deep groove.
12	34	9 days	Usually multiple grooves become obvious around germ disc. In addition, extensive wrinkles and ridges are observed in other regions of embryo and continue to be present throughout Stage 12 and into Stage 13.
13	35	10 days	Germ disc continues to differentiate with multiple grooves around margin becoming obvious. Characteristic is the presence of an apparent groove transecting the germ disc.
14	36	11 days	Six pairs of elongated blocks representing the limbuds are observed. In addition the abdominal region shows some differentiation.
15	37	12 days	The first five pairs of limbbuds are becoming round. The stomodaeum appears and is anterior to the first pair of limbbuds. The other regions of embryo are smooth.
16	38	13 days	As limbbuds elongate into appendages, stomodaeum moves posteriorly to a position between first pair of appendages. Round lateral organs appear laterally to fourth pair of appendages.
17	39	14 days	Final stage before first embryonic moult. If carefully examined, the loosening of the exuvia can eventually be observed in the region between the fourth–sixth pairs of appendages. Appendages are becoming articulated and frequently move. Anterior

TABLE 18–1 *Continued*

Stage	Figure	Time from Fertilization	Brief Description
			end continues to differentiate and stomodaeum is located at the end of a ridge-like structure. Opisthosomal segmentation is obvious. By careful focusing, polygonal plates of extra-embryonic shell (located beneath egg envelope) can be observed.
18	40	15 days	Material released during first embryonic moult is very cloudy and obscures many features of this stage. However, initial segmentation of the green hepatopancreas (occupying most of the embryo's internal space) can be observed.
19	41	17 days	After second embryonic moult, body segmentation of hepatopancreas has become more pronounced and extends over the entire dorsal surface of this organ. Although medium surrounding embryo is clearer than before, the extra-embryonic shell is more opaque and causes difficulties in observing embryo. Lateral organs are oval.
20	42	21 days	After third embryonic moult, the egg envelope breaks away. The process is slow with a crack occurring in the egg envelope and the extra-embryonic shell eventually swelling to maximum size. Infrequently, this swelling can occur earlier in Stage 19. The egg envelope in time falls away.

20	44	22 days	The larva freely moves around within the extra-embryonic shell. If no obstructions are present, larval movements cause rolling around of shell in the Petri dish. As a result, behavioral responses to various tropisms can be studied.
21	45	26 days	The fourth embryonic moult produces the tribolite larva, which generally hatches from extra-embryonic shell in 1–1½ days. The opisthosomal appendages are greatly flattened.
21	46	27 days	Hatched tribolite larva swims freely and frequently in a dorsal down position. Posterior margin of opisthosoma is serrated and sensory filaments can be observed on prosoma. Excellent animal for studying behavioral patterns. Larva does not feed.
First posthatched juvenile	47	59 days	All juvenile stages are similar to adults in appearance and must be fed. The animal is very active and will readily burrow into very fine sand. Length is 6.5 mm; width (across prosoma) is 5.0 mm
Second posthatched juvenile		4½ mo[b]	Length is 9.0 mm; width is 6.5 mm
Third posthatched juvenile		6 mo[b]	Length is 13 mm; width is 9.0 mm
Fourth posthatched juvenile		7½ mo[b]	Length is 18 mm; width is 11 mm
Fifth posthatched juvenile		9 mo[b,c]	Length is 24 mm; width is 16 mm

[a] Magnification of each micrograph is 22 ×.

[b] The information concerning the second through fourth posthatched juvenile stages was obtained from a culture that was fertilized and cultured at 19–22°C. 22–23°C was used for all earlier stages.

[c] Mean moulting time is projected from two animals that have currently moulted.

non and Brown, 1980; Brown and Clapper, 1980; Bannon and Brown, in preparation) and consist of a cortical reaction and at least two granulation cycles. These are designated as "events" and not additional "stages." The yolk block formation (segmentation process) occurring in *Limulus* in Stage 3 (Figure 18–25) is reported by Sekiguchi to occur in Stage 1 of *Tachypleus*. However, the description of nuclei and their appearance on the surface correlates well during the early stages. The swelling of the extra-embryonic shell usually occurs during Stage 20 in *Limulus*, but occurs during Stage 19 in *Tachypleus*. Scholl (1977) has recently completed a careful study of late *Limulus* embryology and demonstrated 9 stages, starting with the appearance of limbbuds (our equivalent of Stage 14). He has described two stages in place of our Stage 20. Further studies are needed to clarify this problem.

ACKNOWLEDGMENTS

The authors thank Gwen Waddington, Sharon Nugent, Susan Frey, and Carol Gostele for expert assistance in the completion of this manuscript, Jim Redmond for close-up photographs and preparation of Figures 18–1-5 through 18–8-10, and David Epel for moral support and encouragement during the final preparations. This study was supported in part by a research grant and a Faculty Improvement Leave from Iowa State University.

FIGURE 18–1 Pair of mature horseshoe crabs, *Limulus polyphemus*. The female (22 cm across prosoma) is larger than the male.

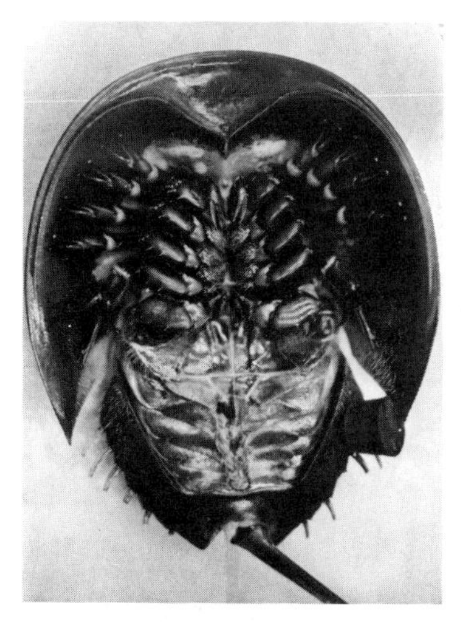

FIGURE 18–2 First four pairs of walking legs in female are structurally similar.

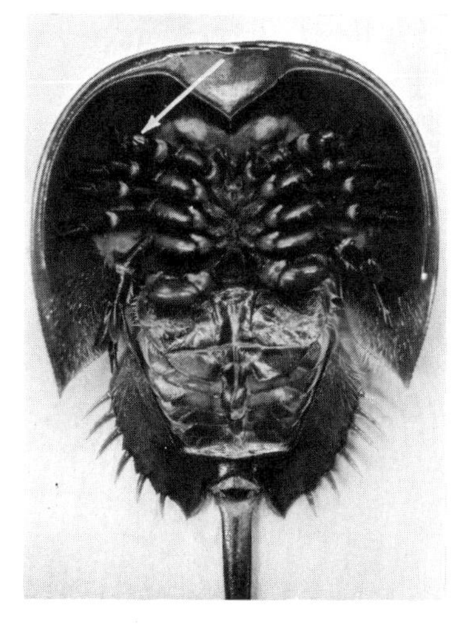

FIGURE 18–3 First pair of walking legs of male are claspers for amplexus.

FIGURE 18–4 570 liters Instant Oceans Aquarium with 10–15 animals.

FIGURE 18–5 Although amplexus occurs in aquaria, we have never observed spawning.

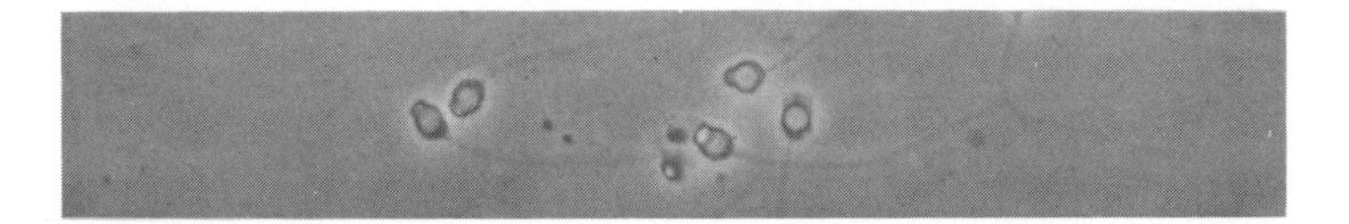

FIGURE 18–6 Spermatozoa have typical primitive morphology: apical acrosome, nucleus, and flagellum. Motility occurs only in close proximity to egg.

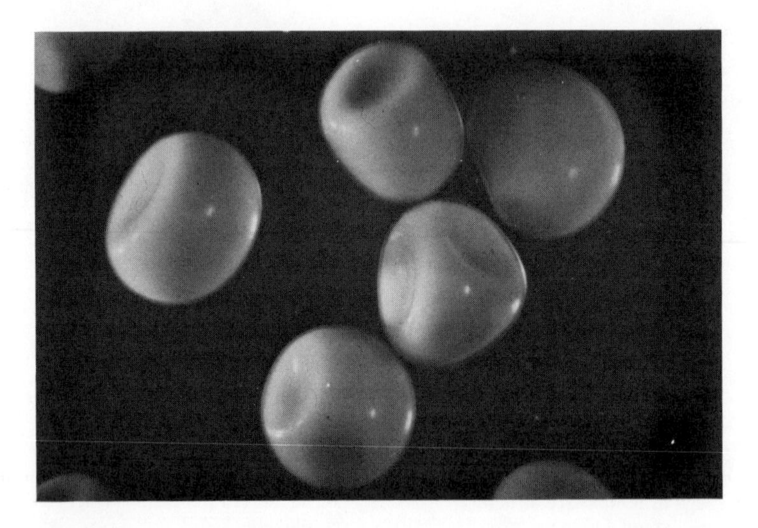

FIGURE 18–7 Unfertilized eggs (1.9 mm in diam.). Indentations are caused by oviductal packing.

FIGURE 18–8 Apparatus for collecting gametes: electrical stimulator, Petri dishes, wooden applicators, flat stone (Figure 18–13), and centrifuge tube and pipette.

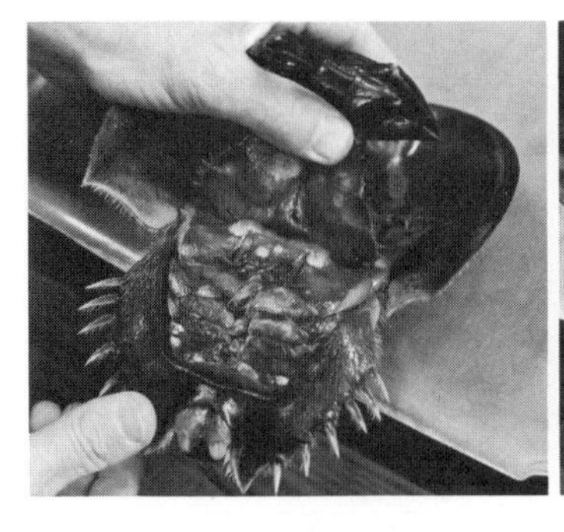

FIGURE 18–9 Positioning of animal before collecting gametes.

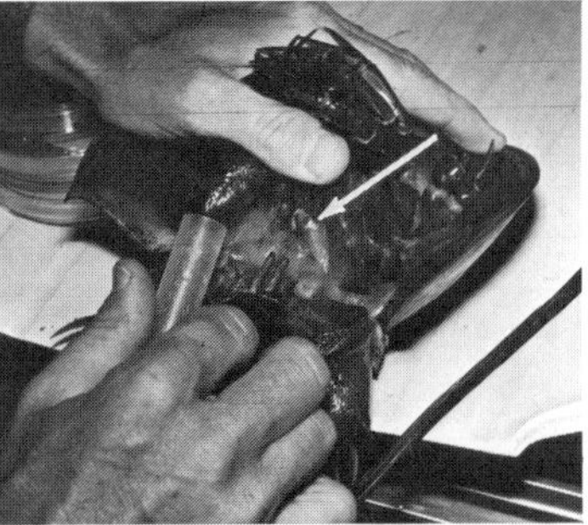

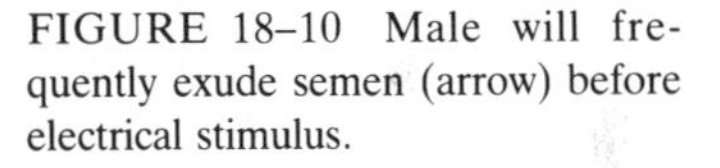

FIGURE 18–10 Male will frequently exude semen (arrow) before electrical stimulus.

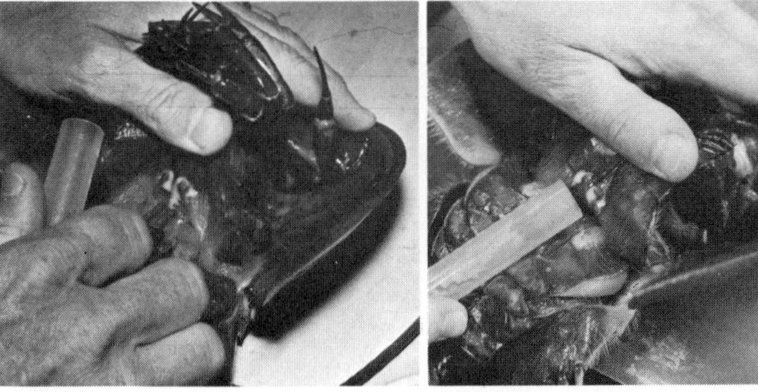

FIGURE 18–11 After electrical stimulation, semen flows copiously from gonopores.

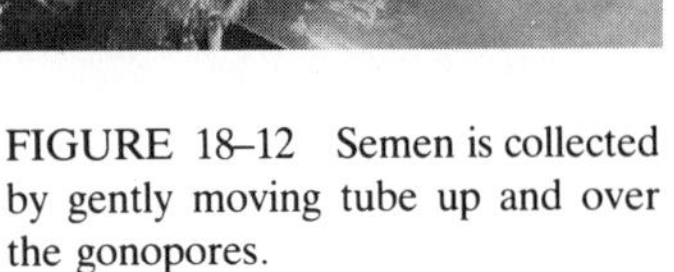

FIGURE 18–12 Semen is collected by gently moving tube up and over the gonopores.

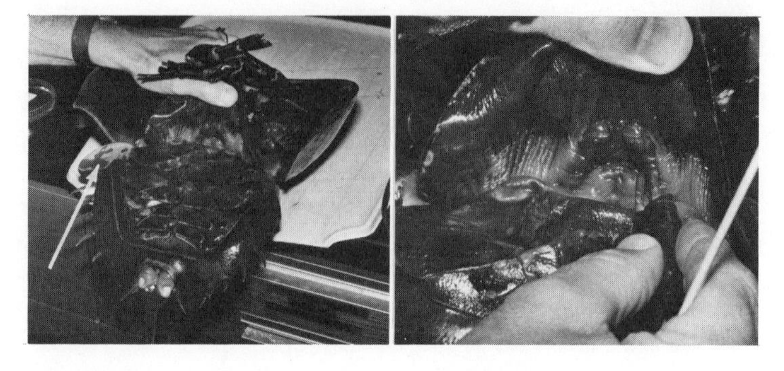

FIGURE 18–13 A flat stone is often used to prevent abdominal flexure of female.

FIGURE 18–14 Stimulus is applied below each gonopore.

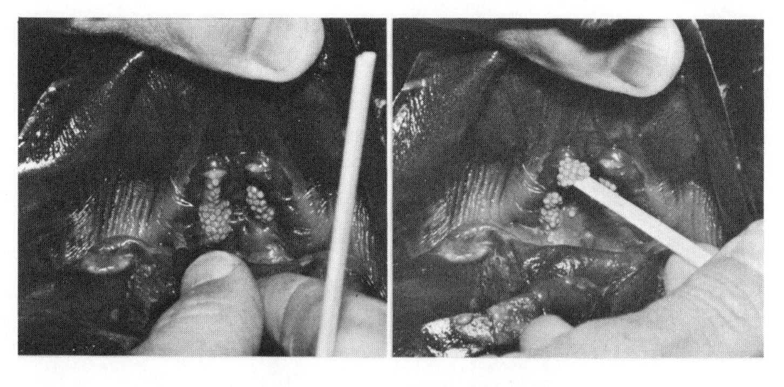

FIGURE 18–15 Short stimuli will produce the desired number of eggs.

FIGURE 18–16 Eggs are collected with wooden applicators or plastic spoons.

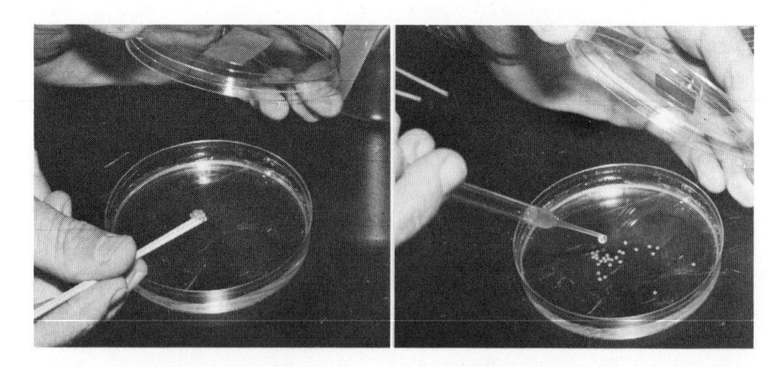

FIGURE 18–17 10–50 eggs are collected and placed in each Petri dish.

FIGURE 18–18 One drop of 1–10 percent sperm suspension is added and dish is swirled.

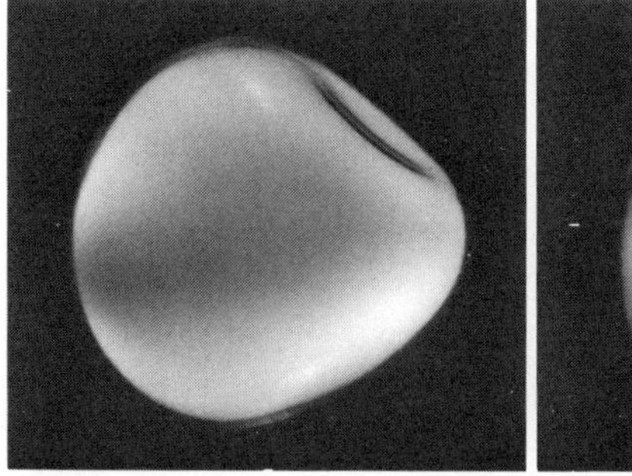

FIGURE 18–19 Stage 1. Unfertilized egg. Indentations are normal.

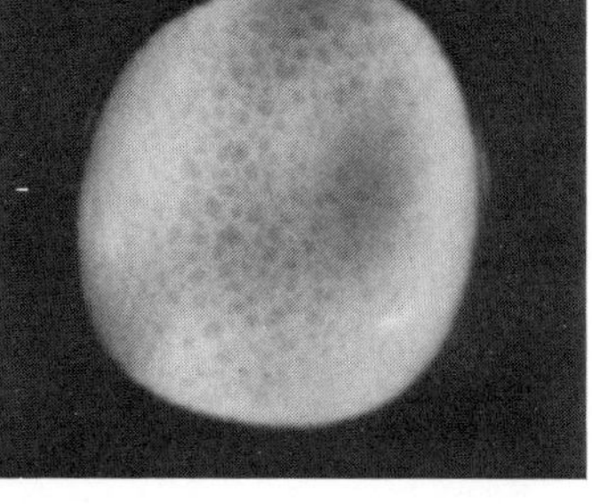

FIGURE 18–20 Stage 1. Pits appear during the cortical reaction.

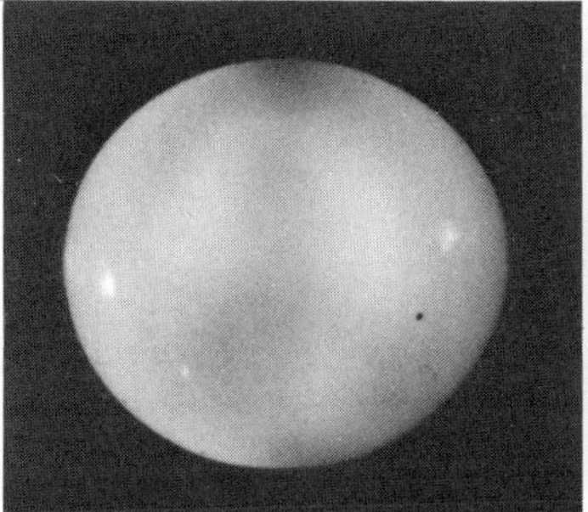

FIGURE 18–21 Stage 1. End of cortical reaction. Surface is smooth.

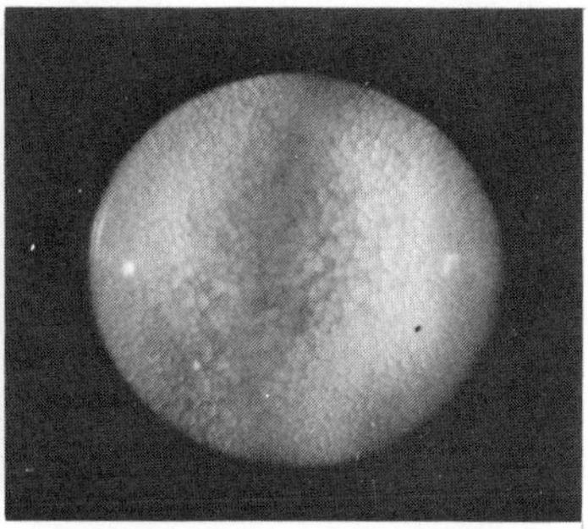

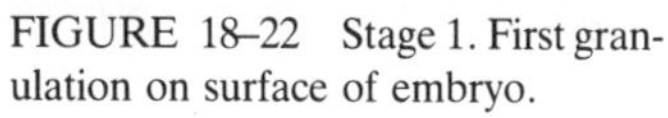

FIGURE 18–22 Stage 1. First granulation on surface of embryo.

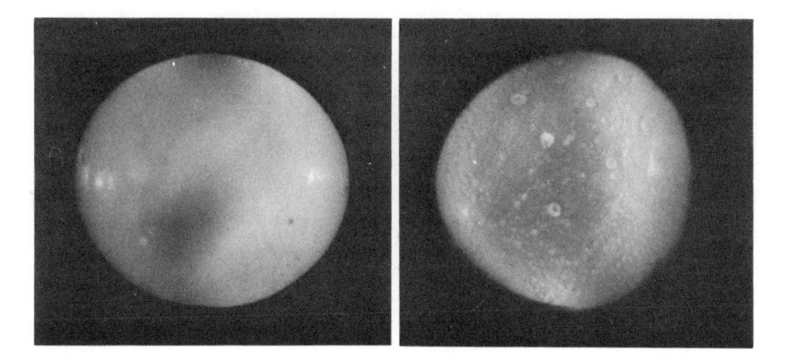

FIGURE 18–23 Stage 1. Smooth surface embryo between first two granulations.

FIGURE 18–24 Stage 2. Large nuclei appear on surface.

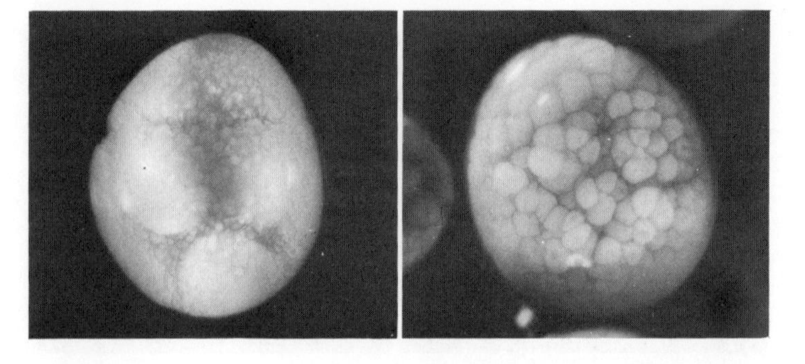

FIGURE 18–25 Stage 3. Yolk block formation.

FIGURE 18–26 Stage 4. Nuclei obvious. Blastomeres are variable in size and shape.

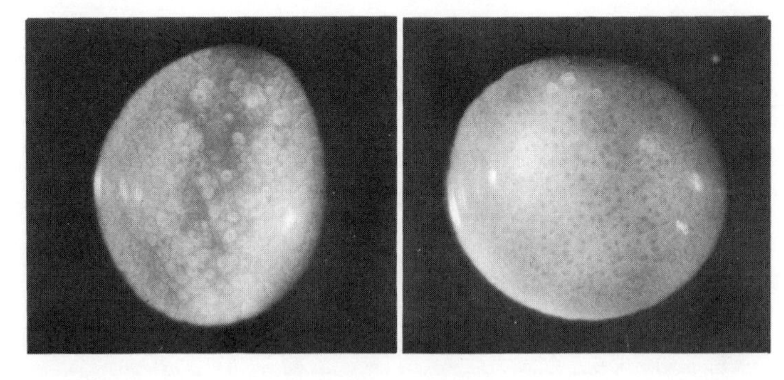

FIGURE 18–27 Stage 5. Size and shape of blastomeres uniform. Surface irregular.

FIGURE 18–28 Stage 6. Blastomeres are regularly arranged. Surface is smooth.

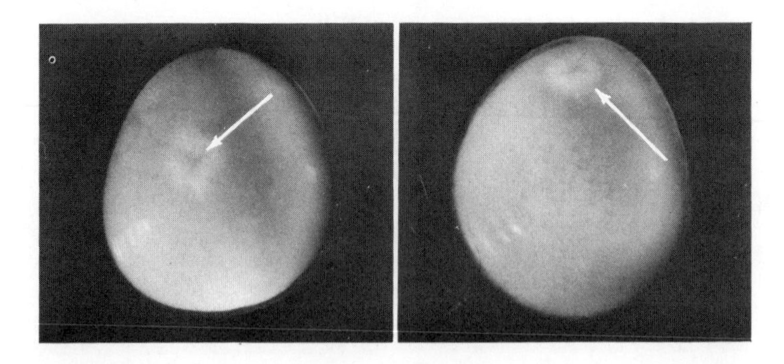

FIGURE 18–29 Stage 7. Irregular indentation on surface can be observed (arrow).

FIGURE 18–30 Stage 8. Area around indentation is now uplifting (arrow).

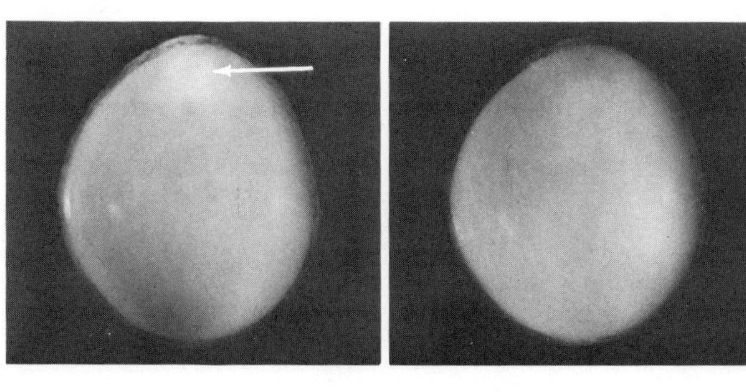

FIGURE 18–31 Stage 9. Germ disc (arrow) is about ¼ diameter of embryo.

FIGURE 18–32 Stage 10. Germ disc (not obvious) is centrally located.

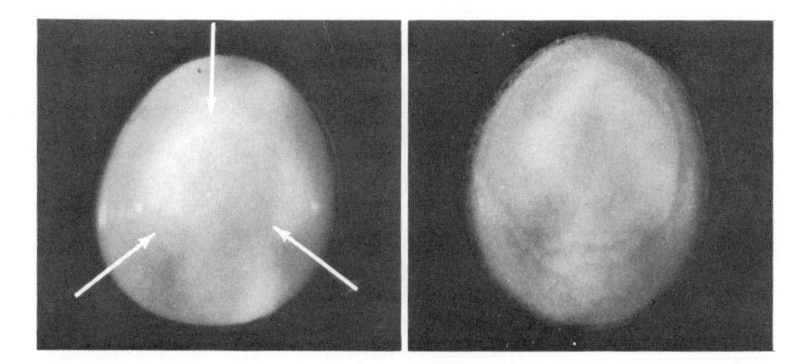

FIGURE 18–33 Stage 11. Large germ disc has formed with definite margin (arrows).

FIGURE 18–34 Stage 12. Germ disc outlined by grooves. Wrinkles present on body.

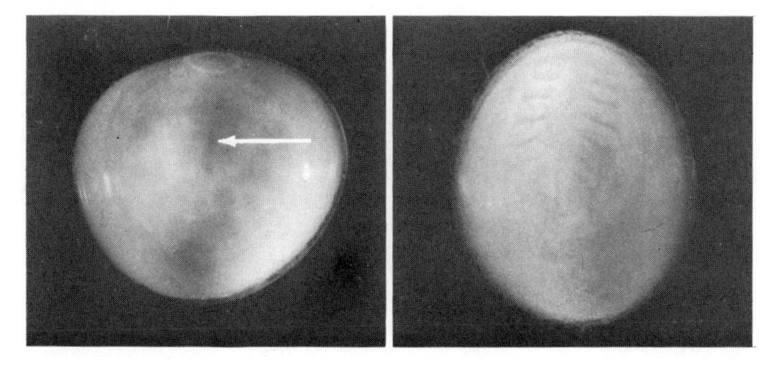

FIGURE 18–35 Stage 13. Groove (arrow) transecting germ disc becomes obvious.

FIGURE 18–36 Stage 14. Paired limbbud blocks are observed.

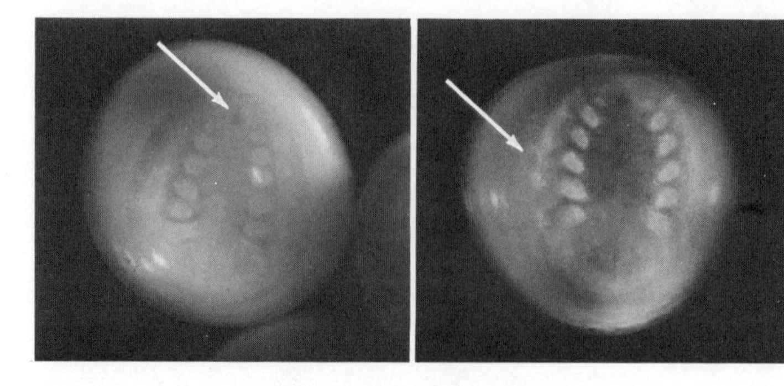

FIGURE 18–37 Stage 15. Limb-buds become round. Stomodaeum appears (arrow).

FIGURE 18–38 Stage 16. Limbs elongated. Lateral organs appear (arrow).

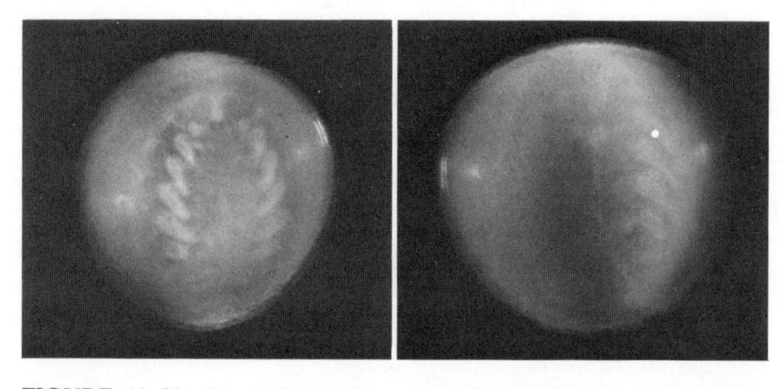

FIGURE 18–39 Stage 17. Limbs articulated. Segmented opisthosoma is obvious.

FIGURE 18–40 Stage 18. After first embryonic moult. Milky medium around embryo.

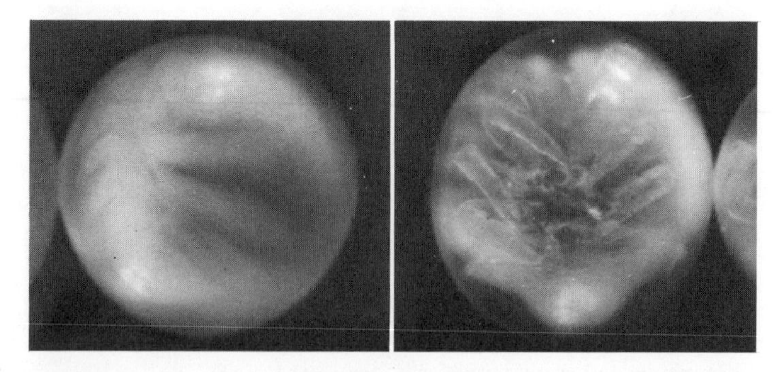

FIGURE 18–41 Stage 19. After second embryonic moult. Hepatopancreas segmented.

FIGURE 18–42 Stage 20. After third embryonic moult. Limb length has doubled.

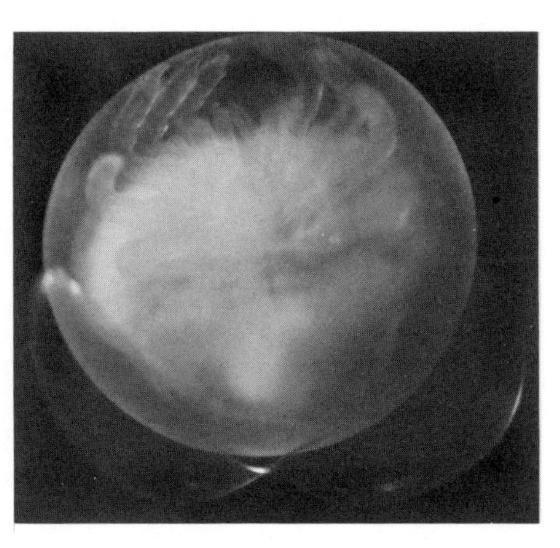

FIGURE 18–43 Stage 20. Egg envelope breaking away while inner extraembryonic shell is expanding.

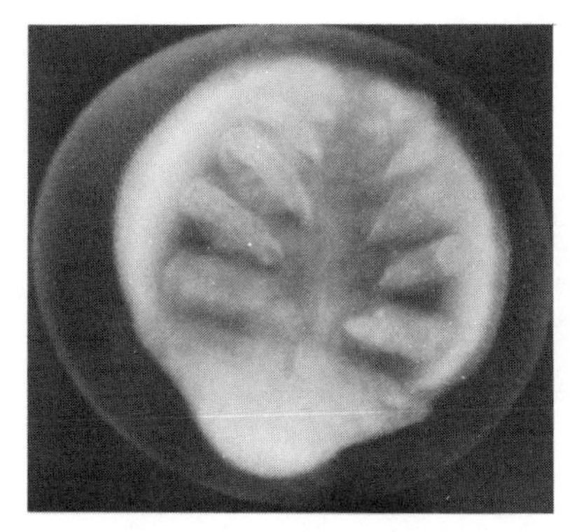

FIGURE 18–44 Stage 20. Larva is active and moves freely within extraembryonic shell.

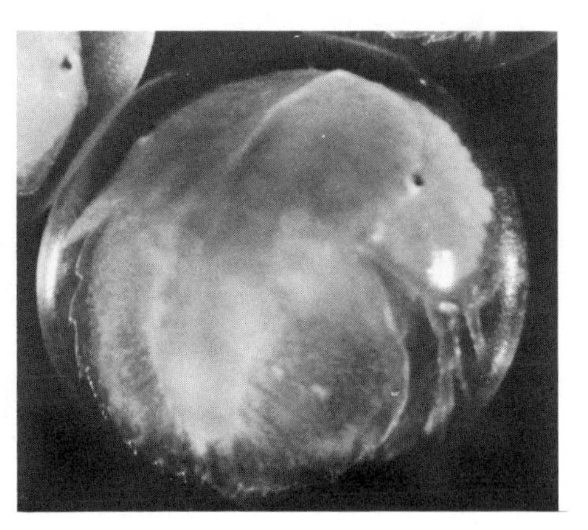

FIGURE 18–45 Stage 21. Fourth and final embryonic moult. Larva is greatly flattened and fills extraembryonic shell. Will hatch in 1–2 days.

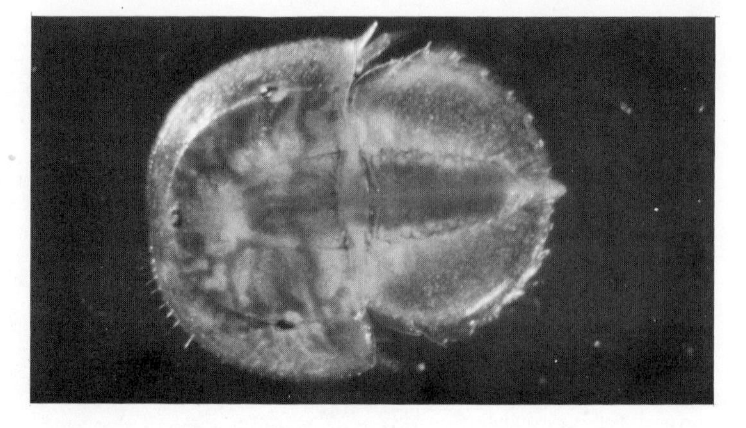

FIGURE 18–46 Stage 21. A hatched larva. Swimming upside down is characteristic.

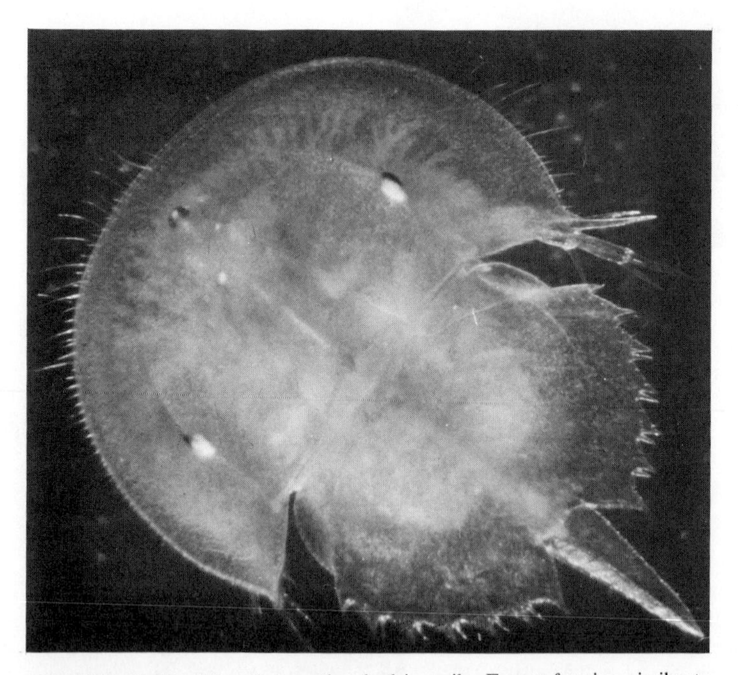

FIGURE 18–47 After first posthatched juvenile. Except for size, similar to adult. Telson is obvious. Must now be fed for development to continue.

19

Laboratory Culture of the Sea Urchin *Lytechinus pictus*

RALPH T. HINEGARDNER *and*
MARY MARGARET ROCHA TUZZI

INTRODUCTION

Sea urchins have been used as experimental laboratory animals for almost 100 years. Their development from egg to pluteus is probably as well studied as any embryo has ever been studied, and there are now more than 5,000 papers on one aspect or another of its early development. By comparison, the stages from pluteus to adult urchin are covered by no more than a handful of papers. The reason for this great difference is simple—the animals have been difficult to raise in the laboratory or even at marine stations. In 1969 a method was devised that makes the laboratory culture of sea urchins a relatively easy task (Hinegardner, 1969). Once the basic hardware is in place and the techniques become familiar, it is now not much more difficult to raise sea urchins than it is to raise an equivalent number of mice.

The procedures that will be described here were developed for the sea urchin *Lytechinus pictus*. This particular species has a number of advantages as a laboratory animal. The adults remain healthy in closed seawater systems, and they are not too large. The eggs are transparent, and the larvae are fairly big. Also, the young urchins are easy to transfer from one container to another without tearing off a significant number of tube feet.

With minor modifications, a number of other species of sea urchins can also be raised using essentially the same procedures as those described here. These include: *Strongylocentrotus purpuratus* and *S. fran-*

ciscanus, Arbacia punculata, Lytechinus variegatus, and probably many others.

We have tried numerous modifications of the procedures that will be described here, and though we may not have found the very best procedure, the methods that will be described work well, and we suggest that anyone starting out to raise sea urchins, particularly *Lytechinus pictus,* follow the descriptions closely.

SEA URCHIN LIFE CYCLE

The life cycle of the sea urchin can be divided into six more or less distinct phases: (1) the fertilized egg; (2) development through blastula and gastrula to pluteus, at which time egg nutrients are usually consumed; (3) growth and development of the feeding pluteus to a fully grown larva; (4) simultaneously with larval growth, the embryonic sea urchin (the rudiment) is developing within the larva; this is almost a separate individual that will ultimately become the ventral half of the newly metamorphosed urchin; (5) metamorphosis from larva to sea urchin; and (6) growth of the young urchin to reproductive maturity. Each of these phases has associated laboratory procedures.

There are a number of references that can be consulted for extensive descriptions of the morphological and biochemical events in sea urchin development. Among these are: Bury, 1895; McBride, 1903; Harvey, 1956; Boolootian, 1966; Hinegardner, 1967; Giudice, 1973; Horstadius, 1973; Czihak, 1975. The following description of the techniques for raising sea urchins in the laboratory assumes that the reader is generally familiar with the basics of larval and adult anatomy.

SPAWNING AND FERTILIZATION

Spawning can be induced by a number of procedures. The two most common ones are dissecting out the gonads or injection of a 0.55 molar KCl solution (Tyler, 1949). This is approximately isotonic to seawater. A few species do not respond to KCl injections and a solution of acetylcholine can be used instead (Hinegardner, 1969). In our experience, KCl injection is the best all-around method. It works on most species, and, if used in moderation, it is nonlethal. We have animals that have been injected every month for almost a year and with no discernable ill effects. The procedure is to inject 0.1 to 1.0 ml of KCl solution into the coelom in the vicinity of the gonads. The amount depends on the size of the animal. The best place for injection is through the peristomal membrane, which surrounds the mouth. Use a thin hypodermic needle (20–22 guage). If injection is done carefully, and if few eggs or sperm are needed, spawning can often be restricted to one or two gonads. For maximum yield, the urchin should be injected at several places. Shaking the animal firmly after injection will often increase spawning yield.

Some urchins tolerate being removed from seawater for extended periods. *L. pictus* cannot, and individuals should be kept out of water for no more than a minute. When the animals are out of water they are covered only by a thin film of seawater, which can become strongly hypertonic from evaporation in a very short time. After KCl injection, the animal is put immediately into a container filled with seawater from the aquarium the animal lives in. The urchins soon crawl up the side to the surface of the water, the gametes stream out and collect around the bottom of the container. Sperm should be collected from the bottom or from the surface of the urchin as it is being spawned and transferred to a test tube that is kept on ice. After the females have completed spawning, the eggs should be filtered through 110-μm nylon mesh and washed by allowing them to settle out of Millipore-filtered seawater (not seawater from the aquarium) three times. This removes substances that are secreted by the adult urchin and which inhibit fertilization.

Sperm that are concentrated by centrifugation will keep for several days on ice. The eggs are usually not viable after about 8 h. This varies from one species to the next. *Lytechinus pictus* eggs should not be kept below 10°C.

Fertilization occurs simply by adding a small amount of sperm to an egg suspension. It is important not to add too much sperm. Large amounts lead to polyspermy and abnormal development. The amount is correct when 10 to 100 sperm can be seen around each egg. Within a minute or so after fertilization, the fertilization membrane arises. The male and female pronuclei fuse in 15 to 20 min, and at 18°C first cleavage is 90 min later. Unless the developing embryos are stirred, they should be at a concentration that leaves more than an egg diameter between embryos on the bottom of the container.

FEEDING AND CARE OF THE LARVAE

At 18° the larvae grow to plutei in about 3 days. They should then begin to be fed. Up to the pluteus, the embryos practically raise themselves if they are given even grudging consideration. After pluteus, a certain amount of care is needed. If the larvae are given an optimum amount of food and care throughout their growth, they will be ready to metamorphose in about 3 wk.

Lytechinus pictus larvae have several requirements. They need the proper food organism, they should not be crowded unless their water is being continuously changed, and they need constant gentle stirring. The last is necessary to keep them from lysing on the surface of the water or getting trapped in the debris on the bottom. *Arbacia* larvae are not as sensitive and do not have to be stirred.

CONTAINERS

The larvae can be grown in almost any sized container, providing they can be stirred. We have found that three different sizes fill most of our needs. These are: polystyrene refrigerator dishes that are 9.5 cm in diameter and 7 cm deep (3¾″ × 2¾″) and hold about 300-ml, 13-cm × 13-cm (5″ × 5″) polystyrene containers which hold 1.5 liters, and large cylindrical jars with a volume of around 8 liters. Care should be taken not to have too many larvae in a given container. During early development, one larva per milliliter is fine. By the time they are mature this should be reduced to one per 10 ml.

Providing the larvae are given reasonable care, about 80 percent of the plutei can be raised to mature larvae. The nonsurvivors seem to die from various developmental defects that are traceable to biological problems, and not to the method.

AGITATION

The larvae of most species of sea urchins require some form of stirring. We have found that a small paddle type stirrer turned at 20 RPM works well. Several different stirrer designs can be used. For large containers, an 8-cm × 5-cm plastic paddle attached to a glass rod and turned by a 20 RPM electric clock type motor makes an inexpensive stirrer. The paddle can be cut from a large polyethylene bottle, two longitudinal slots cut in the paddle and the rod threaded through them. For the 1.5-liter and smaller containers we used a magnetic stirrer capable of stirring a large number of containers simultaneously. One magnetic stirrer will stir 40 1.5-liter containers, all with one motor. The stirring magnets are suspended on a light-weight brass chain that hangs from a motor mounted on top of the apparatus. We have found that bubbling air is a poor method of agitation. Larval spines tend to be damaged and development is abnormal.

SEAWATER

The larvae are cultured in 0.45-μm Millipore-filtered seawater. It is important that the water is filtered, not so much to remove bacteria, but to eliminate algae and protozoa. Some algae, such as the blue greens, can be toxic. The protozoa, particularly ciliates, can grow to become so numerous they starve out the larvae by quickly eating the food. The life of a Millipore filter can be greatly extended if the filter is first covered with a layer of Celite or other form of diatomacious filter aid. We use 3.5 g of filter aid on 125-mm filters and 1 g on 45-mm filters.

FOOD

We have tried a large number of different single-celled algae as well as various other particulate and soluble compounds as possible food (Hinegardner, 1969). Of those we tried, two algae have been particularly useful. So far, soluble mixtures have been toxic, and various particulate nutrients have not worked. The two algae are a rhodamonas designated 3C by Guillard and *Dunalellia tertiolecta. Lytechinus* will not grow on *Dunalellia,* but thrives on a diet of 3C. Some other urchin species, *Arbacia* for example, will grow well on *Dunalellia.* Our algae cultures are not axenic and there are bacteria associated with them that may or may not have an effect on larval growth. As far as we can tell, they have no effect.

The food algae are cultured in a modification of the water enrichment medium of Guillard and Ryther (1962), and has the following composition:

Per Liter of Filtered Seawater

1.0 ml	3 percent sodium nitrate ($NaNO_3$)
0.5 ml	1 percent sodium phosphate ($NaH_2PO_4 \cdot 8H_2O$)
0.5 ml	trace metal solution
1.0 ml	vitamin mix

The trace metal solution is made up in the following way: Dissolve 6.25 g of $FeCl_3 \cdot 6H_2O$ and 8.25 g Na_2EDTA (disodium ethylenedinitrilotetraacetate) in 1 liter of distilled water. We have found that it is not necessary to adjust the pH. Then add:

0.044 g	zinc sulfate ($ZnSO_4 \cdot 7H_2O$)
0.020 g	cobalt chloride ($CoCl_2 \cdot 6H_2O$)
0.360 g	manganese chloride ($MnCl_2 \cdot 4H_2O$)
0.013 g	sodium molybdate ($Na_2MoO_4 \cdot 2H_2O$)

Mold can grow in this solution and it should be kept in the refrigerator.

The vitamin mix is made up from stock solution plus thiamine hydrochloride. The stock solutions are:

Biotin	10 mg in 10 ml distilled water
Vitamin B_{12}	10 mg in 10 ml distilled water

Both of these are kept frozen. The final mix consists of:

Per 100 ml Distilled Water	
1.0 ml	biotin stock
0.1 ml	B_{12} stock
20.0 mg	thiamine HCl

This is distributed among a number of small containers and frozen. The volume per container should be the amount that will be used at one time.

The algae are cultured in 500-ml flasks to which 300 ml of enriched seawater have been added. These are plugged with autoclavable sponge or cotton plugs that have a hole in the center for a bubbling tube. The flasks are autoclaved and stored on the shelf until use. It is best to use polycarbonate flasks. The culture medium extracts silica from glass flasks during autoclaving. This does not harm the algae, but dissolved silica does permit diatoms to grow as a contaminant of the cultures.

Cultures can be grown up at 18°C in front of fluorescent lights. The 3C algae tend to settle to the bottom of the flasks unless bubbling is fairly vigorous. When the algae have reached close to maximum growth, the medium appears to be a brick red to brown color. New cultures should be started at this time. 3C cultures are temperamental and can not be left sitting around for extended periods. The cultures usually die out if they are not subcultured regularly. *Dunalellia* is much hardier.

Algal suspensions should not be transferred directly to the larval cultures, since 3C medium is toxic. Instead, the algae are centrifuged out of the medium at approximately 200 g for 5 min, the medium decanted, and the algae resuspended in Millipore-filtered seawater at a concentration of several times what it was in the culture. This suspension will keep for a day or more at 18°.

The concentration of algae in the larval cultures is important. Too much food will die and begin to rot before the larvae are able to eat it; too little food slows development. Keeping track of the number of algae per culture with a haemocytometer works well but is a tedious and time-consuming approach. An easier method is to use a simple form of light-scattering. We shine the light beam from a penlight type flashlight through the culture. If the light beam is viewed almost head on, the algae can be seen as tiny reddish specks. Once you get used to what the proper algal concentration looks like, feeding can be very fast and the amount of food can be well controlled.

The amount of algae that is used depends on the stage of larval growth, with the earlier stages being fed much less than later stages. The larvae are usually fed once a day and given a little more food than they will consume in 24 h, or, in the case of young cultures, about 3,000 algae per milliliter of seawater.

About every week, or sooner if the larval culture becomes dirty, cloudy, or contaminated in some way, the seawater is siphoned off through 110-μm nylon mesh. The mesh can be cemented over the cut-off end of a 50-ml plastic test tube. The opposite end is plugged with a one-hole stopper with attached tubing. This is used to siphon off the water through the mesh. The siphoning should be fairly slow so that the larvae are not damaged against the mesh. After the water is drawn down to a minimal volume, the larvae are transferred to a clean container of Millipore-filtered seawater. With proper care the larvae will grow to full size in 3 to 4 wk at 18°C.

LARVAL GROWTH

The external features of the developing larvae gradually become more complex as the larvae grow. New spicules and their associated arms arise, and in some species such as *Lytechinus,* dense ciliary bands, called epaulets, form. Three pedicellariae appear, two on the right side and one at the posterior end of the larvae. The pedicellariae will later be incorporated into the anatomy of the urchin after metamorphosis.

During larval growth the embryonic urchin is also developing. In about the seventh day after fertilization, it begins to form out of the union of a small portion of the ectoderm on the left surface of the larva and the middle left hydrocoel (McBride, 1903). The latter arose from the coelomic pouches formed after gastrulation. The developing urchin, while it is in the larva, is called the *rudiment.* The larva serves as a source of nutriment and protection for the growing rudiment. The rudiment is not a little urchin, but is only a portion of the developing ventral half of the urchin and consists primarily of ventral skeleton, the first set of urchin spines, and five tube feet. The rest of the ventral half, as well as almost all the dorsal and internal structures, develop during the metamorphic process. Though most of the larval biomass ends up in the urchin, it does so after passing through a period during early metamorphosis when many of the cells are broken down and the larval material is little more than a lump of protoplasm on the newly metamorphosed urchin. The urchin develops its own mouth, anus, and most of its internal organs.

Larvae stop growing when they are mature, which can be anywhere from 3 wk to more than a month, depending on how well they were fed. Once they reach full size, they are competent to metamorphose. If metamorphosis does not occur, the larva will continue to feed, though at a much slower rate, and can be kept for several months as a progressively degenerating organism. However, they are able to metamorphose normally for only a week or so after they are completely formed.

METAMORPHOSIS

Metamorphosis is not an obligatory phase of the sea urchin life cycle. The larvae must be exposed to a substratum containing bacteria, or to some compound(s) that the bacteria produce (Cameron and Hinegardner, 1974). The easiest way to prepare the necessary bacterial surface is to place Petri dishes in an established aquarium for a few days. We use 9-cm × 2.5-cm (3.5″ × 2″) dishes. When fully developed larvae are placed in these dishes, which could contain water from the aquarium, they begin metamorphosis almost immediately. In about an hour, a swimming larvae is transformed into a little sea urchin-like organism, with five tube feet, five sets of spines, and three pedicellaria in a dorsal lump of larval tissue. The process of metamorphosis is described in Hinegardner (1969), Okazaki (1975), and Cameron and Hinegardner (1974, 1978).

It takes about 5 days for the skeleton and internal organs to form, after that the young urchin will begin to feed. This marks the end of the metamorphic period.

CARE AND FEEDING OF THE YOUNG URCHIN

When the urchins are 5 days old (i.e., 5 days after metamorphosis begins), they are transferred to dishes that have a layer of surface dwelling diatoms growing on the bottom. The urchins are fed this diatom until they are large enough to eat the food being fed the adults. We use three different diameter dishes, depending on the size of the urchins or how many we wish to place in one container. The dishes are made by Tristate Plastic Company, Henderson, Kentucky. These and similar designs can be purchased from a number of plastic container supply companies. The dishes are 15.5 cm in diameter by 6.5 cm deep (6⅛″ × 2⅝″), 18.6 cm × 7.9 cm (7⅜″ × 3⅛″), and 21 cm × 7.9 cm (8¼″ × 3⅛″).

The diatom we use has been identified as a member of the genus *Nitzschia*. It was orginially isolated from one of our aquaria. The diatom is grown up on the dishes in a second modification of the seawater enrichment medium of Guillard and Ryther (1963) and page 295. The composition is:

Per Liter of Filtered Seawater	
3.0 ml	3 percent sodium nitrate
1.0 ml	1 percent sodium phosphate
1.5 ml	3 percent sodium metasilicate ($Na_2SiO_3 \cdot 9H_2O$)
1.0 ml	trace metals
2.0 ml	vitamin mix

The trace metal solution and vitamin mix are those described previously. Each dish is filled to a depth of about 8 mm with this medium and inoculated with the diatom culture. A dish in which diatoms have previously been grown up in is a convenient source of inoculum for a number of dishes. The diatom concentration in the inoculation is enough to allow the dishes to acquire a good layer of diatoms in 3 to 4 days. A probe type sonicator is convenient for knocking the diatoms off of dishes. When the system is working smoothly, it is easier to premix the diatoms and the medium in a beaker and then distribute this to the new dishes. The dishes are incubated in an illuminated environment at 15–18°C for 3 to 4 days. After 6 days unused dishes should be discarded, particularly if the diatoms tend to come off with mild agitation. Though it is convenient to use the culture dishes both for growing urchins and for growing the large numbers of diatoms necessary for inoculating the dishes, a bacterial-free diatom culture should be maintained in sterile autoclaved medium in 250-ml pyrex flasks. This culture is used both to maintain the diatoms under sterile conditions and, should it be necessary, for starting the dishes that will in turn be used to start the dishes the urchins will grow in.

Blue-green algae are the major potential contaminants. Some are toxic to the urchins, and contaminated dishes should be discarded. The diatom culture medium is toxic to the urchins and must be poured off and replaced. Unlike the larvae, fresh seawater is mildly toxic to young urchins, at least that has consistently been our experience. What we do to get around this is to use Millipore-filtered water from the aquaria the adult urchins are growing in. To each liter of this water we add 0.5 ml of the 3 percent sodium metasilicate solution. This keeps the diatoms healthier while the urchins are feeding. This silicate tank water is poured into the urchin culture dishes to a depth of about 4 cm and the urchins transferred. A convenient transfer tool for the newly metamorphosed urchins is a plastic soda straw attached to a dropper bulb. The urchin can be pushed off the bottom with this and drawn up at the same time. They should immediately be transferred to the diatom dishes before they have a chance to adhere to the inside of the straw, where they are sometimes difficult to dislodge. Larger animals can be transferred with the aid of a modified bacterial transfer loop. The loop itself is cut off and the end of the wire bent back into a 1-cm-long V, which is in turn bent so that the V is parallel to the surface of the dish when the handle is being held. The urchins are easily picked up by catching them in the V. *Arbacia* are particularly difficult to remove from a dish without tearing off most of their tube feet. The best way we found to prevent this was to first give the animals a mild electrical shock (about 12 V AC) and then quickly pick them up.

The animals should not be crowded in the dishes. Newly metamorphosed urchins are about a millimeter in diameter, and about 25 can be kept in a single dish. When they are around 8 mm in diameter, no more than four to five should be kept in a 21-cm-diameter dish. A general rule to follow is to maintain them at a population density that just consumes most of the diatoms in a week.

We keep the dishes of young urchins in incubators that are illuminated with fluorescent lights. We have found that about half the lights can be removed from the average commerically available incubator, leaving two or three. This seems to improve general growth conditions. The animals can also be kept out in a room providing it never gets above 23°C or below 10°C. There is however an additional risk here. At least in the building we are in, there are enough stray toxic vapors to eventually kill most of the young urchins if they are kept out in our lab. To get around this problem, we put all dishes in incubators and pump in a slow stream of air from outdoors.

If the urchins are changed to fresh dishes once a week, they will grow from 1 mm in diameter at metamorphosis to 1 cm in 4.5 to 6 mo. Care should be taken not to introduce contaminants into the cultures. If the dishes do become contaminated, the animals should immediately be changed to fresh dishes. Inbred urchins, parthenogenetic individuals, or those carrying deleterious mutations require more intensive care and they usually grow more slowly.

CARE OF ADULT URCHINS

The urchins we raise become sexually mature when they are about 1 cm in diameter, with the males maturing at a slightly smaller size. They should be transferred out of their culture dishes at about 8 mm or whenever they are able to eat the food the adults are being fed. Our adult animals are fed on the blades of the West Coast kelp, *Macrocystis,* which we collect shortly after it has washed up on the local beaches. Each animal is given about 25 cm^2 of algae at a feeding. Urchins will also eat and grow on dried *Macrocystis.* This tends to rot sooner than living pieces, and feeding should be controlled more carefully. *Egregia* and various species of *Ulva* and *Laminaria* can also be used. Most urchins are scavengers, and they will eat a variety of other foods such as hard boiled egg white, lettuce, other leafy vegetables and a number of food combinations that can be made up in agar. One combination we have used for some time consists of:

20 g	agar
5 g	corn meal
2 g	soya flour
1 g	dried yeast
2 g	dried whole egg
1,000 ml	distilled water

The mixture is brought to a boil and poured in thin layers over a smooth surface and allowed to dry. The paper thin sheets can then be cut into appropriate sized pieces. This food will maintain animals for long periods but they do not grow.

Our animals are kept in 20-gal aquaria. Each has a motor-driven filter mounted on the side and a large air stone. The filters we use are made by Eugene G. Danner Mfg., Inc., 1660 Summerfield St., Brooklyn, NY 11227, and are sold in aquarium stores under the name Supreme Aqua-king, Model I. The filter box is half filled with dolomite chips (Aquarium Systems, Inc., Mentor, OH 44094), which are washed when the filter becomes clogged. In our experience, activated charcoal and fiber filters are unnecessary. The filter serves at least three roles. It retains large debris, and, once it is established, it serves as a good biological filter, breaking down the organics that accumulate in the seawater. The filter also keeps the water constantly moving in the aquaria. The air stone does the latter also and is included just in case the filter should stop.

All aquaria are cooled to 16°C, either by being in a low temperature room or by external refrigeration. The aquaria should not drop below

10°C. Cooling is supplied by circulating cooled 50 percent ethanol through a refrigerating unit and then through stainless steel tubing in the tanks. The ethanol solution is transported through tygon or vinyl tubing that attaches to the stainless steel tubing just below the water line. Using this technique, the stainless steel does not rust and seems to have an indefinite life.

We make no attempt to keep the seawater in the aquaria water white, but we do keep it clear. Over time the water becomes brownish, probably from algal pigments. This is harmless. Debris accumulates on the bottom and is cleaned out with a large siphon about once every week or two. The water that is removed is replaced by fresh seawater (unfiltered). If a particular aquarium becomes cloudy from suspended bacteria, we attach a UV aquarium sterilizer to the filter. Within a day or two the water will clear up. The specific gravity should be checked about once a month with an accurate hydrometer and adjusted if necessary. Evaporated water is replaced with deionized glass distilled water. Ordinary building distilled water may be adequate, but it can be toxic. This should be checked before the water is used extensively.

The urchins should not be taken from their aquaria and put in fresh seawater. For reasons we don't understand, this often induces spawning and sometimes kills them. Half new and half old seawater is not harmful.

The particular conditions of your aquarium will determine how many urchins can be maintained in good health. In our experience, a surprisingly large number of animals can exist and grow in a closed system. It is best to start out with 10 to 15 urchins the size of *Lytechinus pictus* in about 80 liters of seawater and gradually increase the number until you feel you have reached a convenient carrying capacity.

If the animals are being used in genetic studies, it is necessary to be able to identify specific individuals. We have tried a number of different marking procedures; none were very successful. The best method is to keep each animal in a separate container. We use plastic boxes that have multiple compartments. One size that is particularly useful is 28-cm long, 16-cm deep, and 4.5-cm high and contains 18 compartments. Boxes similar to this are made by a number of different plastic companies and sold in hardware stores. Before they can be used, the boxes have to have holes drilled in them for circulation. We put five 6-mm-diameter holes in all sides of all compartments.

Basically, maintaining a colony of sea urchins away from the sea requires feeding, removal of debris from aquaria, filtration, temperature control, and assurance that the water is always agitated. This should not be taken to mean no care is needed. The animals should be given the attention a comparable sized mouse colony requires.

References

Abbott, M. B. 1975. Bryozoa. Pp. 155–176 in W. L. Smith, and M. H. Chanley, eds. Culture of Marine Invertebrate Animals. Plenum Press, New York and London.

Abbott, R. T. 1974. American Seashells, the Marine Mollusca of the Atlantic and Pacific Coast of North America. Van Nostrand Reinhold Co., New York. 663 pp.

Abe, Y., and M. Hisada. 1969. On a new rearing method of common jellyfish, *Aurelia aurita*. Bull. Mar. Biol. Stn. Asamushi 13:205–209.

Adelung, D., and A. Ponat. 1977. Studies to establish an optimal diet for the decapod crab *Carcinus maenas* under culture conditions. Mar. Biol. (Berlin) 44:287–292.

Åfzelius, B. A. 1972. Sperm morphology and fertilization biology. Pp. 131–143 in R. A. Beatty and S. Gluecksohn-Waelsch, eds. The Genetics of the Spermatozoa. Edinburgh, August 16–20, 1971.

Ahmad, M. F. 1969. Anaesthetic effects of tricaine methane sulphonate (MS 222 Sandoz) on *Gammarus pulex* (L.) (Amphipoda). Crustaceana (Leiden) 17:197–201.

Åkesson, B. 1970. *Ophryotrocha labronica* as test animal for the study of marine pollution. Helgol. Wiss. Meeresunters. 20:293–303.

Åkesson, B. 1972. Incipient reproductive isolation between geographic populations of *Ophryotrocha labronica* (Polychaeta, Dorvilleidae). Zoöl. Scr. 1:207–210.

Åkesson, B. 1973. Reproduction and larval morphology of five *Ophryotrocha* species (Polychaeta, Dorvilleidae). Zool. Scr. 2:145–155.

Albrechtsen, K. 1979. Experiments in aquaria with the common prawn, *Leander adspersus*. Medd. Dan. Fisk-Havunders. 7:511–527.

Allen, E. J. 1914. On the culture of the plankton diatom *Thalassiosira gravida* Cleve, in artificial sea-water. J. Mar. Biol. Assoc. U.K. 10:417–439.

Allen, E. J., and E. W. Nelson. 1910. On the artificial culture of marine plankton organisms. J. Mar. Biol. Assoc. U.K. 8:421–474.

Allen, H. 1971. Effects of petroleum fractions on the early development of a sea urchin. Mar. Pollut. Bull. 2:138–140.

Allen, R. K. 1976. Common Intertidal Invertebrates of Southern California. Peek Publ., Palo Alto, Calif.

American Society of Zoologists. 1974. The developmental biology of the Cnidaria. A symposium. Am. Zool. 14(2):440–866.

American Society of Zoologists. 1975. Developmental biology of the echinoderms. A symposium. Am. Zool. 15(3):485–775.

American Society of Zoologists. 1976. Spiralian development. A symposium. Am. Zool. 16(3):277–626.

Amiard, J. C. 1976. Les variations de la phototaxie des larves de crustacés sous l'action de divers polluants métalliques: mise au point d'un test de toxicité subléthale. Mar. Biol. (Berlin) 34:239–245.

Andersen, N. M., and J. T. Polhemus. 1976. Water-striders (Hemiptera: Gerridae, Veliidae, etc.). Pp. 187–224 in L. Cheng, ed. Marine Insects. North-Holland Publ. Co., Amsterdam and Oxford. American Elsevier Publ. Co., New York.

Anderson, D. T. 1973. Embryology and Phylogeny in Annelids and Arthropods. Pergamon Press, Oxford, New York, Toronto, Sydney, and Braunschweig. xiv + 495 pp.

André, J. 1963. A propos d'une leçón sur la limule. Ann. Faculté Sci. Univ. Clermont 26:27–38.

Anger, K., and K. K. C. Nair. 1979. Laboratory experiments on the larval development of *Hyas araneus* (Decapoda, Majidae). Helgol. Wiss. Meeresunters. 32:36–54.

Apelt, G. 1969. Fortpflanzungsbiologie, Entwicklungszyklen und vergleichende Frühentwicklung acoeler Turbellarien. Mar. Biol. (Berlin) 4:267–325.

APHA (American Public Health Association). 1976. Standard Methods: For the Examination of Water and Wastewater (see Rand *et al.*, 1976).

Appelgate, A. L. 1968. A cytochemical study of the oocyte and egg of *Ilyanassa obsoleta,* a marine snail. Ph. D. Diss., Emory University.

Aquacop (G. Cuzon, A. Febvre, J. Mélard, G. Parker, G. Fagnoni, J. Calvas, J. M. Griessinger, P. Hatt, G. Poullaouec, J. F. Le Bitoux, J. L. Martin, and A. Michel). 1978. Study of nutritional requirements and growth of *Penaeus merguiensis* in tanks by means of purified and artificial diets. Pp. 225–234 in J. W. Avault, Jr., ed. Proc. Ninth Annu. Meet. World Mariculture Soc. La. State Univ. Div. Continuing Educ., Baton Rouge.

Armstrong, D. A., D. Chippendale, A. W. Knight, and J. E. Colt. 1978. Interaction of ionized and un-ionized ammonia on short-term survival and growth of prawn larvae, *Macrobrachium rosenbergii.* Biol. Bull. (Woods Hole) 154:15–31.

Arndt, W. 1933. Haltung und Aufzucht von Meeresschwämmen. Pp. 443–464 in Abderhalden's Handbuch der Biologischen Arbeitsmethoden, vol. 9, pt. 5. Urban & Schwarzenberg, Berlin and Vienna.

Arneson, A. C., and C. E. Cutress. 1976. Life history of *Carybdea alata* Reynaud, 1830 (Cubomedusae). Pp. 227–236 in G. O. Mackie, ed. Coelenterate Ecology and Behavior. Plenum Press, New York and London.

Arnold, J. M., C. T. Singley, and L. D. Williams-Arnold. 1972. Embryonic development and post-hatching survival of the sepiolid squid, *Euprymna scolopes* under laboratory conditions. Velíger 14:361–364.

Arnold, J. M., W. C. Summers, D. L. Gilbert, R. S. Manalis, N. W. Daw, and R. J. Lasek. 1974. A Guide to Laboratory Use of the Squid *Loligo pealei.* Marine Biological Laboratory, Woods Hole, Mass. 74 pp.

Atkins, D. 1955a. The cyphonautes larvae of the Plymouth area and metamorphosis of *Membranipora membranacea* (L). J. Mar. Biol. Assoc. U.K. 34:441–449.

Atkins, D. 1955b. The ciliary feeding mechanism of the cyphonautes larva (Polyzoa Ectoprocta). J. Mar. Biol. Assoc. U.K. 34:451–566.

Atkinson, J. W. 1968. Organogenesis in normal and lobeless embryos of the marine prosobranch gastropod *Ilyanassa obsoleta.* J. Morphol. 133:339–352.

Atkinson, R. J. A., H. Bailey, and E. Naylor. 1974. Some laboratory methods for recording and displaying temporal patterns of locomotor activity in marine animals. Mar. Behav. Physiol. 3:59–70.

Audesirk, T. E. 1975. Chemoreception in *Aplysia californica.* I. Behavioral localization of distance chemoreceptors used in food-finding. Behav. Biol. 15:45–55.

Audesirk, T. E. 1977. Chemoreception in *Aplysia californica.* III. Evidence for pheromones influencing reproductive behavior. Behav. Biol. 20:235–243.

Auwarter, A. G. 1977. A flow-through system for studying interaction of two toxicants on aquatic organisms. Pp. 90–98 in F. L. Mayer and J. L. Hamelink, eds. Aquatic Toxicology and Hazard Evaluation. American Society for Testing and Materials, Philadelphia. (STP 634)

Avent, R. M. 1975. Evidence for acclimation to hydrostatic pressure in *Uca pugilator* (Crustacea: Decapoda: Ocypodidae). Mar. Biol. (Berlin) 31:193–199.

Bahner, L. H., and D. R. Nimmo. 1975. A salinity controller for flow-through bioassays. Trans. Am. Fish. Soc. 104:388–389.

Bahner, L. H., C. D. Craft, and D. R. Nimmo. 1975. A saltwater flow-through bioassay method with controlled temperature and salinity. Prog. Fish. Cult. 37:126–129.

Baker, L. D., and M. R. Reeve. 1974. Laboratory culture of the lobate ctenophore *Mnemiopsis mccradyi* with notes on feeding and fecundity. Mar. Biol. (Berlin) 26:57–62.

Balazs, G. H., E. Ross, and C. C. Brooks. 1973. Preliminary studies on the preparation and feeding of crustacean diets. Aquaculture 2:369–377.

Balfour, F. M. 1880. A Treatise on Comparative Embryology, vol. I. Macmillan and Co., London. 492 + xxii pp.

Balzer, F. 1933. Zucht- und Versuchsmethoden (Geschlechtsbestimmung) bei Bonellia. Zuchtmethoden bei Thalassema und Echiurus. Pp. 431–442 in Abderhalden's Handbuch der Biologischen Arbeitsmethoden, vol. 9, pt. 5. Urban & Schwarzenberg, Berlin and Vienna.

Bang, F. B. 1979. Ontogeny and phylogeny of response to gram-negative endotoxins among the marine invertebrates. Pp. 109–123 in E. Cohen, ed. Biomedical Applications of the Horseshoe Crab (Limulidae). Alan R. Liss, Inc., New York.

Bannon, G. A., and G. G. Brown, 1980. Vesicle involvement in the egg cortical reaction of the horseshoe crab *Limulus polyphemus* L. Dev. Biol. 76 (in press).

Bardach, J. E., J. H. Ryther, and W. O. McLarney. 1972. Culture of marine gastropods, especially abalone. Pp. 777–785 in J. E. Bardach, J. H. Ryther, and W. O. McLarney. Aquaculture the Farming and Husbandry of Freshwater and Marine Organisms. John Wiley & Sons, New York, London, Sydney, and Toronto. xii + 868 pp.

Barker, M. F. 1979. Breeding and recruitment in a population of the New Zealand starfish *Stichaster australis* (Verrill). J. Exp. Mar. Biol. Ecol. 41:195–211.

Barnes, J. H. 1967. Extraction of cnidarian venom from living tentacle. Pp. 115–129 in F. E. Russell and P. R. Saunders, eds. Animal Toxins. Pergamon Press, New York.

Barnett, A. M. 1974. The feeding ecology of an omnivorous neritic copepod, *Labidocera trispinosa* Esterly. Diss. Abstr. Int. B Sci. Eng. 35:185.

Barr, M. W. 1969. Culturing the marine harpacticoid copepod, *Tisbe furcata* (Baird, 1837). Crustaceana (Leiden) 16:95–97.

Battaglia, B. 1970. Cultivation of marine copepods for genetic and evolutionary research. Helgol. Wiss. Meeresunters. 20:385–392.

Bayne, B. L. 1965. Growth and the delay of metamorphosis of the larvae of *Mytilus edulis* (L.). Ophelia 2:1–47.

Bayne, B. L. 1976. The biology of mussel larvae. Pp. 81–120 in B. L. Bayne, ed. Marine Mussels: Their Ecology and Physiology. Cambridge University Press, Cambridge, London, New York, and Melbourne. xvii + 506 pp.

Bayne, B. L., and R. J. Thompson. 1970. Some physiological consequences of keeping *Mytilus edulis* in the laboratory. Helgol. Wiss. Meeresunters. 20:526–552.

Bayne, B. L., D. R. Livingstone, M. N. Moore, and J. Widdows. 1976. A cytochemical and a biochemical index of stress in *Mytilus edulis* L. Mar. Pollut. Bull. 7:221–224.

Bayne, B. L., M. N. Moore, J. Widdows, D. R. Livingstone, and P. Salkeld. 1979. Measurement of the responses of individuals to environmental stress and pollution: Studies with bivalve molluscs. Philos. Trans. R. Soc. Lond. B Biol. Sci. 286:563–581.

Beard, T. W., J. F. Wickins, and D. R. Arnstein 1977. The breeding and growth of *Penaeus merguiensis* de Man in laboratory recirculation systems. Aquaculture 10:275–289.

Becker, C. D., J. A. Lichatowich, M. J. Schneider, and J. A. Strand. 1973. Regional Survey of Marine Biota for Bioassay Standardization of Oil and Oil Dispersant Chemicals. Am. Pet. Inst. Publ. No. 4167. Washington, D.C. 110 pp.

Beeman, R. D. 1969. The use of succinylcholine and other drugs for anesthetizing or narcotizing gastropod molluscs. Pubbl. Stn. Zool. Napoli 36:267–270.

Beeman, R. D. 1970. An ecological study of *Phyllaplysia taylori* Dall, 1900, Gastropoda: Opisthobranchia, with an emphasis on its reproduction. Vie Milieu Ser. A 21:189–212.

Bellan, G., D. J. Reish, and J. P. Foret. 1972. The sublethal effects of a detergent on the reproduction, development and settlement in the polychaetous annelid *Capitella capitata*. Mar. Biol. (Berlin) 14:183–188.

Bellan-Santini, D., G. Bellan, and D. J. Reish. 1979. Variations de l'influence d'altéragènes suivant le cycle d'activité jour/nuit de certains organismes marins. C.R. Hebd. Seances Acad. Sci. Ser. D Sci. Nat. 288:139–141.

Berg, I., and Å. Granmo. 1976. Procedures for toxicity testing in a continuous flow system. Pp. 75–80 in Lectures Presented at the Second FAO/SIDA Training Course on Marine Pollution in Relation to Protection of Living Resources. FAO/SIDA/TF No. 108, Suppl. 1.

Bergholz, E., and U. Brenning. 1978. Studies on the reproductive cycles of marine nematodes (*Rhabditis marina* and *Prochromadora orleji*). Wiss. Z. Univ. Rostock Math.-Naturwiss. Reihe 27:393–398.

Bernhard, M. 1957. Die Kultur von Seeigellarven (*Arbacia lixula* L.) in künstlichem and natürlichem Meerwasser mit Hilfe von Ionenaustauschsubstanzen und Komplexbildnern. Pubbl. Stn. Zool. Napoli 29:80–95.

Bernhard, M. 1977. Chemical contamination of culture media: Assessment, avoidance and control. Pp. 1459–1499 in O. Kinne, ed. Marine Ecology a Comprehensive, Integrated Treatise on Life in Oceans and Coastal Waters. Vol. III: Cultivation. John Wiley & Sons, Chichester, New York, Brisbane, and Toronto.

Berrill, N. J. 1937. Culture methods for ascidians. Pp. 564–571 in J. G. Needham *et al.*, eds. Culture Methods for Invertebrate Animals. Comstock Publishing Co., Ithaca, N.Y.

Berrill, N. J. 1947. Development and Growth of *Ciona*. J. Mar. Biol. Assoc. U.K. 26(4):616–625.

Berrill, N. J. 1950. The Tunicata with an Account of the British Species. Ray Society, Bernard Quaritch, London. 354 pp.

Betouhim-El, T., and D. Kahan. 1972. *Tisbe pori* n. sp. (Copepoda: Harpacticoida) from the Mediterranean coast of Israel and its cultivation in the laboratory. Mar. Biol. (Berlin) 16:201–209.

Bidwell, J. P. 1976. Water quality and the bioassay. Seascope 6(1/2):1, 6.

Bierbach, M., and D. K. Hofmann. 1973. Experimentelle Untersuchungen zum Stockwachstum und zur Medusenbildung bei dem marinen Hydrozoon *Eirene viridula*. Helgol. Wiss. Meeresunters. 25:63–84.

Bigford, T. E. 1978. Effect of several diets on survival, development time, and growth of laboratory-reared spider crab, *Libinia emarginata,* larvae. U.S. Natl. Mar. Fish. Serv. Fish. Bull. 76:59–64.

Binyon, J. 1972. Physiology of Echinoderms. Pergamon Press, New York. 264 pp.

Birkeland, C. E. 1974. Interactions between a sea pen and seven of its predators. Ecol. Monogr. 44:211–232.

Birkeland, C., F.-S. Chia, and R. R. Strathmann. 1971. Development, substratum selection, delay of metamorphosis and growth in the seastar, *Mediaster aequalis* Stimpson. Biol. Bull. (Woods Hole, Mass.) 141:99–108.

Black, J. H. 1976. Spawning and development of *Bedeva paivae* (Crosse, 1864) (Gastropoda: Muricidae), compiled from notes and observations by Florence V. Murray and G. Prestedge. J. Malacol. Soc. Aust. 3:215–221.

Blake, J. A. 1969. Reproduction and larval development of *Polydora* from northern New England (Polychaeta: Spionidae). Ophelia 7:1–63.

Blanquet, R. 1968. Properties and composition of the nematocyst toxin of the sea anemone, *Aiptasia pallida.* Comp. Biochem. Physiol. 25:893–902.

Blogoslawski, W. J., and C. Brown. 1979. Use of ozone for crustacean disease prevention: A brief review. Pp. 220–230 in D. H. Lewis and J. K. Leong, eds. Proceedings of the Second Biennial Crustacean Health Workshop July 1979. Texas A&M University, College Park. (TAMU–SG–79–114)

Blogoslawski, W. J., M. E. Stewart, and E. W. Rhodes. 1978. Bacterial disinfection in shellfish hatchery disease control. Pp. 589–602 in J. W. Avault, Jr., ed. Proc. Ninth Annu. Meet. World Mariculture Soc. La. State Univ. Div. Continuing Educ., Baton Rouge.

Blundo, R. 1978. The toxic effects of the water soluble fractions of no. 2 fuel oil and of three aromatic hydrocarbons on the behavior and survival of barnacle larvae. Contrib. Mar. Sci. 21:25–37.

Board, P. A., and M. J. Feaver. 1973. Rearing and radiographing the shipworm *Lyrodus pedicellatus.* Mar. Biol. (Berlin) 20:265–268.

Bocquet-Védrine, J. 1974. Chronologie du développement chez *Crinoniscus equitans* Pérez (isopode cryptoniscien). Arch. Zool. Exp. Gen. 115:197–204.

Boghen, A. D., and J. D. Castell. 1979. A recirculating system for small-scale experimental work on juvenile lobsters (*Homarus americanus*). Aquaculture 18:383–387.

Boletzky, S. von. 1974a. Élevage de céphalopodes en aquarium. Vie Milieu Ser. A Biol. Mar. 24:309–340.

Boletzky, S. von. 1974b. The "larvae" of Cephalopoda: A review. Thalassia Jugosl. 10:45–76.

Boletzky, S. von. 1975. Spawning behaviour independent of egg maturity in a cuttlefish (*Sepia officinalis* Linnaeus). Veliger 17:247.

Boletzky, S. von, and M. V. von Boletzky. 1969. First results in rearing *Octopus joubini* Robson 1929. Verh. Naturforsch. Ges. Basel 80:56–61.

Boletzky, S. von, M. V. von Boletzky, and V. Gätzi. 1971. Laboratory rearing of Sepiolinae (Mollusca: Cephalopoda). Mar. Biol. (Berlin) 8:82–87.

Boletzky, S. von, L. Rowe, and L. Aroles. 1973. Spawning and development of the eggs, in the laboratory, of *Illex coindetii* (Mollusca: Cephalopoda). Veliger 15:257–258.

Bookhout, C. G., and J. D. Costlow, Jr. 1970. Nutritional effects of *Artemia* from different locations on larval development of crabs. Helgol. Wiss. Meeresunters. 20:435–442.

Bookhout, C. G., and J. D. Costlow, Jr. 1974a. Crab development and effects of pollutants. Thalassia Jugosl. 10:77–87.

Bookhout, C. G., and J. D. Costlow, Jr. 1974b. Larval development of *Portunus spinicarpus* reared in the laboratory. Bull. Mar. Sci. 24:20–50.

Bookhout, C. G., and J. D. Costlow, Jr. 1975. Effects of mirex on the larval development of blue crab. Water, Air, Soil Pollut. 4:113–126.

Bookhout, C. G., and R. J. Monroe. 1977. Effects of malathion on the development of crabs. Pp. 3–19 in F. J. Vernberg, A. Calabrese, F. P. Thruberg, and W. B. Vernberg, eds. Physiological Responses of Marine Biota to Pollutants. Academic Press, New York.

Bookhout, C. G., J. D. Costlow, Jr., and R. Monroe. 1976. Effects of methoxychlor on larval development of mud-crab and blue crab. Water, Air, Soil Pollut. 5:349–365

Boolootian, R. A., ed. 1966. Physiology of Echinodermata. Wiley Interscience, New York, London, and Sydney. xviii + 822 pp.

Boolootian, R. A., A. C. Giese, A. Farmanfarmaian, and J. Tucker. 1959. Reproductive cycles of five West Coast crabs. Physiol. Zool. 32:213–220.

Borowsky, B. 1980. Reproductive patterns of three intertidal salt-marsh gammaridean amphipods. Mar. Biol. (Berlin) 55:327–334.

Bougis, P., M. C. Corre, and M. Etienne. 1979. Sea urchin larvae as a tool for assessment of the quality of sea-water. Ann. Inst. Oceanogr. (Paris) 55:21–26.

Bowen, S. T. 1962. The genetics of Artemia salina. I. The reproductive cycle. Biol. Bull. (Woods Hole, Mass.) 122:25–32.

Bower, C. E. 1980. Saltwater aquariums: Water quality without water analysis. Freshwater Mar. Aquarium 3 (4):44–46, 59.

Braconnot, J. C. 1963. Étude du cycle annuel des salpes et dolioles en rade de Villefranche-sur-Mer. J. Cons. Cons. Int. Explor. Mer 28:21–36.

Bradley, B. P. 1976. The measurement of temperature tolerance: Verification of an index. Limnol. Oceanogr. 21:596–599.

Bradley, E. A. 1974. Some observations of *Octopus joubini* reared in an inland aquarium. J. Zool. (Lond.) 173:355–368.

Braverman, M. H. 1962. Studies in hydroid differentiation. I. *Podocoryne carnea* culture methods and carbon dioxide induced sexuality. Exp. Cell Res. 26:301–306.

Braverman, M. H. 1971. Studies on hydroid differentiation. VI. Regulation of hydranth formation in *Podocoryne carnea*. J. Exp. Zool. 176:361–381.

Braverman, M. H. 1974. The cellular basis of morphogenesis and morphostasis in hydroids. Oceanogr. Mar. Biol. Annu. Rev. 12:129–221.

Breder, C. M., Jr. 1964. Miniature circulating systems for small laboratory aquariums. Pp. 39–53 in J. R. Clark and R. L. Clark, eds. Sea-water Systems for Experimental Aquariums. U.S. Fish Wildl. Serv. Res. Rep. No. 63. vi + 192 pp.

Breese, W. P., and F. D. Phibbs. 1970. Some observations on the spawning and early development of the butter clam, *Saxidomus giganteus* (Deshayes). Proc. Natl. Shellfish. Assoc. 60:95–98.

Brenko, M. H. 1973. Gonad development, spawning and rearing of *Mytilus* sp. larvae in the laboratory. Stud. Rev. Gen. Fish. Counc. Medit. No. 52:53–65.

Brenko, M. H., and A. Calabrese. 1969. The combined effects of salinity and temperature on larvae of the mussel *Mytilus edulis*. Mar. Biol. (Berlin) 4:224–226.

Brenowitz, A. H. 1975. A review of coelenterate laboratory culture. Pp. 137–143 in W. L. Smith and M. H. Chanley, eds. Culture of Marine Invertebrate Animals. Plenum Press, New York and London.

Brewster, F., and B. L. Nicholson. 1979. In vitro maintenance of amoebocytes from the American oyster (*Crassostrea virginica*). J. Fish. Res. Bd. Can. 36:461–467.

Brick, R. W. 1974. Effects of water quality, antibiotics, phytoplankton and food on survival and development of larvae of *Scylla serrata* (Crustacea: Portunidae). Aquaculture 3:231–244.

Bridges, C. B. 1975. Larval development of *Phyllaplysia taylori* Dall, with a discussion of development in the Anaspidea (Opisthobranchiata: Anaspidea). Ophelia 14:161–184.

Brinckmann, A. 1962. The life cycle of *Merga galleri* sp.n. (Anthomedusae, Pandeidae). Pubbl. Stn. Zool. Napoli 33:1–9.

Brinckmann, A. 1964. Observations on the biology and development of *Staurocladia portmanni* sp. n. (Anthomedusae, Eleutheridae). Can. J. Zool. 42:693–705.

Brock, M. A. 1975. Circannual rhythms. I. Free-running rhythms in growth and development of the marine cnidarian, *Campanularia flexuosa*. Comp. Biochem. Physiol. A: Comp. Physiol. 51:377–383.

Brown, A. C., and P. J. Greenwood. 1978. Use of the sea-urchin *Parechinus angulosus* (Leske) in determining water quality and acceptable pollution levels for the South African marine environment. S. Afr. J. Sci. 74:102–103.

Brown, A., Jr., J. McVey, B. S. Middleditch, and A. L. Lawrence. 1979. Maturation of white shrimp (*Penaeus setiferus*) in captivity. Pp. 435–444 in J. W. Avault, Jr., ed. Proc. Tenth Annu. Meet. World Mariculture Soc. La. State Univ. Div. Continuing Educ., Baton Rouge.

Brown, C. H. 1975. Structural Materials in Animals. John Wiley & Sons, New York and Toronto. 448 pp.

Brown, F. A., Jr. 1950. Studies on the physiology of *Uca* red chromatophores. Biol. Bull. (Woods Hole, Mass.) 98:218–226.

Brown, G. G. 1976. Scanning electron–microscopical and other observations of sperm fertilization reactions in *Limulus polyphemus* L. (Merostomata: Xiphosura). J. Cell Sci. 22:547–562.

Brown, G. G., and D. L. Clapper. 1980. Cortical reaction in inseminated eggs of the horseshoe crab, *Limulus polyphemus* L. Dev. Biol. 76:410–417.

Brown, G. G., and W. J. Humphreys. 1971. Sperm-egg interactions of *Limulus polyphemus* with scanning electron microscopy. J. Cell Biol. 51:904–907.

Brown, G. G., and J. R. Knouse. 1973. Effects of sperm concentration, sperm aging, and other variables on fertilization in the horseshoe crab, *Limulus polyphemus* L. Biol. Bull. (Woods Hole, Mass.) 144:462–470.

Brown, S. C. 1969. The structure and function of the digestive system of the mud snail *Nassarius obsoletus* (Say). Malacologia 9:447–500.

Brownell, W. N. 1978. Reproduction, laboratory culture, and growth of *Strombus gigas, S. costatus* and *S. pugilus* in Los Roques, Venezuela. Bull. Mar. Sci. Gulf Caribb. 27:668–680.

Brusca, R. C. 1980. Common Intertidal Invertebrates of the Gulf of California, 2d ed. University of Arizona Press. 513 pp.

Brusca, G. J., and R. C. Brusca. 1978. A Naturalist's Seashore Guide. Common Marine Life of the Northern California Coast and Adjacent Shore. Mad River Press, Eureka, Calif. 205 pp.

Buchanan, D. V., M. J. Myers, and R. S. Caldwell. 1975. Improved flowing water apparatus for the culture of brachyuran crab larvae. J. Fish. Res. Bd. Can. 32:1880–1883.

Buikema, A. L., Jr., and E. F. Benfield. 1979. Use of macroinvertebrate life history information in toxicity tests. J. Fish. Res. Bd. Can. 36:321–328.

Buikema, A. L., Jr., and J. Cairns, Jr., eds. 1980. Aquatic Invertebrate Bioassays. American Society for Testing and Materials, Philadelphia. 209 pp. (STP 715)

Bulla, L. A., and T. C. Cheng, eds. 1978. Invertebrate Models for Biomedical Research. Comparative Pathobiology, vol. 4. Plenum Publ. Corp., New York and London. 163 pp.

Bulnheim, H. P. 1978. Interaction between genetic, external and parasitic factors in sex determination of the crustacean amphipod *Gammarus duebeni*. Helgol. Wiss. Meeresunters. 31:1–33.

Bunde, T. A., and M. Fried. 1978. The uptake of dissolved free fatty acids from seawater by a marine filter feeder, *Crassostrea virginica.* Comp. Biochem. Physiol. A: Comp. Physiol. 60:139–144.

Burton, D. T. 1977. General test conditions and procedures for chlorine toxicity tests with estuarine and marine macroinvertebrates and fish. Chesapeake Sci. 18:130–136.

Bury, M. 1895. The metamorphosis of echinoderms. Q. J. Microsc. Sci. 38:45–137.

Butler, P. A., P. Doudoroff, M. A. Fontaine, M. Fujiya, R. Lange, R. Lloyd, D. J. Reish, J. B. Sprague, and M. Swedmark. 1977. Manual of methods in aquatic environment research. Part 4: Bases for selecting biological tests to evaluate marine pollution. FAO Fish. Tech. Pap. No. 164. viii + 31 pp.

Buu-Hoi, N. P., and Pham-Huu Chanh. 1970. Effect of various types of carcinogens on the hatching of *Artemia salina* eggs. J. Natl. Cancer Inst. 44:795–799.

Cable, W. D., and W. S. Landers. 1974. Development of eggs and embryos of the surf clam, *Spisula solidissima,* in synthetic seawater. U.S. Natl. Mar. Fish. Serv. Fish. Bull. 72:247–249.

Calabrese, A. 1969. *Mulinia lateralis:* Molluscan fruit fly? Proc. Natl. Shellfish. Assoc. 59:65–66.

Calabrese, A., and H. C. Davis. 1970. Tolerances and requirements of embryos and larvae of bivalve molluscs. Helgol. Wiss. Meeresunters. 20:553–564.

Calabrese, A., and E. W. Rhodes. 1974. Culture of *Mulinia lateralis* and *Crepidula fornicata* embryos and larvae for studies of pollution effects. Thalassia Jugosl. 10:89–102.

Calabrese, A., J. R. MacInnes, D. A. Nelson, and J. E. Miller. 1977. Survival and growth of bivalve larvae under heavy-metal stress. Mar. Biol. (Berlin) 41:179–184.

Calder, D. R. 1973. Laboratory observations on the life history of *Rhopilema verrilli* (Scyphozoa: Rhizostomeae). Mar. Biol. (Berlin) 21:109–114.

Calton, G. J., J. W. Burnett, and L. M. Staling. 1978. The Maryland blue crab: An experimental animal for cardiotoxicological investigations. Pp. 27–33 in P. N. Kaul and C. J. Sindermann, eds. Drugs and Food from the Sea. Myth or Reality? University of Oklahoma, Norman. xi + 448 pp.

Cameron, R. A., and R. T. Hinegardner. 1974. Initiation of metamorphosis in laboratory cultured sea urchins. Biol. Bull. (Woods Hole, Mass.) 146:335–342.

Cameron, R. A., and R. T. Hinegardner. 1978. Early events in sea urchin metamorphosis, description and analysis. J. Morphol. 157:21–32.

Campbell, R. D. 1966. Colony growth and pattern in the two-tentacled hydroid *Proboscidactyla flavicirrata.* Biol. Bull. (Woods Hole, Mass.) 135:96–104.

Campillo, A. 1976. Données pratiques sur l'élevage au laboratoire des larves de *Palaemon serratus* (Pennant). Rev. Trav. Inst. Peches Marit. 39:395–405.

Capo, T. R., S. E. Perritt, and C. J. Berg, Jr. 1979. New developments in the mariculture of *Aplysia californica.* Biol. Bull. (Woods Hole, Mass.) 157:360.

Cardwell, R. D., and C. E. Woelke. 1979. Marine Water Quality Compendium for Washington State. Vol. I. Introduction. Department of Fisheries, State of Washington, Olympia. vi + 75 pp.

Cardwell, R. D., C. E. Woelke, M. I. Carr, and E. Sanborn. 1977. Appraisal of a reference toxicant for estimating the quality of oyster larvae. Bull. Environ. Contam. Toxicol. 18:719–725.

Cardwell, R. D., C. E. Woelke, M. I. Carr, and E. Sanborn. 1978. Variation in toxicity tests of bivalve mollusc larvae as a function of termination technique. Bull. Environ. Contam. Toxicol. 20:128–134.

Cardwell, R. D., C. E. Woelke, M. I. Carr, and E. W. Sanborn. 1979. Toxic Substance and Water Quality Effects on Larval Marine Organisms. Wash. Dep. Fish. Tech. Rep. No. 45. vi + 71 pp.

Carefoot, T. H. 1967. Growth and nutrition of *Aplysia punctata* feeding on a variety of marine algae. J. Mar. Biol. Assoc. U.K. 47:565–589.

Carefoot, T. H. 1970. A comparison of absorption and utilization of food energy in two species of tropical *Aplysia.* J. Exp. Mar. Biol. Ecol. 5:47–62.

Carefoot, T. H. 1979. Artificial diets for sea hares. Can. J. Zool. 57:2271–2273.

Carefoot, T. H. 1980. Studies on the nutrition and feeding preferences of *Aplysia*: Development of an artificial diet. J. Exp. Mar. Biol. Ecol. 42:241–252.

Cargo, D. G. 1975. Comments on the laboratory culture of Scyphozoa. Pp. 145–154 in W. L. Smith and M. H. Chanley, eds. Culture of Marine Invertebrate Animals. Plenum Press, New York and London.

Carr, R. S., and D. J. Reish. 1977. The effect of petroleum hydrocarbons on the survival and life history of polychaetous annelids. Pp. 168–173 in D. A. Wolfe, ed. Rates and Effects of Petroleum Hydrocarbons in Marine Organisms and Ecosystems. Pergamon Press, London.

Carriker, M. R. 1961. Interrelation of functional morphology, behavior, and autecology in early stages of the bivalve *Mercenaria mercenaria.* J. Elisha Mitchell Soc. 77:168–241.

Castagna, M. 1975. Culture of the bay scallop, *Argopecten irradians,* in Virginia. NOAA (Natl. Ocean. Atmos. Adm.) Mar. Fish. Rev. 37(1):19–24.

Castilla, J. C., and J. Cancino. 1976. Spawning behaviour and egg capsules of *Concholepas concholepas* (Mollusca: Gastropoda: Muricidae). Mar. Biol. (Berlin) 37:255–263.

Castille, F. L., Jr., and A. L. Lawrence. 1979. The role of bacteria in the uptake of hexoses from seawater by postlarval penaeid shrimp. Comp. Biochem. Physiol. A: Comp. Physiol. 64:41–48.

Castro, W. E., P. B. Zielinski, and P. A. Sandifer. 1975. Performance characteristics of air lift pumps of short length and small diameter. Pp. 451–461 in J. W. Avault, Jr., ed. Proc. Sixth Annu. Meet. World Mariculture Soc. La. State Univ. Div. Continuing Educ., Baton Rouge.

Cather, J. N. 1963. A time schedule of the meiotic and early mitotic stages of *Ilyanassa.* Caryologia 16:663–670.

Cather, J. N. 1967. Cellular interactions in the development of the shell gland of the gastropod, *Ilyanassa.* J. Exp. Zool. 166:205–224.

Cather, J. N. 1971. Cellular interactions in the regulation of development in annelids and molluscs. Pp. 67–125 in M. Abercrombie, J. Brachet, and T. J. King, eds. Advances in Morphogenesis, vol. 9. Academic Press, New York.

Cavanaugh, G. M., ed. 1975. Formulae and Methods VI of the Marine Biological Laboratory Chemical Room. Marine Biological Laboratory, Woods Hole, Mass. 84 pp.

Chandler, J. H., Jr., and S. K. Partridge. 1975. A solenoid-actuated chemical-metering apparatus for use in flow-through toxicity tests. Prog. Fish Cult. 37:93–95.

Chang, E. S., and J. D. O'Connor. 1977. Secretion of ∝-ecdysone by crab Y-organs *in vitro.* Proc. Natl. Acad. Sci. 74:615–618.

Chanley, M. H., and O. W. Terry. 1974. Inexpensive modular habitats for juvenile lobsters (*Homarus americanus*). Aquaculture 4:89–92.

Chanley, P. 1969. Larval development in the class bivalvia. Pp. 475–481 in Proc. Symp. Mollusca—Part III. Mar. Biol. Assoc. India.

Chanley, P. 1975. Laboratory cultivation of assorted bivalve mollusks. Pp. 297–318 in W. L. Smith and M. H. Chanley, eds. Culture of Marine Invertebrate Animals. Plenum Press, New York and London.

Chanley, P., and M. Castagna. 1971. Larval development of the stout razor clam, *Tagelus plebeius* Solander (Solecurtidae: Bivalvia). Chesapeake Sci. 12:167–172.

Chanley, P., and M. Chanley. 1970. Larval development of the commensal clam, *Montacuta percompressa* Dall. Proc. Malacol. Soc. Lond. 39:59–67.

Chen, C.-H., J. W. Greenawalt, and A. L. Lehninger. 1974. Biochemical and ultrastructural aspects of Ca^{2+} transport by mitochondria of the hepatopancreas of the blue crab *Callinectes sapidus*. J. Cell Biol. 61:301–315.

Cheng, L., ed. 1976. Marine Insects. Elsevier North-Holland, New York. 582 pp.

Chia, F. 1976. Sea anemone reproduction, patterns and adaptive radiations. Pp. 261–270 in G. O. Mackie, ed. Coelenterate Ecology and Behavior. Plenum Press, New York and London.

Chia, F., and L. R. Bickell. 1978. Mechanisms of larval attachment and the induction of settlement and metamorphosis in coelenterates: A review. Pp. 1–12 in F. Chia and M. E. Rice, eds. Settlement and Metamorphosis of Marine Invertebrate Larvae. Elsevier, New York and Oxford.

Chia, F., and B. J. Crawford. 1973. Some observations on gametogenesis, larval development and substratum selection of the sea pen *Ptilosarcus guerneyi*. Mar. Biol. (Berlin) 23:73–82.

Chia, F. S., and R. Koss. 1978. Development and metamorphosis of the planktotrophic larvae of *Rostanga pulchra* (Mollusca: Nudibranchia). Mar. Biol. (Berlin) 46:109–119.

Chia, F. S., and M. E. Rice, eds. 1978. Settlement and Metamorphosis of Marine Invertebrate Larvae. Proceedings of a Symposium, Toronto, Dec. 1977. Elsevier North-Holland, New York. xii + 290 pp.

Chia, F., and M. A. Rostron. 1970. Some aspects of the reproductive biology of *Actinia equina* (Cnidaria: Anthozoa). J. Mar. Biol. Assoc. U.K. 50:253–264.

Chittleborough, R. G. 1976. Breeding of *Panulirus longipes cygnus* George under natural and controlled conditions. Aust. J. Mar. Freshwater Res. 27:499–516.

Chitwood, B. G., and D. G. Murphy. 1964. Observations on two marine monhysterids—their classification, cultivation, and behavior. Trans. Am. Microsc. Soc. 83:311–329.

Choe, S. 1966. On the eggs, rearing, habits of the fry, and growth of some Cephalopoda. Bull. Mar. Sci. Gulf Caribb. 16:330–348.

Choe, S., and Y. Ohshima. 1963. Rearing of cuttlefishes and squids. Nature (Lond.) 197:307.

Christiansen, M. E. 1971. Larval development of *Hyas araneus* (Linnaeus) with and without antibiotics (Decapoda, Brachyura, Majidae). Crustaceana (Leiden) 21:307–315.

Christiansen, M. E. 1973. The complete larval development of *Hyas araneus* (Linnaeus) and *Hyas coaractatus* Leach (Decapoda, Brachyura, Majidae) reared in the laboratory. Norw. J. Zool. 21:63–89.

Christiansen, M. E., and Won Tack Yong. 1976. Feeding experiments on the larvae of the fiddler crab *Uca pugilator* (Brachyura, Ocypodidae), reared in the laboratory. Aquaculture 8:91–98.

Christiansen, M. E., J. D. Costlow, Jr., and R. J. Monroe. 1977. Effects of the juvenile hormone mimic ZR–512 (Altozar) on larval development of the mud-crab *Rhithropanopeus harrisii* at various cyclic temperatures. Mar. Biol. (Berlin) 39:281–288.

Chu, G. W. T. C., and E. P. Ryan. 1960. A technique for maintenance of the snail *Littorina pintado* in the laboratory. J. Parasitol. 46:249.

Clapper, D. L., and G. G. Brown. 1980, Sperm motility in the horseshoe crab, *Limulus polyphemus* L. I. Sperm behavior near eggs and motility initiation by egg extracts. Dev. Biol. 76:341–449.

Clark, R. C., Jr., and J. S. Finley. 1974. Tidal aquarium for laboratory studies of environmental effects on marine organisms. Prog. Fish Cult. 36:134–137.

Claus, C., F. Benijts, and P. Sorgeloos. 1977. Comparative study of different geographical strains of the brine shrimp, *Artemia salina*. Pp. 91–105 in E. Jaspers and G. Persoone, eds. Fundamental and Applied Research on the Brine Shrimp *Artemia salina* (L.) in Belgium. Europ. Mariculture Soc. Spec. Publ. No. 2. Bredene, Belgium. vi + 110 pp.

Clegg, J. S. 1964. The control of emergence and metabolism by external osmotic pressure and the role of free glycerol in developing cysts of *Artemia salina.* J. Exp. Biol. 41:879–892.

Clement, A. C. 1952. Experimental studies on germinal localization in *Ilyanassa.* I. The role of the polar lobe in determination of the cleavage pattern and its influence in later development. J. Exp. Zool. 121:593–626.

Clement, A. C. 1956. Experimental studies on germinal localization in *Ilyanassa.* II. The development of isolated blastomeres. J. Exp. Zool. 132:427–446.

Clement, A. C. 1962. Development of *Ilyanassa* following removal of the D macromere at successive cleavage stages. J. Exp. Zool. 149:193–216.

Clement, A. C. 1967. The embryonic value of the micromeres in *Ilyanassa obsoleta,* as determined by deletion experiment. I. The first quartet cells. J. Exp. Zool. 166:77–88.

Clement, A. C. 1971. *Ilyanassa.* Pp. 188–214 in G. Reverberi, ed. Experimental Embryology of Marine and Fresh-water Invertebrates. North-Holland Publ. Co., Amsterdam–London. American Elsevier Publ. Co., New York.

Clement, A. 1976. Cell determination and organogenesis in molluscan development: A reappraisal based on deletion experiments in *Ilyanassa.* Am. Zool. 16:447–453.

Cloney, R. A. 1961. Observations on the mechanism of tail resorption in ascidians. Am. Zool. 1:67–88.

Cloney, R. A. 1978. Ascidian metamorphosis: Review and analysis. Pp. 255–282 in F. S. Chia and M. E. Rice, eds., Settlement and Metamorphosis of Marine Invertebrate Larvae. Elsevier North-Holland Biomedical Press, New York.

Clutter, R. I., and G. H. Theilacker. 1971. Ecological efficiency of a pelagic mysid shrimp; estimates from growth, energy budget, and mortality studies. U.S. Natl. Mar. Fish. Serv. Fish. Bull. 69:93–115.

Coe, W. R. 1937. Methods for the laboratory culture of nemerteans. Pp. 162–165 in J. G. Needham *et al.,* eds. Culture Methods for Invertebrate Animals. Comstock Publ. Co., Ithaca, N.Y.

Coffaro, K. A. 1979. Transplantation Immunity in the Sea Urchin. Ph.D. Thesis, University of California, Santa Cruz.

Coffaro, K., and R. H. Hinegardner. 1977. Immune response in the sea urchin *Lytechinus pictus.* Science 197:1389–1390.

Coffin, H. G. 1958. The laboratory culture of *Pagurus samuelis* (Stimpson) (Crustacea, Decapoda). Walla Walla Coll. Publ. No. 22:1–5.

Cohen, E., F. B. Bang, J. Levin, J. J. Marchalonis, T. G. Pistole, R. A. Prendergast, C. Shuster, Jr., and S. W. Watson, eds. 1979. Biomedical Applications of the Horseshoe Crab (Limulidae). Proceedings of a Symposium Held at the Marine Biological Laboratory, Woods Hole, Massachusetts, October 1978. Alan R. Liss, New York. xx + 688 pp.

Coleman, D. E. 1979. Chamber for culturing microorganisms. Prog. Fish Cult. 41:110.

Collier, J. R. 1965. Morphogenetic significance of biochemical patterns in "Mosaic Embryos." Pp. 203–241 in R. Weber, ed. Biochemical Aspects of Development, vol. 1. Academic Press, New York.

Collier, J. R. 1966. The transcription of genetic information in the spiralian embryo. Pp. 39–58 in A. A. Moscona and A. Monroy, eds. Current Topics in Developmental Biology, vol. 1. Academic Press, New York.

Collier, J. R. 1975. Nucleic acid synthesis in the normal and lobeless embryo of *Ilyanassa obsoleta.* Exp. Cell Res. 95:254–262.

Collier, J. R. 1976. Nucleic acid chemistry of the *Ilyanassa* embryo. Am. Zool. 16:483–500.

Collier, J. R. 1977. Rates of RNA synthesis in the normal and lobeless embryo of *Ilyanassa obsoleta.* Exp. Cell Res. 106:390–394.

Collier, M. M. 1975. A Cytochemical, Ultrastructural, and Autoradiographic Study of Oogenesis in the Marine Mud Snail *Ilyanassa obsoleta* (Say). Ph.D. Diss., City University of New York.

Collins, M. T., J. B. Gratzek, D. L. Dawe, and T. G. Nemetz. 1976. Effects of antibacterial agents on nitrification in an aquatic recirculating system. J. Fish. Res. Bd. Can. 33:215–218.

Colón-Urban, R., P. J. Cheung, G. D. Ruggieri, and R. F. Nigrelli. 1979. Observations on the development and maintenance of the deep sea barnacle, *Octolasmis aymonini geryonophila* (Pilsbry). Int. J. Invertebr. Reprod. 1:245–252.

Comely, C. A. 1972. Larval culture of the scallop *Pecten maximus* (L.). J. Cons. Cons. Int. Explor. Mer 34:365–378.

Conger, K. A., M. L. Swift, J. B. Reeves III, and S. Lakshmanan. 1978. Shell growth of unfed oysters in the laboratory: A sublethal bioassay system for pollutants. Life Sci. 22:245–254.

Conklin, D. E., and L. Provasoli. 1978. Biphasic particulate media for the culture of filter-feeders. Biol. Bull. (Woods Hole, Mass.) 154:47–54.

Conklin, D. E., M. J. Goldblatt, C. E. Bordner, N. A. Baum, and T. B. McCormick. 1978. Artificial diets for the lobster, *Homarus americanus:* A revaluation. Pp. 243–250 in J. W. Avault, Jr., ed. Proc. Ninth Annu. Meet. World Mariculture Soc. La. State Univ. Div. Continuing Educ., Baton Rouge.

Connell, A. D., and D. D. Airey. 1979. Life-cycle bioassays using two estuarine amphipods, *Grandidierella lutosa* and *G. lignorum,* to determine detrimental sublethal levels of marine pollutants. S. Afr. J. Sci. 75:313–314.

Connell, J. H. 1970. A predator-prey system in the marine intertidal region. I. *Balanus glandula* and several predatory species of *Thais.* Ecol. Monogr. 40:49–78.

Connor, P. M., and K. W. Wilson. 1972. A continuous-flow apparatus for assessing the toxicity of substances to marine animals. J. Exp. Mar. Biol. Ecol. 9:209–215.

Conover, R. J., and C. M. Lalli. 1972. Feeding and growth in *Clione limacina* (Phipps), a pteropod mollusc. J. Exp. Mar. Biol. Ecol. 9:279–302.

Conrad, G. W., and D. C. Williams. 1974. Polar lobe formation and cytokinesis in fertilized eggs of *Ilyanassa obsoleta.* II. Large bleb formation caused by high concentrations of exogenous calcium ions. Dev. Biol. 37:280–294.

Conrad, G. W., D. C. Williams, F. R. Turner, K. M. Newrock, and R. A. Raff. 1973. Microfilaments in the polar lobe constriction of fertilized eggs of *Ilyanassa obsoleta.* J. Cell Biol. 59:228–233.

Conrad, G. W., A. E. Kammer, and G. F. Athey. 1977. Membrane potential of fertilized eggs of *Ilyanassa obsoleta* during polar lobe formation and cytokinesis. Dev. Biol. 57:215–220.

Cook, D. W. 1972. A circulating seawater system for experimental studies with crabs. Prog. Fish Cult. 34:61–62.

Cook, H. L. 1969. A method of rearing penaeid shrimp larvae for experimental purposes. FAO Fish. Rep. No. 57:709–715.

Cook, P. L., and P. J. Chimonides. 1978. Observations on living colonies of *Selenaria* (Bryozoa, Cheilostomata). I. Cah. Biol. Mar. 19:147–158.

Coombs, R. F. 1972. Device for detecting and measuring activity of large marine crustaceans. N.Z. J. Mar. Freshwater Res. 6:194–205.

Cori, C. J. 1938. Narkose und Anästhesie wirbelloser Tiere des Süss- und Meerwassers. Pp. 542–584 in Abderhalden's Handbuch der Biologischen Arbeitsmethoden, vol. 9, pt. 6. Urban & Schwarzenberg, Berlin and Vienna.

Corkett, C. J. 1967. Technique for rearing marine calanoid copepods in laboratory conditions. Nature (Lond.) 216:58–59.

Corkett, C. J. 1968. La reproduction en laboratoire des copépodes marins *Acartia clausi* Giesbrecht et *Idya furcata* (Baird). Pelagos 10:77–90.

Corkett, C. J. 1970. Techniques for breeding and rearing marine calanoid copepods. Helgol. Wiss. Meeresunters. 20:318–324.

Corkett, C. J., and D. L. Urry. 1968. Observations on the keeping of adult female *Pseudocalanus elongatus* under laboratory conditions. J. Mar. Biol. Assoc. U.K. 48:97–105.

Cormier, S. M., and D. A. Hessinger. 1980. Sensory receptor of *Physalia* nematocytes. J. Ultrastruc. Res. (in press).

Costello, D. P., and C. Henley. 1971. Methods for Obtaining and Handling Marine Eggs and Embryos, 2d ed. Marine Biological Laboratory, Woods Hole, Mass. xvii + 247 pp.

Costello, T. J., J. H. Hudson, J. L. Dupuy, and S. Rivkin. 1973. Larval culture of the calico scallop, *Argopecten gibbus*. Proc. Natl. Shellfish. Assoc. 63:72–76.

Costlow, J. D., Jr. 1968. Metamorphosis in crustaceans. Pp. 3–41 in W. Etkin and L. I. Gilbert, eds. Metamorphosis. A Problem in Developmental Biology. Appleton-Century-Crofts, New York. 459 pp.

Costlow, J. D., Jr., and C. G. Bookhout. 1957. Larval development of Balanus eburneus in the laboratory. Biol. Bull. (Woods Hole, Mass.) 112:313–324.

Costlow, J. D., Jr., and C. G. Bookhout. 1959. The larval development of *Callinectes sapidus* Rathbun reared in the laboratory. Biol. Bull. (Woods Hole, Mass.) 116:373–396.

Costlow, J. D., Jr., and C. G. Bookhout. 1960a. A method for developing brachyuran eggs *in vitro*. Limnol. Oceanogr. 5:212–215.

Costlow, J. D., Jr., and C. G. Bookhout. 1960b. The complete larval development of *Sesarma cinereum* (Bosc) reared in the laboratory. Biol. Bull. (Woods Hole, Mass.) 118:203–214.

Costlow, J. D., Jr., and C. G. Bookhout. 1962. The larval development of *Sesarma reticulatum* Say reared in the laboratory. Crustaceana 4:281–294.

Costlow, J. D., Jr., and C. G. Bookhout. 1965. The effect of environmental factors on larval development of crabs. Pp. 77–86 in C. M. Tarzwell, ed. Biological Problems in Water Pollution. Third Seminar 1962. Division of Water Supply and Pollution Control, U.S. Public Health Service, Cincinnati. (PHS Publ. 999–WP–25)

Costlow, J. D., Jr., C. G. Bookhout, and R. Monroe. 1960. The effect of salinity and temperature on larval development of *Sesarma cinereum* (Bosc) reared in the laboratory. Biol. Bull. (Woods Hole, Mass.) 118:183–202.

Costlow, J. D., Jr., C. G. Bookhout, and R. Monroe. 1966. Studies on the larval development of the crab, *Rhithropanopeus harrisii* Gould. I. The effect of salinity and temperature on larval development. Physiol. Zool. 39:81–100.

Couch, J. A., and L. Courtney. 1977. Interaction of chemical pollutants and virus in a crustacean: A novel bioassay system. Ann. N.Y. Acad. Sci. 298:497–504.

Courtright, R. C., W. P. Breese, and H. Krueger. 1971. Formulation of a synthetic seawater for bioassays with *Mytilus edulis* embryos. Water Res. 5:877–888.

Cox, G. V., B. L. Olla, F. A. Cross, T. W. Duke, G. La Roche, and R. J. Livingston. 1974. Marine Bioassays Workshop Proceedings 1974. Marine Technology Society, Washington, D.C. xviii + 308 pp.

Crampton, H. E. 1896. Experimental studies on gastropod development. Arch. Entwicklungsmech. Org. 3:1–19.

Crane, J. 1975. Fiddler Crabs of the World, Ocypodidae: Genus *Uca*. Princeton University Press, Princeton. 660 pp.

Crecelius, E. A. 1979. Measurements of oxidants in ozonized seawater and some biological reactions. J. Fish. Res. Bd. Can. 36:1006–1008.

Cripe, C. R. 1979. An automated device (AGARS) for studying avoidance of pollutant gradients by aquatic organisms. J. Fish. Res. Bd. Can. 36:11–16.

Cripps, R. A., and D. J. Reish. 1973. The effect of environmental stress on the activity of malate dehydrogenase and lactate dehydrogenase in *Neanthes arenaceodentata* (Annelida: Polychaeta). Comp. Biochem. Physiol. 46B:123–133.

Crisp, D. J. 1974. Factors influencing the settlement of marine invertebrate larvae. Pp. 177–266 in P. T. Grant and A. M. Mackie, eds. Chemoreception in Marine Organisms. Academic Press, New York.

Crisp, D. J., and P. A. Davies. 1955. Observations *in vivo* on the breeding of *Elminius modestus* grown on glass slides. J. Mar. Biol. Assoc. U.K. 34:357–380.

Crow, T., and J. F. Harrigan. 1979. Reduced behavioral variability in laboratory-reared *Hermissenda crassicornis* (Eschscholtz, 1831) (Opisthobranchia: Nudibranchia). Brain Res. 173:179–184.

Crowell, J. 1965. The fine structure of the polar lobe of *Ilyanassa obsoleta.* Acta Embryol. Morphol. Exp. 7:225–234.

Crowell, P. 1957. Differential responses of growth zones to nutritive level, age, and temperature in the colonial hydroid *Campanularia.* J. Exp. Zool. 134:63–90.

Crowell, P., and C. R. Wyttenbach. 1957. Factors affecting terminal growth in the hydroid *Campanularia.* Biol. Bull. (Woods Hole, Mass.) 113:233–244.

Crowell, S. 1953. The regression-replacement cycle of hydranths of *Obelia* and *Campanularia.* Physiol. Zool. 26:319–327.

Crowell, S. 1967. Coelenterates. Pp. 257–264 in F. H. Wilt and N. K. Wessels, eds. Methods in Developmental Biology. Thomas Y. Crowell Co., New York.

Cruz, L. J., G. Corpuz, and B. M. Olivera. 1978. Mating, spawning, development and feeding habits of *Conus geographus* in captivity. Nautilus 92:150–152.

Culliney, J. L. 1975. Comparative larval development of the shipworms *Bankia gouldi* and *Teredo navalis.* Mar. Biol. (Berlin) 29:245–251.

Culliney, J. L., P. J. Boyle, and R. D. Turner. 1975. New approaches and techinques for studying bivalve larvae. Pp. 257–271 in W. L. Smith and M. H. Chanley, eds. Culture of Marine Invertebrate Animals. Plenum Press, New York and London.

Czihak, G., ed. 1975. The Sea Urchin Embryo. Biochemistry and Morphogenesis. Springer-Verlag, Berlin, Heidelberg, and New York. xix + 700 pp.

Dagg, M. J. 1976. Complete carbon and nitrogen budgets for the carnivorous amphipod, *Calliopius laeviusculus* (Krøoyer). Int. Rev. Gesamten Hydrobiol. 61:297–357.

D'Agostino, A. 1975. Antibiotics in cultures of invertebrates. Pp. 109–133 in W. L. Smith and M. H. Chanley, eds. Culture of Marine Invertebrate Animals. Plenum Press, New York and London.

D'Agostino, A., and C. Finney. 1974. The effect of copper and cadmium on the development of *Tigriopus japonicus.* Pp. 445–463 in F. J. Vernberg, and W. B. Vernberg eds. Pollution and Physiology of Marine Organisms. Academic Press, New York, San Francisco, and London.

Dallot, S. 1968. Observations préliminaires sur la reproduction en élevage du chaetognathe planctonique *Sagitta setosa* Müller. Rapp. P.-V. Reun. Cons. Int. Explor. Mer 19:521–523.

D'Asaro, C. N. 1970. Egg capsules of prosobranch mollusks from south Florida and the Bahamas and notes on spawning in the laboratory. Bull. Mar. Sci. Gulf Caribb. 20:414–440.

D'Asaro, C. N. 1973. Cultured lugworms—a new source of bait for marine sport fishermen. Am. Fish. Farmer, pp. 4–7, September.

Davenport, J. 1977. A study of the effects of copper applied continuously and discontinuously to specimens of *Mytilus edulis* (L.) exposed to steady and fluctuating salinity levels. J. Mar. Biol. Assoc. U.K. 57:63–74.

Davis, C. C. 1965. A study of the hatching process in aquatic invertebrates. XIV. An examination of hatching in *Palaemonetes vulgaris* (Say). Crustaceana (Leiden) 8:233–238.

Davis, H. C. 1961. Effects of some pesticides on eggs and larvae of oysters. (*Crassostrea virginica*) and clams (*Venus mercenaria*). Commer. Fish. Rev. 23(12):8–23.

Davis, H. C., and P. E. Chanley. 1956. Effects of some dissolved substances on bivalve larvae. Proc. Natl. Shellfish. Assoc. 46:59–74.

Davis, J. C. 1978. Disruption of precopulatory behavior in the amphipod *Anisogammarus pugettensis* upon exposure to bleached kraft pulpmill effluent. Water Res. 12:273–275.

Davis, J. D. 1977. Nutrition for Scaphopoda. Pp. 103–105 in M. Rechcigl, Jr., ed. CRC Handbook Series in Nutrition and Food. Section G: Diets, Culture Media, Food Supplements. Volume II: Food Habits of, and Diets for Invertebrates and Vertebrates—Zoo Diets. CRC Press, Cleveland.

Davis, L. V. 1971. Growth and development of colonial hydroids. Pp. 16–36 in H. M. Lenhoff, L. Muscatine, and L. V. Davis, eds. Experimental Coelenterate Biology. University of Hawaii Press, Honolulu.

Davis, W. R., and D. J. Reish. 1975. The effect of reduced dissolved oxygen concentration on the growth and production of oocytes in the polychaetous annelid *Neanthes arenaceodentata*. Rev. Int. Oceanogr. Med. 37–38:3–16.

Dawson, C. E. 1957. Studies on the marking of commercial shrimp with biological stains. Spec. Sci. Rep.-Fish. No. 231. U.S. Department of the Interior, Washington, D.C. 24 pp.

Dawson, E. Y. 1966. Seashore plants of Northern California. University of California Press, Berkeley. 103 pp.

Day, J. H. 1973. New Polychaeta from Beaufort with a Key to All Species Recorded from North Carolina. NOAA Tech. Rep. NMFS Circ. No. 375. 140 pp.

Day, R., L. Day, and J. A. Blake. 1979. Reproduction and larval development of *Polydora giardi* Mesnil (Polychaeta: Spionidae). Biol. Bull. (Woods Hole, Mass.) 156:20–30.

Dayton, P. K., G. A. Robilliard, R. T. Paine, and L. B. Dayton. 1974. Biological accommodation in the benthic community at McMurdo Sound, Antarctica. Ecol. Monogr. 44:105–128.

Dean, D. 1977. Diets and culture media: Annelida, Echiura, and Sipuncula. Pp. 117–129 in M. Rechcigl, Jr., ed. CRC Handbook Series in Nutrition and Food. Section G: Diets, Culture Media, Food Supplements. Volume II: Food Habits of, and Diets for Invertebrates and Vertebrates—Zoo Diets. CRC Press, Cleveland.

Dean, D., and M. Mazurkiewicz. 1975. Methods of culturing polychaetes. Pp. 177–197 in W. L. Smith and M. H. Chanley, eds. Culture of Marine Invertebrate Animals. Plenum Press, New York and London.

de Angelis, E. 1977a. The problem of bacterial contamination during sea urchin culture. Pubbl. Stn. Zool. Napoli 40:1–25.

de Angelis, E. 1977b. The effect of antibiotics on sea urchin egg culture. Pubbl. Stn. Zool. Napoli 40:26–33.

DeCoursey, P. J., and W. B. Vernberg. 1972. Effects of mercury on survival, metabolism and behavior of larval *Uca pugliator*. Oikos 23:241–247.

de Graaf, F. 1973. Marine Aquarium Guide. Pet Library, Ltd., Sternco Industries, Inc., Harrison, N.J. 284 pp.

DeManche, J. M., P. L. Donaghay, W. P. Breese, and L. F. Small. 1975. Residual toxicity of ozonized seawater to oyster larvae. Publ. Oreg. State Univ. Sea Grant Program. 7 pp. (ORESU-T-75-003)

Dempster, R. P. 1953. The use of larval and adult brine shrimp in aquarium fish culture. Calif. Fish Game 39:355–364.

Deshimaru, O., and K. Kuroki. 1979. Requirement of prawn for dietary thiamine, pyridoxine, and choline chloride. Bull. Jpn. Soc. Sci. Fish. 45:363–367.

Devaney, D. M., and L. G. Eldredge, eds. 1977. Reef and Shore Fauna of Hawaii. Section 1: Protozoa through Ctenophora. Bernice P. Bishop Mus. Spec. Publ. 64(1):xii + 278.

Dexter, D. M. 1972. Molting and growth in laboratory reared phyllosomes of the California spiny lobster, *Panulirus interruptus*. Calif. Fish Game 58:107–115.

Dix, T. G. 1969. Larval life span of the echinoid *Evechinus chloroticus* (Val.). N.Z. J. Mar. Freshwater Res. 3:13–16.

Dix, T. G. 1970. Biology of *Evechinus chlorotecus* (Echinoides: Echinometridae) from different localities. 2. Movement. N.Z. J. Mar. Freshwater Res. 4:267–277.

Dix, T. G. 1976. Laboratory rearing of larval *Ostrea angasi* in Tasmania, Australia. J. Malacol. Soc. Aust. 3:209–214.

Dobkin, S. 1969. Abbreviated larval development in caridean shrimps and its significance in the artificial culture of these animals. FAO Fish. Rep. No. 57:935–946.

Dodd, J. M. 1957. Artificial fertilisation, larval development and metamorphosis in *Patella vulgata* L. and *Patella coerulea* L. Pubbl. Stn. Zool. Napoli 29:172–186.

Donahue, W. H., R. T. Wang, M. Welch, and J. A. C. Nicol. 1977. Effects of water-soluble components of petroleum oils and aromatic hydrocarbons on barnacle larvae. Environ. Pollut. 13:187–202.

Dougherty, E. C., ed. 1959. Axenic culture of invertebrate Metazoa: A goal. Ann. N.Y. Acad. Sci. 77:25–405.

Dowds, R. E. 1979. References for the Identification of Marine Invertebrates on the Southern Atlantic Coast of the United States. NOAA Tech. Rep. NMFS No. 729. iv + 37 pp.

Dries, M., and D. Adelung. 1976. Neue Ergebnisse über die Aufzucht von *Carcinus maenas* im Laboratorium. Mar. Biol. (Berlin) 38:17–24.

Droop, M. R., and J. M. Scott. 1978. Steady-state energetics of a planktonic herbivore. J. Mar. Biol. Assoc. U.K. 58:749–772.

Ďuračková, Z., V. Betina, B. Horníková, and P. Nemec. 1977. Toxicity of mycotoxins and other fungal metabolites to *Artemia salina* larvae. Zentralbl. Bakteriol. Parasitenkd. Infektionskr. Hyg. Zweite Naturwiss. Abt. Allg. Landwirtsch. Tech. Mikrobiol. 132:294–299.

Eales, N. B. 1960. Revision of the world species of *Aplysia* (Gastropoda, Opisthobranchia). Br. Mus. (Nat. Hist.) Bull. Zool. 5:267–400.

Ebert, T. A. 1965. A technique for the individual marking of sea urchins. Ecology 46:193–194.

Ebstein, B. S., M. D. Rosenthal, and R. L. DeHaan. 1965. Cells from isolated blastomeres of *Ilyanassa obsoleta* in tissue culture. Exp. Cell Res. 40:174–177.

Eckelbarger, K. J. 1975. Developmental studies of the post-settling stages of *Sabellaria vulgaris* (Polychaeta: Sabellariidae). Mar. Biol. (Berlin) 30:137–149.

Edds, K. T. 1977. Dynamic aspects of filopodial formation by reorganization of microfilaments. J. Cell Biol. 73:479–491.

Edmonson, C. H. 1946. Reef and shore fauna of Hawaii. Bernice P. Bishop Mus. Spec. Publ. 22:1–381.

Ehrmann, R. L., and G. O. Gey. 1956. The growth of cells on a transparent rat-tail collagen. J. Natl. Cancer Inst. 16:1375–1403.

Eisig, H. 1898. Zur Entwicklungsgeschichte der Capitelliden. Zool. Stan. Neapel Mitt. 13:1–292.

Eisler, R. 1979. Behavioural responses of marine poikilotherms to pollutants. Philos. Trans. R. Soc. Lond. B Biol. Sci. 286:507–521.

Environmental Research Laboratory, Gulf Breeze–Narragansett–Corvallis. 1976. Bioassay Procedures for the Ocean Disposal Permit Program. U.S. Environmental Protection Agency. EPA–600/9–76–010. 109 pp.

Epifanio, C. E. 1976. Culture of bivalve mollusks in recirculating systems: Nutritional requirements. Pp. 173–194 in K. S. Price, Jr., W. N. Shaw, and K. S. Danberg, eds. Proc. First Int. Conf. Aquaculture Nutr. Coll. Mar. Stud. University of Delaware, Newark.

Epifanio, C. E., and C. A. Mootz. 1976. Growth of oysters in a recirculating maricultural system. Proc. Natl. Shellfish. Assoc. 65:32–37.

Ewald, J. J. 1965. The laboratory rearing of pink shrimp, *Penaeus duodarum* Burkenroad. Bull. Mar. Sci. Gulf Caribb. 15:436–449.

Fannaly, M. T. 1978. A method for tagging immature blue crabs (*Callinectes sapidus* Rathbun). Northeast. Gulf Sci. 2:124–126.

Farke, H., and E. M. Berghuis. 1979. Spawning, larval development and migration behaviour of *Arenicola marina* in the laboratory. Neth. J. Sea Res. 13:512–528.

Fauchald, K., and P. A. Junars. 1979. The diet of worms: A study of polychaete feeding guilds. Oceanogr. Mar. Biol. Annu. Rev. 17:193–284.

Feder, H. M. 1955. The use of vital stains in marking Pacific Coast starfish. Calif. Fish Game 41:245–246.

Federighi, H. 1937. Culture methods for *Urosalpinx cinerea.* Pp. 532–536 in J. G. Needham *et al.,* eds. Culture Methods for Invertebrate Animals. Comstock Publ. Co., Ithaca, N.Y.

Fell, P. E. 1967. Sponges. Pp. 265–276 in F. H. Wilt and N. K. Wessels, eds. Methods in Developmental Biology. Thomas Y. Crowell Co., New York.

Fell, P. E. 1976. Analysis of reproduction in sponge populations: An overview with specific information on the reproduction of *Haliclona loosanoffi.* Pp. 51–67 in F. W. Harrison and R. R. Cowden, eds. Aspects of Sponge Biology. Academic Press, New York, San Francisco, and London. xiii + 354 pp.

Fell, P. E. 1977. Diets and culture media for Porifera. Pp. 3–6 in M. Rechcigl, Jr., ed. CRC Handbook Series in Nutrition and Food. Section G: Diets, Culture Media, Food Supplements. Volume II: Food Habits of, and Diets for Invertebrates and Vertebrates—Zoo Diets. CRC Press, Cleveland.

Fenaux, R. 1976. Cycle vital d'un appendiculaire *Oikopleura dioica* Fol. 1872 description et chronologie. Ann. Inst. Oceanogr. 52:89–101.

Fenaux, R., and G. Gorsky. 1979. Techniques d'élevage des appendiculaires. Ann. Inst. Oceanogr. 55:195–200.

Ferraris, J. D. 1978. Neurosecretion in selected Nemertina: A histological study. Zoomorphologie 91:275–287.

Figueiredo, M. J. de, and M. H. Vilela. 1972. On the artificial culture of *Nephrops norvegicus* reared from the egg. Aquaculture 1:173–180.

Fincham, A. A. 1977. Larval development of British prawns and shrimps (Crustacea: Decapoda: Natantia). 1. Laboratory methods and a review of *Palaemon* (*Paleander*) *elegans* Rathke 1837. Bull. Br. Mus. (Nat. Hist.) Zool. 32:1–28.

Fingerman, M. 1976. Animal Diversity, 2d ed. Holt, Rinehart and Winston, New York. 250 pp.

Fingerman, M., and S. W. Fingerman. 1977. Antagonistic actions of dopine and 5-hydroxytryptamine on color changes in the fiddler crab, *Uca pugilator.* Comp. Biochem. Physiol. 58C:121–127.

Fingerman, M., R. A. Krasnow, and S. W. Fingerman. 1971. Separation, assay, and properties of the distal retinal pigment light-adapting and dark-adapting hormones in the eyestalks of the prawn *Palaemonetes vulgaris*. Physiol. Zool. 44:119–128.

Fisher, W. K. 1952. The sipunculid worms of California and Baja California. Proc. U.S. Nat. Mus. 102:371–450.

Fisher, W. S., and R. T. Nelson. 1978. Application of antibiotics in the cultivation of Dungeness crab, *Cancer magister*. J. Fish. Res. Bd. Can. 35:1343–1349.

Fisher, W. S., E. H. Nilson, J. F. Steenbergen, and D. V. Lightner. 1978. Microbial diseases of cultured lobsters: A review. Aquaculture 14:115–140.

Flores, E. E. C., S. Igarashi, T. Mikami, and K. Kobayashi. 1976. Studies on squid behavior in relation to fishing. I. On the handling of squid, *Todarodes pacificus* Steenstrup, for behavioral study. Bull. Fac. Fish. Hokkaido Univ. 27:145–151.

Flores, E. E. C., S. Igarashi, and T. Mikami. 1977. Studies on squid behavior in relation to fishing. II. On the survival of squid, *Todarodes pacificus* Steenstrup, in experimental aquarium. Bull. Fac. Fish. Hokkaido Univ. 28:137–142.

Flower, N. E., A. J. Geddes, and K. M. Rudall. 1969. Ultrastructure of the fibrous protein from the egg capsules of the whelk, *Buccinum undatum*. J. Ultrastruct. Res. 26:262–273.

Foret, J. P. 1972. Étude des Effets à Long Terme de Quelques Détergents (Issus de la Petroleochemie) sur la Séquence du Développement de Deux Espèces de Polychétes Sedentaires: *Scolelepis fuliginosa* (Claparède) et *Capitella capitata* (Fabricus). Ph.D. thesis, Universite de Marseille, Luminy, France. 125 pp.

Foret-Montardo, P. 1970. Étude d'action des produits de base entrant dans la composition des détergents issus de la pétroléchimie vis-à-vis de quelques invertébrés benthique marina. Tethys 2:567–614.

Forster, J. R. M. 1973. Studies on compounded diets for prawns. Pp. 389–402 in J. W. Avault, Jr., ed. Proc. Third Annu. Meet. World Mariculture Soc. La. State Univ. Div. Continuing Educ., Baton Rouge.

Forsythe, J. W., and R. T. Hanlon. 1980. A closed marine culture system for rearing *Octopus joubini* and other large-egged benthic octopods. Lab. Anim. 14:137–142.

Foster, N. R. 1973. Occurrence of *Vallentinia gabriellae* (Hydrozoa: Olindiadidae) in coastal Yucatan, with notes on its biology and laboratory culture. Proc. Acad. Nat. Sci. Phila. 125(4):69–74.

Fotheringham, N., S. L. Brunenmeister, and P. Menefee. 1980. Beachcomber's Guide to Gulf Coast Marine Life. Gulf Publishing Co., Houston, Tex. xi + 124 pp.

Føyn, B., and I. Gjøen. 1954. Studies on the serpulid Pomatoceros triqueter L. I. Observations on the life history. Nytt Mag. Zool. (Oslo) 2:73–81.

Frank, J. R., S. D. Sulkin, and R. P. Morgan II. 1975. Biochemical changes during larval development of the xanthid crab *Rhithropanopeus harrisii*. I. Protein, total lipid, alkaline phosphatase, and glutamic oxaloacetic transaminase. Mar. Biol. (Berlin) 32:105–111.

Frank, P. W. 1965. The biodemography of an intertidal snail population. Ecology 46:831–844.

Franz, D. R. 1971. Development and metamorphosis of the gastropod *Acteocina canaliculata* (Say). Trans. Am. Microsc. Soc. 90:174–182.

Franz, D. R. 1975. Opisthobranch culture. Pp. 245–256 in W. L. Smith and M. H. Chanley, eds. Culture of Marine Invertebrate Animals. Plenum Press, New York and London.

French, K. A. 1979. Laboratory culture of embryonic and juvenile Limulus. Pp. 61–71 in E. Cohen *et al.*, eds. Biomedical Applications of the Horseshoe Crab (Limulidae). Alan R. Liss, New York.

Fridberger, A., T. Fridberger, and L.-G. Lundin. 1979. Cultivation of sea urchins of five different species under strict artificial conditions. Zoon 7:149–151.

Fulton, C. 1960. Culture of a colonial hydroid under controlled conditions. Science 132:473–474.

Fuseler, J. W. 1973. Repetitive procurement of mature gametes from individual sea stars and sea urchins. J. Cell Biol. 57:879–881.

Gabbott, P. A., D. A. Jones, and D. H. Nichols. 1976. Studies on the design and acceptability of micro-encapsulated diets for marine particle feeders. II. Bivalve molluscs. Pp. 127–141 in G. Persoone and E. Jaspers, eds. Proc. 10th Europ. Symp. Mar. Biol., vol. 1. Universa Press, Welteren, Belgium.

Gabe, S. H. 1975. Reproduction in the giant octopus of the North Pacific, *Octopus dofleini martini*. Veliger 18:146–150.

Gallagher, M. L., and W. D. Brown. 1976. Comparison of artificial and natural sea water as culture media for juvenile lobsters. Aquaculture 9:87–90.

Gallagher, M. L., D. E. Conklin, and W. D. Brown. 1976. The effects of pelletized protein diets on growth, molting and survival of juvenile lobsters. Pp. 363–378 in J. W. Avault, Jr., ed. Proc. Seventh Annu. Meet. World Mariculture Soc. La. State Univ. Div. Continuing Educ., Baton Rouge.

Galtsoff, P. S., W. W. Anderson, A. H. Banner, F. M. Bayer, E. H. Behre, P. A. Butler, F. A. Chace, Jr., A. C. Chandler, B. G. Chitwood, A. H. Clark, W. R. Coe, P. S. Conger, G. A. Cooper, C. C. Davis, E. S. Deevey, Jr., E. Deichmann, H. W. Graham, G. Gunter, O. Hartman, J. W. Hedgpeth, D. P. Henry, L. H. Hyman, R. Lasker, D. F. Leipper, M. J. Lindner, G. H. Lowery, Jr., S. A. Lynch, H. W. Manter, H. A. Marmer, H. B. Moore, R. J. Newman, R. C. Osburn, F. L. Parker, F. B. Phleger, E. L. Pierce, W. A. Price, H. A. Rehder, L. R. Rivas, G. A. Rounsefell, W. L. Schmitt, M. Sears, C. M. Shigley, W. S. Shoemaker, F. G. Walton Smith, V. Sprague, W. R. Taylor, R. F. Thorne, J. Q. Tierney, R. W. Timm, W. L. Tressler, W. G. Van Name, G. L. Voss, R. H. Williams, and C. E. Zobell. 1954. Gulf of Mexico—Its Origin, Waters, and Marine Life. U.S. Fish Wildl. Serv. Fish. Bull. No. 89. xiv + 604 pp.

Gamble, J. C. 1969. An anesthetic for *Corophium volutator* (Pallas) and *Marinogammarus obtustatus* (Dahl), Crustacea, Amphipoda. Experientia 25:539–540.

Geduldig, D., and T. Hoekman. 1979. A source of large axons for neurophysiology: The North Atlantic squid *Illex illecebrosus*. Can. J. Physiol. Pharmacol. 57:912–916.

Gentile, J. H., and S. L. Sosnowski. 1978. Methods for the culture and short term bioassay of the calanoid copepod (*Acartia tonsa*). Pp. 28–58 in Mimeographed Report in EPA–600/9–78/010, Bioassay Procedures for the Ocean Disposal Permit Program. U.S. EPA Environmental Research Laboratory, Gulf Breeze, Fla.

Gentile, J. H., S. Sosnowski, and J. Cardin. 1974. Marine zooplankton. Pp. 144–155 in G. V. Cox *et al.*, eds. Marine Bioassays Workshop Proceedings 1974. Marine Technology Society, Washington, D.C.

George, J. D. 1976. The culture of benthic polychaetes and harpacticoid copepods on agar. Pp. 143–159 in G. Persoone and E. Jaspers, eds. Proc. 10th Europ. Symp. Mar. Biol., vol. 1. Universa Press, Welteren, Belgium.

Gerin, Y. 1976. Origin and evolution of some organelles during oogenesis in the mud snail *Ilyanassa obsoleta*. I. The yoke platelets. Acta Embryol. Exp. 1:15–26.

Gerlach, S. A., and M. Schrage. 1971. Life cycles in marine meiobenthos. Experiments at various temperatures with *Monhystera disjuncta* and *Theristus pertenuis* (Nematoda). Mar. Biol. (Berlin) 9:274–280.

Gerould, J. H. 1913. The sipunculids of the eastern coast of North America. Proc. U.S. Nat. Mus. 44:373–437.

Gibson, V. R., and G. D. Grice. 1977. The developmental stages of *Labidocera aestiva* Wheeler, 1900 (Copepoda, Calanoida). Crustaceana (Leiden) 32:7–20.

Giesa, S. 1966. Die Embryonalentwicklung von *Monocelis fusca* Oersted (Turbellaria, Proseriata). Z. Morphol. Oekol. Tiere 57:137–230.

Giese, A. C., and J. S. Pearse, eds. 1974. Reproduction of Marine Invertebrates. Vol. I: Acoelomate and Pseudocoelomate Metazoans. Academic Press, New York, San Francisco, and London. xi + 546 pp.

Giese, A. C., and J. S. Pearse, eds. 1975a. Reproduction of Marine Invertebrates. Vol. II: Entoprocts and Lesser Coelomates. Academic Press, New York, San Francisco, and London. xiii + 344 pp.

Giese, A. C., and J. S. Pearse, eds. 1975b. Reproduction of Marine Invertebrates. Vol. III: Annelids and Echiurans. Academic Press, New York, San Francisco, and London. xii + 343 pp.

Giese, A. C., and J. S. Pearse, eds. 1977. Reproduction of Marine Invertebrates. Vol. IV: Molluscs: Gastropods and Cephalopods. Academic Press, New York, San Francisco, and London. xii + 369 pp.

Giese, A. C., and J. S. Pearse, eds. 1979. Reproduction of Marine Invertebrates. Vol V: Molluscs: Pelecypods and Lesser Classes. Academic Press, New York, London, Toronto, Sydney, and San Francisco. xvi + 369 pp.

Gilgan, M. W., and B. G. Burns. 1976. The successful induction of molting in the adult male lobster (*Homarus americanus*) with a slow-release form of ecdysterone. Steroids 27:571–580.

Gilgan, M. W., and B. G. Burns. 1977. On the reduced sensitivity of the adult male lobster (*Homarus americanus*) to ecdysterone at reduced temperatures. Comp. Biochem. Physiol. 58A:33–36.

Ginn, T. C., and J. M. O'Connor. 1978. Response of the estuarine amphipod *Gammarus daiberi* to chlorinated power plant effluent. Estuarine Coastal Mar. Sci. 6:459–469.

Giudice, G. 1973. Developmental Biology of the Sea Urchin Embryo. Academic Press, New York. 469 pp.

Glass, G. E. 1973. Bioassay Techniques and Environmental Chemistry. Ann Arbor Science Publishers, Ann Arbor. xi + 499 pp.

Goerke, H. 1971a. Die Ernährungsweise der *Nereis*-Arten (Polychaeta, Nereidae) der deutschen Küsten. Veroeff. Inst. Meeresforsch. Bremerhaven 13:1–50.

Goerke, H. 1971b. *Nereis fucata* (Polychaeta, Nereidae) als Kommensale von *Eupagurus bernhardus* (Crustacea, Paguridae), Entwicklung einer Population und Verhalten der Art. Veroeff. Inst. Meeresforsch. Bremerhaven 13:79–118.

Goerke, H. 1979. *Nereis virens* (Polychaeta) in marine pollution research: Culture methods and oral administration of a polychlorinated biphenyl. Veroeff. Inst. Meeresforsch. Bremerhaven 17:151–161.

Goldberg, R. 1978. Some effects of gas-supersaturated seawater in *Spisula solidissima* and *Argopecten irradians*. Aquaculture 14:281–287.

Gomot, L. 1972. The organotypic culture of invertebrates other than insects. Pp. 41–136 in C. Vago, ed. Invertebrate Tissue Culture, vol. II. Academic Press, New York and London. xiv + 415 pp.

Gontcharoff, M. 1959. Rearing of certain nemerteans (genus *Lineus*). Ann. N.Y. Acad. Sci. 77:93–95.

Gonzalez, J. G., P. P. Yevich, J. H. Gentile, and N. F. Lackie. 1975. Problems associated with culture of marine copepods. Pp. 199–207 in W. L. Smith and M. H. Chanley, eds. Culture of Marine Invertebrate Animals. Plenum Press, New York and London.

Goodbody, I. 1977. Diets and culture media for ascidians and other tunicates. Pp. 251–254 in M. Rechcigl, Jr., ed. CRC Handbook Series in Nutrition and Food. Section G: Diets, Culture Media, Food Supplements. Volume II: Food Habits of, and Diets for Invertebrates and Vertebrates—Zoo Diets. CRC Press, Cleveland.

Goodwin, L., W. Shaul, and C. Budd. 1979. Larval development of the geoduck clam (*Panope generosa,* Gould). Proc. Natl. Shellfish. Assoc. 69:73–76.

Gopalakrishnan, K. 1973. Developmental and growth studies of the euphausiid Nematoscelis difficilis (Crustacea) based on rearing. Bull. Scripps Inst. Oceanogr. Univ. Calif. 20:1–87.

Gore, R. H. 1968. The larval development of the commensal crab *Polyonx gibbesi* Haig, 1956 (Crustacea: Decapoda). Biol. Bull. (Woods Hole, Mass.) 135:111–129.

Goreau, T. F. 1961. On the relation of calcification to primary productivity in reef building organisms. Pp. 269–286 in H. M. Lenhoff and W. F. Loomis, eds. The Biology of Hydra and of Some Other Coelenterates: 1961. University of Miami Press, Coral Gables.

Gosner, K. L. 1971. Guide to Identification of Marine and Estuarine Invertebrates. Cape Hatteras to the Bay of Fundy. John Wiley & Sons, New York. xix + 693 pp.

Gosner, K. L. 1979. A Field Guide to the Atlantic Seashore. Invertebrates and Seaweeds of the Atlantic Coast from the Bay of Fundy to Cape Hatteras. Houghton Mifflin Co., Boston. xvi + 329 pp.

Goswami, S. C. 1977. Laboratory culture of a harpacticoid copepod *Laophonte setosa* (Boeck). Pp. 563–570 in Proceedings of the Symposium on Warm Water Zooplankton. Spec. Publ. Natl. Inst. Oceanogr. (Goa, India). xi + 722 pp.

Gould, M. C. 1967. Echiuroid worms: *Urechis*. Pp. 163–171 in F. H. Wilt and N. K. Wessels, eds. Methods in Developmental Biology. Thomas Y. Crowell Co., New York.

Gould-Somero, M. 1975. Echiura. Pp. 277–311 in A. C. Giese and J. S. Pearse, eds. Reproduction of Marine Invertebrates. Vol. III: Annelids and Echiurans. Academic Press, New York, San Francisco, and London.

Granade, H. R., P. C. Cheng, and N. J. Doorenbos. 1976. Ciguatera I: Brine shrimp (*Artemia salina* L.) larval assay for ciguatera toxins. J. Pharmaceut. Sci. 65:1414–1415.

Grassle, J. F., and J. P. Grassle. 1974. Opportunistic life histories and genetic systems in marine benthic polychaetes. J. Mar. Res. 32:253–284.

Grassle, J. F., and J. P. Grassle. 1975. Sibling species in the marine pollution indicator *Capitella* (Polychaeta). Science 192:567–569.

Grave, B. H. 1937a. *Bugula flabellata* and *B. turrita*. Pp. 178–179 in J. G. Needham *et al.,* eds. Culture Methods for Invertebrate Animals. Comstock Publishing Co., Ithaca, N.Y.

Grave, B. H. 1937b. *Hydroides hexagonus*. Pp. 185–187 in J. G. Needham *et al.*, eds. Culture Methods for Invertebrate Animals. Comstock Publishing Co., Ithaca, N.Y.

Grave, C. 1937. Notes on the culture of eight species of ascidians. Pp. 560–564 in J. G. Needham *et al.*, eds. Culture Methods for Invertebrate Animals. Comstock Publishing Co., Ithaca, N.Y.

Greve, W. 1968. The "planktonkreisel," a new device for culturing zooplankton. Mar. Biol. (Berlin) 1:201–203.

Greve, W. 1970. Cultivation experiments on North Sea ctenophores. Helgol. Wiss. Meeresunters. 20:304–317.

Greve, W. 1972. Ökologische Untersuchungen an *Pleurobrachia pileus*. 2. Laboratoriumsuntersuchungen. Helgol. Wiss. Meeresunters. 23:141–164.

Greve, W. 1975. The "Meteor planktonküvette": A device for the maintenance of macrozooplankton aboard ships. Aquaculture 6:77–82.

Griffiths, R. J. I. 1979. The reproductive season and larval development of the barnacle *Tetraclita serrata* Darwin. Trans. R. Soc. S. Afr. 44:97–111.

Grimpe, G. 1933. Pflege, Behandlung und Zucht der Cephalopoden für zoologische und physiologische Zwecke. Pp. 331–402 in Abderhalden's Handbuch der Biologischen Arbeitsmethoden, vol. 9, pt. 5. Urban & Schwarzenberg, Berlin and Vienna.

Groat, C. S., C. R. Thomas, and K. Schurr. 1980. Improved culture of *Aurelia aurita* scyphistomae for bioassay and research. Ohio J. Sci. 80:83–87.

Grodhaus, G., and B. Keh. 1958. The marine, dermatitis-producing cercaria of *Austrobilharzia variglandis* in California (Trematoda: Schistosomatidae). J. Parasitol. 44:633–638.

Gruffydd, Ll. D., and A. R. Beaumont. 1972. A method for rearing *Pecten maximus* larvae in the laboratory. Mar. Biol. (Berlin) 15:350–355.

Guberlet, J. E. 1937. A method for rearing *Nereis agassizi* and *N. procera*. Pp. 184–185 in J. G. Needham *et al.*, eds. Culture Methods for Invertebrate Animals. Comstock Publishing Co., Ithaca, N.Y.

Guérin, J. P. 1971. Modalités d'élevage et description des stades larvaires de *Polyophthalmus pictus* Dujardin (Annélide Polychète). Vie Milieu A 22:143–151.

Guérin, J. P. 1973. Premières données sur la longevité, le rhythme de ponte et la fécondité de *Scolelepis* cf. *fuliginosa* (polychète, spionidé) en élavage. Mar. Biol. (Berlin) 19:27–40.

Guérin, J. P., and J. P. Reys. 1978. Influence d'une temperature elevée sur le rhythme de ponte et la fécondité des populations méditerranéenes de *Scolelepis fuliginosa* (Annélide: Polychaeta) en élevage au laboratoire. Pp. 341–348 in D. S. McLusky and A. J. Berry, eds. Physiology and Behaviour of Marine Organisms. Pergamon Press, New York and Oxford.

Guillard, R. R. L. 1975. Culture of phytoplankton for feeding marine invertebrates. Pp. 29–60 in W. L. Smith and M. H. Chanley, eds. Culture of Marine Invertebrate Animals. Plenum Press, New York and London.

Guillard, R. R. L., and J. H. Ryther. 1962. Studies on marine planktonic diatoms. I. *Cyclotella nana* Hustedt and *Detonula confervacea* (Cleve) Gran. Can. J. Microbiol. 8:229–239.

Hagström, B. E., and S. Lönning. 1973. The sea urchin egg as a testing object in toxicology. Acta Pharmacol. Toxicol. 32(Suppl. 1):49 pp.

Hall, L. W., Jr., and A. L. Buikema, Jr. 1977. Rearing larval grass shrimp in the laboratory. Prog. Fish Cult. 39:129–131.

Hamabe, M. 1963. Spawning experiments of the common squid, *Ommastrephes sloani pacificus* Steenstrup, in an indoor aquarium. Bull. Jpn. Soc. Sci. Fish. 29:930–934. (in Japanese with English summary)

Hamada, T., and S. Mikami. 1977. A fundamental assumption on the habitat condition of *Nautilus* and its application to the rearing of *N. macromphalus*. Sci. Pap. Coll. Gen. Educ. Univ. Tokyo 27:31–39.

Hamond, R. 1974. The culture, experimental taxonomy, and comparative morphology of the planktonic stages of Norfolk autolytoids (Polychaeta: Syllidae: Autolytinae). Zool. J. Linn. Soc. 54:299–320.

Hanlon, R. T. 1977. Laboratory rearing of the Atlantic reef octopus, *Octopus briareus* Robson, and its potential for mariculture. Pp. 471–482 in J. W. Avault, Jr., ed. Proc. Eighth Annu. Meet. World Mariculture Soc. La. State Univ. Div. Continuing Educ., Baton Rouge.

Hanlon, R. T. 1979. Aspects of the biology of the squid *Loligo* (*Doryteuthis*) *plei* in captivity. Diss. Abstr. Int. B Sci. Eng. 39:5707.

Hanlon, R. T., R. F. Hixon, and W. H. Hulet. 1978. Laboratory maintenance of wild-caught loliginid squids. Pp. 20.1–20.13 in N. Balch, T. Amaratunga, and R. K. O'Dor, eds. Proceedings of the Workshop on the Squid *Illex illecebrosus*. Can. Fish. Mar. Serv. Tech. Rep. No. 833.

Hanlon, R. T., R. F. Hixon, W. H. Hulet, and W. T. Yang. 1979. Rearing experiments on the California market squid *Loligo opalescens* Berry, 1911. Veliger 21:428–431.

Haq, S. M. 1972. Breeding of *Euterpina acutifrons*, a harpacticid copepod, with special reference to dimorphic males. Mar. Biol. (Berlin) 15:221–235.

Hardy, B. L. S. 1978. A method for rearing sand-dwelling harpacticoid copepods in experimental conditions. J. Exp. Mar. Biol. Ecol. 34:143–149.

Harrigan, J. F., and D. L. Alkon. 1978a. Laboratory cultivation of *Haminoea solitaria* (Say, 1822) and *Elysia chlorotica* (Gould, 1870). Veliger 21:299–305.

Harrigan, J. F., and D. L. Alkon. 1978b. Larval rearing, metamorphosis, growth and reproduction of the eolid nudibranch *Hermissenda crassicornis* (Eschscholtz, 1831) (Gastropoda: Opisthobranchia). Biol. Bull. (Woods Hole, Mass.) 154: 430–439.

Harris, L. G. 1975. Studies on the life history of two coral-eating nudibranchs of the genus *Phestilla*. Biol. Bull. (Woods Hole, Mass.) 149:539–550.

Harris, R. P. 1977. Some aspects of the biology of the harpacticoid copepod, *Scottolana canadensis* (Willey), maintained in laboratory cutures. Chesapeake Sci. 18:245–252.

Harris, R. P., and G. A. Pafenhöfer. 1976. Feeding, growth, and reproduction of the marine planktonic copepod *Temora longicornis* Müller. J. Mar. Biol. Assoc. U.K. 56:675–690.

Harrison, S. E., W. R. Lillie, E. Passah, J. Loch, J. C. MacLeod, and J. F. Klaverkamp. 1975. A Nodular System for Aquatic Toxicity Studies. Can. Fish. Mar. Serv. Tech. Rep. No. 592. iv + 15 pp.

Hartline, H. K. 1969. Visual receptors and retinal interaction. Science 164:270–278.

Hartman, M. C. 1977. A mass rearing system for the culture of brachyuran crab larvae. Pp. 147–155 in J. W. Avault, Jr., ed. Proc. Eighth Annu. Meet. World Mariculture Soc. La. State Univ. Div. Continuing Educ., Baton Rouge.

Hartman, M. C., and G. R. Letterman. 1978. An evaluation of three species of diatoms as food for *Cancer magister* larvae. Pp. 271–276 in J. W. Avault, Jr., ed. Proc. Ninth Annu. Meet. World Mariculture Soc. La. State Univ. Div. Continuing Educ., Baton Rouge.

Hartman, M. C., C. E. Epifanio, G. Pruder, and R. Srna. 1974. Farming the artificial sea: Growth of clams in a recirculating seawater system. Proc. Gulf Caribb. Fish. Inst. 26:59–74.

Hartman, O. 1969. Atlas of Sedentariate Polychaetous Annelids from California. Allan Hancock Foundation, University of Southern California, Los Angeles. 828 pp.

Hartnoll, R. G., and S. M. Smith. 1978. Pair formation and the reproductive cycle in *Gammarus duebeni*. J. Nat. Hist. 12:501–511.

Harvey, E. B. 1956. The American Arbacia and Other Sea Urchins. Princeton University Press, Princeton, N.J. 298 pp.

Harwig, J., and P. M. Scott. 1971. Brine shrimp (*Artemia salina* L.) larvae as a screening system for fungal toxins. Appl. Microbiol. 21:1011–1016.

Hauenschild, C. 1954. Genetische und entwicklungsphysiologische Untersuchungen über Intersexualität und Gewebeverträglichkeit bei *Hydractinia echinata* Flemm. (Hydroz. Bougainvill.).Wilhelm Roux' Arch. Entwicklungsmech. Organ. 147:1–41.

Hauenschild, C. 1962. Die Zucht mariner Wirbelloser im Laboratorium (Methoden und Anwendung). Kieler Meeresforsch. 18 (3, Sonderheft): 28–37.

Hauenschild, C. 1970. Die Zucht von niederen marinen Wirbellosen und ihre Anwendung in der experimentellen Zoologie. Helgol. Wiss. Meeresunters. 20:249–263.

Hauenschild, C. 1972. Invertebrates. Pp. 216–235 in C. Schlieper, ed. Research Methods in Marine Biology. University of Washington Press, Seattle.

Heald, D. 1978. A successful marking method for the saucer scallop *Amusium balloti* (Bernardi) Aust. J. Mar. Freshwater Res. 29:845–851.

Heinle, D. R. 1969. Culture of calanoid copepods in synthetic sea water. J. Fish Res. Bd. Can. 26:150–153.

Heinle, D. R. 1970. Population dynamics of exploited cultures of calanoid copepods. Helgol. Wiss. Meeresunters. 20:360–372.

Heinle, D. R., R. P. Harris, J. F. Ustach, and D. A. Flemer. 1977. Detritus as food for estuarine copepods. Mar. Biol. (Berlin) 40:341–353.

Heip, C., N. Smol, and V. Absillis. 1978. Influence of temperature on the reproductive period of *Oncholaimus oxyuris* (Nematoda: Oncholaimidae). Mar. Biol. (Berlin) 45:255–260.

Heitkamp, U. 1972. Entwicklungsdauer und Lebenszyklen von *Mesostoma productum* (O. Schmidt, 1848) (Turbellaria, Neorhabdocoela). Oecologia (Berlin) 10:59–68.

Helm, M. M. 1977. Mixed algal feeding of *Ostrea edulis* larvae with *Isochrysis galbana* and *Tetraselmis suecica*. J. Mar. Biol. Assoc. U.K. 57:1019–1029.

Helm, M. M., and B. E. Spencer. 1972. The importance of the rate of aeration in hatchery cultures of the larvae of *Ostrea edulis* L. J. Cons. Cons. Int. Explor. Mer 34:244–255.

Henderson, J. A., and J. S. Lucas. 1971. Larval development and metamorphosis of *Acanthaster planci* (Asteroidea). Nature (Lond.) 232:655–657.

Heron, A. C. 1972. Population ecology of a colonizing species: The pelagic tunicate *Thalia democratica*. I. Individual growth rate and generation time. Oecologia (Berlin) 10:269–293.

Herpin, R. 1926. Recherches biologiques sur la reproduction et le développement de quelques annélides polychétes. Soc. Sci. Nat. Ouest Fr. Bull. Sêr. 4. No. 5:1–250.

Herring, J. L. 1961. The genus *Halobates* (Hemiptera: Gerridae). Pac. Insects 3:223–305.

Herrnkind, W. F., and R. McLean. 1971. Field studies of homing, mass emigration, and orientation of the spiny lobster, *Panulirus argus*. Pp. 359–377 in H. E. Adler, ed. Orientation: Sensory Basis. Ann. N.Y. Acad. Sci. 188:1–408.

Hessinger, D. A., and H. M. Lenhoff. 1976. Mechanism of hemolysis induced by nematocyst venom: Roles of phospholipase A and direct lytic factor. Arch. Biochem. Biophys. 173:603–613.

Hessinger, D. A., H. M. Lenhoff, and L. Kahan. 1973. Haemolytic, phospholipase A and nerve-affecting activities of sea anemone nematocyst venom. Nature New Biol. 241:125–127.

Heusner, A. A., and J. T. Enright. 1966. Long-term activity recording in small aquatic animals. Science 154:532–533.

Hidu, H. 1975. Culture of American and European oysters. Pp. 283–295 in W. L. Smith and M. H. Chanley, eds. Culture of Marine Invertebrate Animals. Plenum Press, New York and London.

Hill, R. W. 1976. Comparative Physiology of Animals: An Environmental Approach. Harper & Row, New York. 672 pp.

Hill, S. D., and J. N. Cather. 1969. A simple method for the laboratory culture of the marine coelenterate *Cassiopeia*. Turtox News 47:259–260.

Hillman, R. E. 1978. Invertebrate mucus: Model systems for studying diseases in man. Pp. 17–25 in P. N. Kaul and C. J. Sindermann, eds. Drugs and Food from the Sea. Myth or Reality? University of Oklahoma, Norman. xi + 448 pp.

Hinegardner, R. T. 1967. Echinoderms. Pp. 139–155 in F. H. Wilt and N. K. Wessells, eds. Methods in Developmental Biology. Thomas Y. Crowell Co., New York.

Hinegardner, R. T. 1969. Growth and development of the laboratory cultured sea urchin. Biol. Bull. (Woods Hole, Mass.) 137:465–475.

Hinegardner, R. T. 1975a. Morphology and genetics of sea urchin development. Am. Zool. 15:679–689.

Hinegardner, R. T. 1975b. Care and handling of sea urchin eggs, embryos, and adults (principally North American species). Pp. 10–25 in G. Czihak and R. Peter, eds. The

Sea Urchin Embryo, Biochemistry and Morphogenesis. Springer-Verlag, Berlin, Heidelberg, and New York.

Hino, A., and R. Hirano. 1976. Ecological studies on the mechanism of bisexual reproduction in the rotifer *Brachionus plicatilis*. I. General aspects of bisexual reproduction inducing factors. Bull. Jpn. Soc. Sci. Fish. 42:1093–1099.

Hino, A., and R. Hirano. 1977. Ecological studies on the mechanism of bisexual reproduction in the rotifer *Brachionus plicatilis*. II. Effects of cumulative parthenogenetic generation on the frequency of bisexual reproduction. Bull. Jpn. Soc. Sci. Fish. 43:1147–1155.

Hinton, S. 1969. Seashore Life of Southern California an Introduction to the Animal Life of California Beaches South of Santa Barbara. Calif. Nat. Hist. Guides No. 26. University of California Press, Berkeley and Los Angeles. 181 pp.

Hirano, R. 1962. Mass rearing of barnacle larvae. Bull. Mar. Biol. Stn. Asamushi 11:77–80.

Hirata, H. 1974. An attempt to apply an experimental microcosm for the mass culture of marine rotifer, *Brachionus plicatilis* Müller. Mem. Fac. Fish. Kagoshima Univ. 23:163–172.

Hirata, H. 1975. An introduction to the rearing methods of prawn, *Penaeus japonicus* Bate, in Japan. Mem. Fac. Fish. Kagoshima Univ. 24:7–12.

Hirata, H., M. Marchiori, and A. Shinomiya. 1978. Rearing of prawn *Penaeus japonicus* with reference to ecological succession. Mem. Fac. Fish. Kagoshima Univ. 27:295–303.

Hirayama, K. 1966. Influence of nitrate accumulated in culturing water on *Octopus vulgaris*. Bull. Jpn. Soc. Sci. Fish. 32:105–111. (in Japanese with English summary)

Hirota, J. 1972. Laboratory culture and metabolism of the planktonic ctenophore, *Pleurobrachia bachei* A. Agassiz. Pp. 465–484 in A. Y. Takenouti *et al.*, eds. Biological Oceanography of the North Pacific Ocean. Idemitsu Shoten, Tokyo. xiii + 626 pp.

Hirota, J. 1974. Quantitative natural history of *Pleurobrachia bachei* in La Jolla bight. Natl. Oceanic Atmos. Adm. Fish. Bull. 72:295–335.

Hofmann, D. K., R. Neumann, and K. Henne. 1978. Strobilation, budding and initiation of scyphistoma morphogenesis in the rhizostome *Cassiopea andromeda* (Cnidaria: Scyphozoa). Mar. Biol. (Berlin) 47:161–176.

Holland, C. A., and D. M. Skinner. 1976. Interactions between molting and regeneration in the land crab. Biol. Bull. (Woods Hole, Mass.) 150:222–240.

Hopkins, S. H. 1944. The external morphology of the third and fourth zoeal stages of the blue crab, *Callinectes sapidus* Rathbun. Biol. Bull. (Woods Hole, Mass.) 87:145–152.

Hoppenheit, M. 1975. Zur Dynamik exploitierter Populationen von *Tisbe holothuriae* (Copepoda, Harpacticoida). I. Methoden, Verlauf der Populationsentwicklung und Einfluss der Wassererneuerung. Helgol. Wiss. Meeresunters. 27:235–253.

Hopper, B. E., and R. C. Cefalù. 1973. Free-living marine nematodes from Biscayne Bay, Florida. VII. Enoplidae: *Enoplus* species in Biscayne Bay with observations on culture and bionomics of *E. paralittoralis* Wieser, 1953. Proc. Helminthol. Soc. Wash. 40:275–280.

Hopper, B. E., and S. P. Meyers. 1966. Observations on the bionomics of the marine nematode, *Metoncholaimus* sp. Nature (Lond.) 209:899–900.

Horstadius, S. 1973. Experimental Embryology of Echinoderms. Oxford University Press, London. 192 pp.

Hudinaga, M. 1942. Reproduction, development and rearing of *Penaeus japonicus* Bate. Jpn. J. Zool. 10:305–393.

Hughes, J. T., J. J. Sullivan, and R. Shleser. 1972. Enhancement of lobster growth. Science 177:1110–1111.

Hughes, J. T., R. A. Shleser, and G. Tchobanoglous. 1974. A rearing tank for lobster larvae and other aquatic species. Prog. Fish Cult. 36:129–132.

Hughes, J. T. *et al.* 1975. Lobster culture. Pp. 221–227 in W. L. Smith and M. H. Chanley, eds. Culture of Marine Invertebrate Animals. Plenum Press, New York and London.

Huguenin, J. E. 1976. Heat exchangers for use in the culturing of marine organisms. Chesapeake Sci. 17:61–64.

Hulet, W. H., M. R. Villoch, R. F. Hixon, and R. T. Hanlon. 1979. Fin damage in captured and reared squids. Lab. Anim. Sci. 29:528–533.

Hunt, S. 1966. Carbohydrate and amino acid composition of the egg capsule of the whelk *Buccinum undatum.* Nature (Lond.) 210:436.

Hurley, A. C. 1976. Feeding behavior, food consumption, growth, and respiration of the squid *Loligo opalescens* raised in the laboratory. U.S. Natl. Mar. Fish. Serv. Fish. Bull. 74:176–182.

Hurst, A. 1967. The egg masses and veligers of thirty Northeast Pacific optisthobranchs. Veliger 9:255–288.

Hyman, L. H. 1940. The Invertebrates. Vol. I: Protozoa through Ctenophora. McGraw-Hill, New York and London. x + 726 pp.

Hyman, L. H. 1951. The Invertebrates. Vol. II: Platyhelminthes and Rhynchocoela. The Acoelomate Bilateria. McGraw-Hill, New York and London. vii + 550 pp.

Hyman, L. H. 1959. The Invertebrates. Vol V: Smaller Coelomate Groups, Chaetognatha, Hemichordata, Pogonophora, Phoronida, Ectoprocta, Brachiopoda, Sipunculida. The Coelomate Bilateria. McGraw-Hill, New York and London. vii + 783 pp.

Ide, H. 1973. Effects of ACTH on melanophores and iridophores isolated from bullfrog tadpoles. Gen. Comp. Endocrinol. 21:390–397.

Ikegami, S., N. Honji, and M. Yoshida. 1978. Light-controlled production of spawning-inducing substance in jellyfish ovary. Nature (Lond.) 272:611–612.

Ikegami, S., K. Kawada, Y. Kimura, and A. Suzuki. 1979. Rapid and convenient procedure for the detection of inhibitors at DNA-synthesis using starfish oocytes and sea-urchin embryos. Agr. Biol. Chem. 43:161–166.

Ingle, R. W., and P. F. Clark. 1977. A laboratory module for rearing crab larvae. Crustaceana (Leiden) 32:220–222.

Itami, K., Y. Izawa, S. Maeda, and K. Nakai. 1963. Notes on the laboratory culture of the octopus larvae. Bull. Jpn. Soc. Sci. Fish. 29:514–520. (in Japanese with English summary)

Ivanovici, A. M. 1979. Adenylate energy charge: Potential value as a tool for rapid determination of toxicity effects. Pp. 241–255 in P. T. S. Wong, P. V. Hodson, A. J. Niimi, V. Cairns, and U. Borgmann, eds. Proceedings of the Fifth Annual Toxicity Workshop November 7–9, 1978, Hamilton, Ontario. Can. Fish. Mar. Serv. Tech. Rep. 862.

Ivker, F. B. 1972. A hierarchy of histo-incompatibility in *Hydractinia echinata.* Biol. Bull. (Woods Hole, Mass.) 143:162–174.

Iwanoff, P. P. 1933. Die embryonale Entwicklung von *Limulus molluccanus.* Zool. Jahrb. Anat. Ont. 56:163–348.

Jahan-Parwar, B. 1972. Behavioral and electrophysiological studies on chemoreception in *Aplysia.* Am. Zool. 12:529–537.

Jebram, D. 1977a. Experimental techniques and culture methods. Pp. 273–306 in R. M. Woollacott and R. L. Zimmer, eds. Biology of Bryozoans. Academic Press, New York, San Francisco, and London.

Jebram, D. 1977b. Culture media and diets for Bryozoa. Pp. 77–92 in M. Rechcigl, Jr., ed. CRC Handbook Series in Nutrition and Food. Section G: Diets, Culture Media,

Food Supplements. Volume II: Food Habits of, and Diets for Invertebrates and Vertebrates—Zoo Diets. CRC Press, Inc., Cleveland.
Jebram, D. 1980. Laboratory diets and qualitative nutritional requirements for bryozoans. Zool. Anz. 205:333–344.
Jefferts, K. B., P. K. Bergman, and H. F. Fiscus. 1963. A coded wire identification system for macroorganisms. Nature 198:460–462.
Jegla, T. C., and J. D. Costlow. 1979. The *Limulus* bioassay for ecdysteroids. Biol. Bull. (Woods Hole, Mass.) 156:103–114.
Jegla, T. C., J. D. Costlow, and J. Alspaugh. 1972. Effects of edysones and some synthetic analogs on horseshoe crab larvae. Gen. Comp. Endocrinol. 19:159–166.
Jenner, C. E. 1956a. The timing of reproductive cessation in geographically separated populations of *Nassarius obsoletus*. Biol. Bull. (Woods Hole, Mass.) 111:292.
Jenner, C. E. 1956b. A striking behavioral change leading to the formation of extensive aggregations in a population of *Nassarius obsoletus*. Biol. Bull. (Woods Hole, Mass.) 111:291–292.
Jenner, C. E. 1956c. The occurrence of a crystalline style in the marine snail, *Nassarius obsoletus*. Biol. Bull. (Woods Hole, Mass.) 111:304.
Jenner, C. E. 1957. Schooling behavior in mud snails in Barnstable Harbor leading to the formation of massive aggregations at the completion of seasonal reproduction. Biol. Bull. (Woods Hole, Mass.) 113:328–329.
Jenner, C. E. 1958. An attempted analysis of schooling behavior in the marine snail *Nassarius obsoletus*. Biol. Bull. (Woods Hole, Mass.) 115:337–338.
Jenner, C. E. 1959. Aggregation and schooling in the marine snail, *Nassarius obsoletus*. Biol. Bull. (Woods Hole, Mass.) 117:397.
Jenner, M. G., and C. E. Jenner. 1979. Pseudohermaphroditism in *Ilyanassa obseleta* (Mullusca: Neogastropoda). Science 205:1407–1409.
Johnson, L. L., and J. M. Shick. 1977. Effects of fluctuating temperature and immersion on asexual reproduction in the intertidal sea anemone *Haliplanella luciae* (Verrill) in laboratory culture. J. Exp. Mar. Biol. Ecol. 28:141–149.
Johnson, P. T. 1968. An Annotated Bibliography of Pathology in Invertebrates Other Than Insects. Burgess Publishing Co., Minneapolis, Minn. xiii + 322 pp.
Johnson, S. K. 1978. Handbook of Shrimp Diseases. Sea Grant College Program, Texas Agricultural Extension Service, Texas A&M University, College Station. (TAMU–SG–75–603. Rev.) 23 pp.
Joll, L. M. 1976. Mating, egg-laying and hatching of *Octopus tetricus* (Mollusca: Cephalopoda) in the laboratory. Mar. Biol. (Berlin) 36:326–333.
Jones, D. A., J. G. Munford, and P. A. Gabbott. 1974. Microcapsules as artificial food particles for aquatic filter feeders. Nature (London) 247:233–235.
Jones, D. A., T. H. Möller, R. J. Campbell, J. G. Munford, and P. A. Gabbott. 1976. Studies on the design and acceptability of micro-encapsulated diets for marine particle feeders. I. Crustacea. Pp. 229–239 in G. Persoone and E. Jaspers, eds. Proc. 10th Europ. Symp. Mar. Biol., vol. 1. Universa Press, Welteren, Belgium.
Jones, D. A., A. Kanazawa, and S. Abdel Rahman. 1979. Studies on the presentation of artificial diets for rearing the larvae of *Penaeus japonicus* Bate. Aquaculture 17:33–43.
Jørgensen, C. B. 1976. August Pütter, August Krogh, and modern ideas on the use of dissolved organic matter in aquatic environments. Biol. Rev. Camb. Philos. Soc. 51:291–328.
Jørgensen, N. O. G. 1979. Uptake of L-valine and other amino acids by the polychaete *Nereis virens*. Mar. Biol. (Berlin) 52:45–52.
Josephson, R. K., and W. E. Schwab. 1979. Unpublished observations.

Joyce, E. A., Jr. 1972. A Partial Bibliography of Oysters, with Annotations. Florida Department of Natural Resources, Special Scientific Report No. 34. 846 pp.

Kakinuma, Y. 1967. Development of a scyphozoan, *Dactylometra pacifica* Goette. Bull. Mar. Biol. Stn. Asamushi 13:29–33.

Kalyanasundaram, N., and S. S. Ganti. 1975. Rearing of barnacle larvae in the laboratory. Bull. Dep. Mar. Sci. Univ. Cochin 7:761–767.

Kalyanasundaram, N., and S. S. Ganti. 1976. Some factors influencing the settlement of barnacle larvae in the laboratory. J. Mar. Biol. Assoc. India 16:455–461.

Kamemoto, F. I. 1976. Neuroendocrinology of osmoregulation in decapod Crustacea. Am. Zool. 16:141–150.

Kanatani, H. 1969. Induction of spawning and oocyte maturation by L-methyladenine in starfishes. Exp. Cell Res. 57:333–337.

Kanatani, H. 1973. Maturation-inducing substance in starfishes. Int. Rev. Cytol. 35:253–298.

Kanatani, H., and H. Shirai. 1971. Chemical structural requirements for induction of oocyte maturation and spawning in starfishes. Dev. Growth Differ. 13:53–64.

Kandel, E. R. 1979. Behavioral Biology of *Aplysia.* A Contribution to the Comparative Study of Opisthobranch Molluscs. W. H. Freeman and Company, San Francisco. xiii + 463 pp.

Kanie, Y., S. Mikami, T. Yamada, H. Hirano, and T. Hamada. 1979. Shell growth of *Nautilus macromphalus* in captivity. Venus Jpn. J. Malacol. 38:129–134. (in Japanese with English summary)

Kan-no, H. 1976. Recent advances in abalone culture in Japan. Pp. 195–211 in K. S. Price, Jr., W. N. Shaw, and K. S. Danberg, eds. Proc. First Int. Conf. Aquaculture Nutr. Coll. Mar. Stud. Univ. Del., Newark.

Kaplan, H. M. 1969. Anesthesia in invertebrates. Fed. Proc. 28:1557–1569.

Karande, A. A., and S. S. Pendsey. 1969. Field and laboratory observations on *Teredo furcifera* M., a test organism for the bioassessment of toxic compounds. Proc. Indian Acad. Sci. Sect. B 70:223–231.

Karande, A. A., and M. K. Thomas. 1971. Laboratory rearing of *Balanus amphitrite communis* (D.). Curr. Sci. (Bangalore) 40:109–110.

Karbe, L. 1972. Marine Hydroiden als Testorganismen zur Prüfung der Toxizität von Abwasserstoffen. Die Wirkung von Schwermetallen auf Kolonien von *Eirene viridula.* Mar. Biol. (Berlin) 12:316–328.

Karnofsky, D. A., and E. B. Simmel. 1963. Effects of growth-inhibiting chemicals on the sand-dollar embryo, *Echinarachnius parma.* Prog. Exp. Tumor Res. 3:254–295.

Karp, R. D., and W. H. Hildemann. 1976. Specific allograft reactivity in the sea star *Dermasterias imbricata.* Transplantation 22:434–439.

Kato, K. 1940. On the development of some Japanese polyclads. Jpn. J. Zool. 8:537–573.

Katona, S. K. 1970. Growth characteristics of the copepods *Eurytemora affinis* and *E. herdmani* in laboratory cultures. Helgol. Wiss. Meeresunters. 20:373–384.

Katona, S. K., and C. F. Moodie. 1969. Breeding of *Pseudocalanus elongatus* in the laboratory. J. Mar. Biol. Assoc. U.K. 49:743–747.

Kawamoto, N. 1978. Progress report of the study of *Nautilus macromphalus* in captivity. Proc. Jpn. Acad. Ser. B 54:87–91.

Kay, D. G., and A. E. Brafield. 1973. The energy relations of the polychaete *Neanthes* (=*Nereis*) *virens* (Sars). J. Anim. Ecol. 42:673–692.

Kay, E. A. 1979. Hawaiian Marine Shells. Reef and Shore Fauna of Hawaii. Section 4: Mollusca. Bernice P. Bishop Mus. Spec. Publ. 64 (4). xviii + 653 pp.

Keller, R. 1977. Comparative electrophoretic studies of crustacean neurosecretary hyperglycemia and melanophore-stimulating hormones from isolated sinus glands. J. Comp. Physiol. 122:359–373.

Kempf, S. C., and A. O. D. Willows. 1977. Laboratory culture of the nudibranch *Tritonia diomedea* Bergh (Tritoniidae: Opisthobranchia) and some aspects of its behavioral development. J. Exp. Mar. Biol. Ecol. 30:261–276.

Ketchum, B. H., and A. C. Redfield. 1938. A method for maintaining a continuous supply of marine diatoms by culture. Biol. Bull. (Woods Hole, Mass.) 75:165–169.

Kidder, G. M., A. J. Clark, and P. A. Gerdes. 1977. Polyadenylic acid in *Ilyanassa*. Localization and size distribution of newly-synthesized Poly(A) in embryonic and larval stages. Differentiation 9:77–84.

Kikuchi, S., and N. Uki. 1974a. Technical study on artificial spawning of abalone, genus *Haliotis*. I. Relation between water temperature and advancing sexual maturity of *Haliotis discus hannai* Ino. Bull. Tohoku Reg. Fish. Res. Lab. 49:69–78. (in Japanese with English summary)

Kikuchi, S., and N. Uki. 1974b. Technical study on artificial spawning of abalone, genus *Haliotis*. II. Effect of irradiated sea water, with ultraviolet rays, on inducing spawning. Bull. Tohoku Reg. Fish. Res. Lab. 49:79–86. (in Japanese with English summary)

Kimeldorf, D. J., and R. W. Fortner. 1971. The prompt detection of ionizing radiations by a marine coelenterate. Radiat. Res. 46:52–63.

King, J. M., and S. Spotte. 1974. Marine Aquariums in the Research Laboratory. Aquarium Systems, Inc., Eastlake, Ohio. 39 pp.

Kinghorn, A. D., K. K. Harjes, and N. J. Doorenbos. 1977. Screening procedure for phorbol esters using brine shrimp (*Artemia salina*) larvae. J. Pharmaceut. Sci. 66:1362–1363.

Kinghorn, A. D., F. H. Jawad, and N. J. Doorenbos. 1978. Structure-activity relationship of grayanotoxin derivatives using a tetrodotoxin-antagonized spasmotic response of brine shrimp larvae (*Artemia salina*). Toxicon 16:227–234.

Kingsley, J. S. 1892. The embryology of *Limulus*. J. Morphol. 7:35–68.

Kingsley, J. S. 1893. The embryology of *Limulus*, Part II. J. Morphol. 8:195–268.

Kinne, O. 1976. Cultivation of marine organisms: Water-quality management and technology. Pp. 19–300 in O. Kinne, ed. Marine Ecology a Comprehensive, Integrated Treatise on Life in Oceans and Coastal Waters. Vol. III: Cultivation. John Wiley & Sons, London, Chichester, New York, Sydney, Brisbane, and Toronto.

Kinne, O., ed. 1976–1977. Marine Ecology a Comprehensive, Integrated Treatise on Life in Oceans and Coastal Waters. Vol. III: Cultivation. John Wiley & Sons, London, Chichester, New York, Sydney, Brisbane, and Toronto. Part 1, xiii + 577 pp. Part 2, xv + 579–1293 pp. Part 3, ix + 1295–1521 pp.

Kinne, O. 1977. Cultivation of animals. Research cultivation. Pp. 579–1293 in O. Kinne, ed. Marine Ecology a Comprehensive, Integrated Treatise on Life in Oceans and Coastal Waters. Vol. III: Cultivation. John Wiley & Sons, London, Chichester, New York, Sydney, Brisbane, and Toronto.

Kinne, O., ed. 1980. Diseases of Marine Animals. Vol. I: General Aspects. Protozoa to Gastropoda. John Wiley & Sons, Chichester, New York, Brisbane, and Toronto. xiii + 466 pp.

Kinne, O., and H. P. Bulnheim, eds. 1970. International symposium "Cultivation of marine organisms and its importance for marine biology." Helgol. Wiss. Meeresunters. 20:1–721.

Kinne, O., and G. A. Paffenhöfer. 1965. Hydranth structure and digestion rate as a function of temperature and salinity in *Clava multicornis* (Cnidaria, Hydrozoa). Helgol. Wiss. Meeresunters. 12:329–341.

Kishinouye, K. 1892. On the development of *Limulus longispina*. J. Coll. Sci. Imp. Univ. Tokyo 5:53–100.

Kittaka, J. 1976. Food and growth of penaeid shrimp. Pp. 249–285 in K. S. Price, Jr., W. N. Shaw, and K. S. Danberg, eds. Proc. First Int. Conf. Aquaculture Nutr. Coll. Mar. Stud. Univ. Del., Newark.

Kleinholz, L. H. 1976. Crustacean neurosecretory hormones and physiological specificity. Am. Zool. 16:151–166.

Knowlton, R. E. 1973. Larval development of the snapping shrimp *Alpheus heterochaelis* Say, reared in the laboratory. J. Nat. Hist. 7:273–306.

Knudsen, J. W. 1966. Biological Techniques. Harper & Row, New York. 525 pp.

Kobayashi, N. 1971. Fertilized sea urchin eggs as an indicatory material for marine pollution bioassay, preliminary experiments. Publ. Seto Mar. Biol. Lab. 18:379–406, 421–424.

Kobayashi, N. 1974. Marine pollution bioassay by sea urchin eggs, an attempt to enhance accuracy. Publ. Seto Mar. Biol. Lab. 21:377–391, 411–432.

Kobayashi, N. 1977. Preliminary experiments with sea urchin pluteus and metamorphosis in marine pollution bioassay. Publ. Seto Mar. Biol. Lab. 24:9–21.

Koike, Y. 1978. Biological and ecological studies on the propagation of the ormer, *Haliotis tuberculata* Linnaeus. I. Larval development and growth of juveniles. Mer (Tokyo) 16:124–145.

Komaki, Y. 1966. Technical notes on keeping euphausiids live in the laboratory, with a review of experimental studies on euphausiids. Inf. Bull. Planktonol. Jpn. No. 13:95–105.

Korschelt, E. 1931. Art und Dauer der ungeschlecht lichen Protpflanzung bei *Ctenodrilus*. Zool. Ang. Leipzig 93:227–238.

Kozloff, E. N. 1969. Monoxenic cultivation of an acoel turbellarian, *Parotocelis luteola* Kozloff. J. Exp. Mar. Biol. Ecol. 3:224–230.

Kozloff, E. N. 1973. Seashore Life of Puget Sound, the Strait of Georgia, and the San Juan Archipelago. University of Washington Press, Seattle. 282 pp.

Kriegstein, A. R. 1977a. Stages in the posthatching development of *Aplysia californica*. J. Exp. Zool. 199:275–288.

Kriegstein, A. R. 1977b. Development of the nervous system of *Aplysia californica*. Proc. Natl. Acad. Sci. 74:375–378.

Kriegstein, A. R., V. Castellucci, and E. R. Kandel. 1974. Metamorphosis of *Aplysia californica* in laboratory culture. Proc. Natl. Acad. Sci. USA 71:3654–3658.

Kropach, C. 1979. Observations on the potential of *Limulus* aquaculture in Israel. Pp. 103–106 in E. Cohen, ed. Biomedical Applications of the Horseshoe Crab (Limulidae). Alan R. Liss, New York.

Kuenen, D. J. 1939. Systematical and physiological notes on the brine shrimp, *Artemia*. Arch. Neerl. Zool. 3:365–449.

Kukinuma, Y. 1966. Life cycle of a hydrozoan, *Sarsia tubulosa* (Sars.). Bull. Mar. Biol. Stn. Asamushi Jpn. 12:207–210.

Kumé, M., and K. Dan. 1968. Invertebrate Embryology. NOLIT Publishing House, Belgrade. xvi + 605 pp. (translation of Musekitsui Dobutsu Hasseigaku, 1957)

Kupfermann, I. 1967. Stimulation of egg-laying: Possible neuroendocrine function of bag cells of abdominal ganglion of *Aplysia californica*. Nature 216:814–815.

Kupfermann, I. 1970. Stimulation of egg laying by extracts of neuroendocrine cells (bag cells) of abdominal ganglion of *Aplysia*. J. Neurophysiol. 33:877–881.

Kupfermann, I. 1972. Studies of the neurosecretory control of egg laying in *Aplysia*. Am. Zool. 12:513–519.

La Barbera, M. 1975. Larval and post-larval development of the giant clams *Tridacna maxima* and *Tridacna squamosa* (Bivalvia: Tridacnidae). Malacologia 15:69–79.

Laird, C. E., and P. A. Haefner, Jr. 1976. Effects of intrinsic and environmental factors on oxygen consumption in the blue crab, *Callinectes sapidus* Rathbun. J. Exp. Mar. Biol. Ecol. 2:171–178.

Lalli, C. M., and R. J. Conover. 1973. Reproduction and development of *Paedocline doliiformis,* and a comparison with *Clione limacina* (Opisthobranchia: Gymnosomata). Mar. Biol. (Berlin) 19:13–22.

Lambert, D. T. 1977. An investigation into the events following hormonal stimulation of pigment granule translocation in chromatophores of the decapod crustaceans *Uca pugilator* and *Palaemonetes pugio.* Diss. Abstr. Int. B Sci. Eng. 38:1591.

Landau, M., and A. D'Agostino. 1977. Enhancement of laboratory cultures of the barnacle *Balanus eburneus* Gould using antibiotics. Crustaceana (Leiden) 33:223–224.

Landau, M., and A. D'Agostino. 1978. Culture of the barnacle *Balanus eburneus* Gould in artificial seawaters. Crustaceana (Leiden) 34:315–318.

Landau, M., C. M. Finney, and A. D'Agostino. 1979. The barnacle, *Balanus eburneus* Gould (Cirripedia) conditioned to spawn in the laboratory. Crustaceana (Leiden) 37:241–246.

Landers, W. S. 1976. Reproduction and early development of the ocean quahog, *Arctica islandica,* in the laboratory. Nautilus 90:88–92.

Lang, K. 1948. Contribution to the ecology of *Priapulus caudatus* Lam. Ark. Zool. 41A(5):1–12.

LaRoche, G., R. Eisler, and C. M. Tarzwell. 1970. Bioassay procedures for oil and oil dispersant toxicity evaluation. J. Water Pollut. Control Fed. 42:1982–1989.

LaRoe, E. T. 1971. The culture and maintenance of the loliginid squids *Sepioteuthis sepioidea* and *Doryteuthis plei.* Mar. Biol. (Berlin) 9:9–25.

LaRoe, E. T. 1973. Laboratory culture of squid. Fed. Proc. 32:2212–2214.

Lasker, R., and G. H. Theilacker. 1965. Maintenance of euphausiid shrimps in the laboratory. Limnol. Oceanogr. 10:287–288.

Latigan, M. J. 1976. Some aspects of the breeding biology of *Charonia lampas pustulata* (Euthyme, 1889) and *Mayena australasia gemmifera* (Euthyme, 1889) under aquarium conditions (Gastropoda: Prosobranchiata). Ann. Cape Prov. Mus. Nat. Hist. 11:47–55.

Laughlin, R. 1979. Water, sodium and chloride fluxes in the early developmental stages of the horseshoe crab, *Limulus polyphemus.* Am. Zool. 19:958. (Abstr.)

Laughlin, R. B., Jr., L. G. L. Young, and J. M. Neff. 1978. A long-term study of the effects of water-soluble fractions of No. 2 fuel oil on the survival, development rate, and growth of the mud crab *Rhithropanopeus harrisii.* Mar. Biol. (Berlin) 47:87–95.

Lawler, A. R., and S. L. Shepard. 1978. Procedures for eradication of hydrozoan pests in closed-system mysid culture. Gulf Res. Rep. 6:177–178.

Lawson, T. J., and G. D. Grice. 1970. The developmental stages of *Centropages typicus* Krøyer (Copepoda, Calanoida). Crustaceana (Leiden) 18:187–208.

Leahy, P. S., T. C. Tutschulte, R. J. Britten, and E. H. Davidson. 1978. A large-scale laboratory maintenance system for gravid purple sea urchins (*Strongylocentrotus purpuratus*). J. Exp. Zool. 204:369–380.

Lebour, V. 1927. Studies of the Plymouth Brachyura. I. The rearing of crabs in captivity with a description of the larval stages of *Inachus dorsettensis, Macropodia longirostris* and *Maia squinado.* J. Mar. Biol. Assoc. U.K. 14:795–821.

Lebour, M. V. 1928. The larval stages of Plymouth Brachyura. Proc. Zool. Soc. Lond. 1928:473–560.

Le Douarin, N. 1971. Organ culture methods. Pp. 41–114 in C. Vago, ed. Invertebrate Tissue Culture, vol. I. Academic Press, New York and London. xiv + 441 pp.

Lee, J. J., and W. A. Muller. 1975. Culture of salt marsh microorganisms and micrometazoa. Pp. 87–107 in W. L. Smith and M. H. Chanley, eds. Culture of Marine Invertebrate Animals. Plenum Press, New York and London.

Lee, J. J., J. H. Tietjen, R. J. Stone, W. A. Muller, J. Rullman, and M. McEnery. 1970. The cultivation and physiological ecology of members of salt marsh epiphytic communities. Helgol. Wiss. Meeresunters. 20:136–156.

Lee, J. J., J. H. Tietjen, and J. R. Garrison. 1976. Seasonal switching in the nutritional requirements of *Nitocra typica,* a harpacticoid copepod from salt marsh aufwuchs communities. Trans. Am. Microsc. Soc. 95:628–637.

Lee, W. Y. 1977. Some laboratory cultured crustaceans for marine pollution studies. Mar. Pollut. Bull. 8:258–259.

Lee, W. Y., and J. A. C. Nicol. 1978. The effect of naphthalene on survival and activity of the amphipod *Parhyale.* Bull. Environ. Contam. Toxicol. 20:233–240.

Leibovitz, L. 1978. Shellfish diseases. NOAA (Natl. Ocean. Atmos. Adm.) Mar. Fish. Rev. 40(3):61–64.

Leibovitz, L., T. R. Meyers, R. Elston, and P. Chanley. 1977. Necrotic exfoliative dermatitis of captive squid (*Loligo pealei*). J. Invertebr. Pathol. 30:369–376.

Leighton, D. L. 1972. Laboratory observations on the early growth of the abalone, *Haliotis sorenseni,* and the effect of temperature on larval development and settling success. U.S. Natl. Mar. Fish. Serv. Fish. Bull. 70:373–381.

Lenhoff, H. M. 1971. Principles of coelenterate culture methods. Pp. 9–15 in H. M. Lenhoff, L. Muscatine, and L. V. Davis, eds. Experimental Coelenterate Biology. University of Hawaii Press, Honolulu.

Lenhoff, H. M., and R. D. Brown. 1970. Mass culture in hydra: Improved method and applications to other invertebrates. Lab. Anim. 4:139–154.

Lenhoff, H. M., and W. Heagy. 1977. Aquatic invertebrates: Model systems for studying the receptor-activation and evolution of receptor proteins Annu. Rev. Pharm. Toxicol. 16:243–258.

Le Pennec, M., and D. Prieur. 1977. Les antibiotiques dans les élevages de larves de bivalves marins. Aquaculture 12:15–30.

Le Roux, A. 1973. Observations sur le développement larvaire de *Nyctiphanes couchii* (Crustacea: Euphausiacea) au laboratoire. Mar. Biol. (Berlin) 22:159–166.

Le Roux, S., and A. Lucas. 1978. Techniques d'étude des effets des hydrocarbures sur la physiologie des larves de moules. Rev. Int. Oceanogr. Med. 49:75–79.

Levine, G., and T. L. Meade. 1976. The effects of disease treatment on nitrification in closed system aquaculture. Pp. 483–493 in J. W. Avault, Jr., ed. Proc. Seventh Annu. Meet. World Mariculture Soc. La. State Univ. Div. Continuing Educ., Baton Rouge.

Lewis, A. G. 1967. An enrichment solution for culturing the early developmental stages of the planktonic copepod *Euchaeta japonica* Marukawa. Limnol. Oceanogr. 12:147–148.

Lewis, C. A. 1975. Some observations on factors affecting embryonic and larval growth of *Pollicipes polymerus* (Cirripedia: Lepadomorpha) *in vitro.* Mar. Biol. (Berlin) 32:127–139.

Lewis, E. G., and P. A. Haefner, Jr. 1976. Oxygen consumption of the blue crab, *Callinectes sapidus* Rathbun, from proecdysis to postecdysis. Comp. Biochem. Physiol. 54A:55–60.

Lickey, M. E., R. L. Emigh, and F. R. Randle. 1970. A recirculating seawater aquarium system for inland laboratories. Mar. Biol. (Berlin) 7:149–152.

Lightner, D. V., B. R. Salser, and R. S. Wheeler. 1974. Gas-bubble disease in the brown shrimp (*Penaeus aztecus*). Aquaculture 4:81–84.

Lillie, W. R., and J. F. Klaverkamp. 1977. A System for Regulating and Recording pH of Solutions in Aquatic Flow-Through Toxicity Tests. Can. Fish. Mar. Serv. Tech. Rep. No. 710. v + 8 pp.

Lin, D. C., and D. A. Hessinger. 1979. Possible involvement of red cell membrane proteins in the hemolytic action of Portuguese Man-of-War toxin. Biochim. Biophys. Res. Commun. 91:761–769.

Little, G. 1969. The larval development of the shrimp, *Palaemon macrodactylus* Rathbun reared in the laboratory, and the effect of eyestalk extirpation on development. Crustaceana (Leiden) 17:69–87.

Lockwood, S. 1870. The horse-foot crab. Am. Nat. 4:257–274.

Lockwood, A. P. M. 1967. Aspects of the Physiology of Crustacea. W. H. Freeman and Co., San Francisco. 328 pp.

Loeb, M. J. 1973. The effect of light on strobilation in the Chesapeake Bay sea nettle *Chrysaora quinquecirrha*. Mar. Biol. (Berlin) 20:144–147.

Lönning, S. 1977. The sea urchin egg as a test object in oil pollution studies. Rapp. P.-V. Reun. Cons. Int. Explor. Mer 171:186–188.

Lönning, S., and B. E. Hagström. 1975. The effects of crude oils and the dispersant Corexit 8666 on sea urchin gametes and embryos. Norw. J. Zool. 23:121–129.

Loomis, W. F. 1954. Environmental factors controlling growth in hydra. J. Exp. Zool. 126:223–234.

Loosanoff, V. L. 1954. New advances in the study of bivalve larvae. Am. Sci. 42:607–624.

Loosanoff, V. L., and H. C. Davis. 1963. Rearing of bivalve mollusks. Adv. Mar. Biol. 1:1–136.

Lough, R. G. 1974. A re-evaluation of the combined effects of temperature and salinity on survival and growth of *Mytilus edulis* larvae using response surface techniques. Proc. Natl. Shellfish. Assoc. 64:73–76.

Lubet, P., M. A. Mannevy, and M. Mathieu. 1978. Analyse expérimentale en cultures d'organes, de l'action de DDT sur la gamétogenèse de la moule (*Mytilus edulis* L.) (mollusque lamellibranche). Bull. Soc. Zool. Fr. 103:283–288.

Lubzens, E., R. Fishler, and V. Berdugo-White. 1980. Induction of sexual reproduction and resting egg production in *Brachionus plicatilis* reared in sea water. Hydrobiologia 73:55–58.

Lucas, A. 1976. Remarques méthodologiques sur l'emploi des larves de moule comme tests biologiques. Haliotis 5:126–132.

Lucas, J. S., and M. M. Jones. 1976. Hybrid crown-of-thorns starfish (*Acanthaster planci* × *A. brevispinus*) reared to maturity in the laboratory. Nature (Lond.) 263:409–412.

Lüdemann, D., and H. Neumann. 1961. Studien über die Verwendung von *Artemia salina* L. als Testtier zum Nachweis von Kontaktinsektiziden. Z. Angew. Zool. 48:325–332.

Lui, C. W., and J. D. O'Connor. 1977. Biosynthesis of crustacean lipovitellin. III. The incorporation of labeled amino acids into the purified lipovitellin of the crab *Pachygrapsus crassipes*. J. Exp. Zool. 199:105–108.

Lumare, F. 1976. Research on the reproduction and culture of the shrimp *Penaeus kerathurus* in Italy. Stud. Rev. Gen. Fish. Counc. Medit. No. 55:35–48.

Lund, W., and R. Lockwood. 1970. Sonic tag for decapod crustaceans. J. Fish. Res. Bd. Can. 27:1147–1151.

Lyes, M. C. 1979. The reproductive behaviour of *Gammarus duebeni* (Lilljeborg), and the inhibitory effect of a surface active agent. Mar. Behav. Physiol. 6:47–55.

Maciorowski, A. F. 1975. An inexpensive macroinvertebrate bioassay table for use in continuous-flow toxicity tests. Bull. Environ. Contam. Toxicol. 13:420–423.

Maciorowski, H. D., R. M. Clarke, and E. Scherer. 1977. The use of avoidance-preference bioassays with aquatic invertebrates. Pp. 49–58 in Proceedings of the 3rd Aquatic Toxicity Workshop, Halifax, N.S., 1976. Environ. Prot. Serv. Tech. Rep. (EPS-5AR-77-1)

Mackie, G. O. 1966. Growth of the hydroid *Tubularia* in culture. Pages 397–410 in W. J. Rees, ed. The Cnidaria and Their Evolution. Academic Press, New York.

Mackie, G. O., and D. A. Boag. 1963. Fishing, feeding and digestion in siphonophores. Pubbl. Stn. Zool. Napoli 33:178–196.

MacLean, S. A., A. C. Longwell, and W. J. Blogoslawski. 1973. Effects of ozone-treated seawater on the spawned, fertilized, meiotic, and cleaving eggs of the commercial American oyster. Mutat. Res. 21:283–285.

Magarelli, P. C., Jr., B. Hunter, D. V. Lightner, and L. B. Colvin. 1979. Black death: An ascorbic acid deficiency disease in penaeid shrimp. Comp. Biochem. Physiol. A: Comp. Physiol. 63:103–108.

Maki, A. W. 1977. Modifications of continuous-flow toxicity test methods for small aquatic ganisms. Prog. Fish Cult. 39:172–174.

Maramorosch, K. 1976. Invertebrate Tissue Culture. Research Applications. Academic Press, New York. 393 pp.

Martin, Y. 1977. Modifications de la microflore bactérienne liées à l'utilisation d'antibiotiques dans les élevages expérimentaux des larves de *Mytilus galloprovincialis* Lmk (mollusque bivalve). Première partie. Evolution d'un élevage et charactères des bactéries associées. Rev. Int. Oceanogr. Med. 65/66:17–27.

Marullo, F., D. A. Emiliani, C. W. Caillouet, and S. H. Clark. 1976. A vinyl streamer tag for shrimp (*Penaeus* spp.). Trans. Am. Fish. Soc. 105:658–663.

Mason, C. F. 1977. Diets and culture media: Gastropoda. Pp. 95–102 in M. Rechcigl, Jr., ed. CRC Handbook Series in Nutrition and Food. Section G: Diets, Culture Media, Food Supplements. Volume II: Food Habits of, and Diets for Invertebrates and Vertebrates—Zoo Diets. CRC Press, Cleveland.

Mather, J. A. 1978. Mating behavior of *Octopus joubini* Robson. Veliger 21:265–267.

Mazur, J. E., and J. W. Miller. 1971. A description of the complete metamorphosis of the sea urchin *Lytechinus variegatus* cultured in synthetic sea water. Ohio J. Sci. 71:30–36.

Mazurkiewicz, M. 1975. Larval development and habits of *Laeonereis culveri* (Webster) (Polychaeta: Nereidae). Biol. Bull. (Woods Hole, Mass.) 149:186–204.

McBride, E. W. 1903. The development of *Echinus esculentus*, together with some points in the development of *E. miliaris* and *E. acutus*. Philos. Trans. R. Soc. Lond. Ser. B, 195:285–327.

McCammon, H. M. 1972. Establishing and maintaining articulate brachiopods in aquaria. J. Geol. Educ. 20:139–142.

McCammon, H. M. 1975. Maintenance of some marine filter feeders on beef heart extract. Pp. 15–27 in W. L. Smith and M. H. Chanley, eds. Culture of Marine Invertebrate Animals. Plenum Press, New York and London.

McCann-Collier, M. 1977. An unusual cytoplasmic organelle in oocytes of *Illyanassa obsoleta*. J. Morphol. 153:119–128.

McKillup, S. C. 1979. A technique for the isolation of microscopic algae suitable for feeding to specific invertebrate larvae. Aquaculture 16:361–362.

McLaren, I. A. 1976. Inheritance of demographic and production parameters in the marine copepod *Eurytemora herdmani*. Biol. Bull. (Woods Hole, Mass.) 151:200–213.

Mearns, A. J., P. S. Oshida, M. J. Sherwood, D. A. Young, and D. J. Reish. 1976. Chromium effects on coastal organisms. J. Water Pollut. Control Fed. 48:1929–1939.

Meixner, R. 1966. Eine Methode zur Aufzucht von *Crangon crangon* (L.) (Crust. Decap. Natantia). Arch. Fischereiwiss. 17:1–4.

Menzies, R. J. 1972. Experimental interbreeding between geographically separated populations of the marine wood-boring isopod *Limnoria tripunctata* with preliminary indications of hybrid vigor. Mar. Biol. (Berlin) 17:149–157.

Menzies, R. J., and D. Frankenberg. 1966. Handbook on the Common Marine Isopod Crustacea of Georgia. University of Georgia Press, Athens. 93 pp.

Meyer, R. M. 1974. Marking fishes and invertebrates. IV. A nonpermanent tag for king crabs, *Paralithodes camtschatica,* and tanner crabs, *Chionoecetes bairdi.* Mar. Fish. Rev. 367:14–16.

Meyers, S. P. 1980. Water stable extruded diets and feeding of invertebrates. J. Aquariculture 1:41–46.

Meyers, S. P., and Z. P. Zein-Eldin. 1973. Binders and pellet stability in development of crustacean diets. Pp. 351–364 in J. W. Avault, Jr., ed. Proc. Third Annu. Meet. World Mariculture Soc. La. State Univ. Div. Continuing Educ., Baton Rouge.

Meyers, S. P., W. A. Feder, and King Mon Tsue. 1963. Nutritional relationships among certain filamentous fungi and a marine nematode. Science 141:520–522.

Meyers, S. P., D. P. Butler, and W. H. Hastings. 1972. Alginates as binders for crustacean rations. Prog. Fish Cult. 34:9–12.

Michael, A. D., and B. Brown. 1978. Effects of laboratory procedure on fuel oil toxicity. Environ. Pollut. 15:277–287.

Michael, A. S., C. G. Thompson, and M. Abramovitz. 1956. *Artemia salina* as a test organism for bioassay. Science 123:464.

Mikami, S., and T. Okutani. 1977. Preliminary observations on maneuvering, feeding, copulating and spawning behaviors of *Nautilus macromphalus* in captivity. Venus Jpn. J. Malacol. 36:29–41.

Mikulich, L. V., and L. P. Kozak. 1972. Experimental rearing of Pacific Ocean squid under artificial conditions. Ecology (Engl. Transl. Ekologiya) 2:266–268.

Milkman, R. 1967. Genetic and developmental studies on *Botryllus schlosseri.* Biol. Bull. (Woods Hole, Mass.) 132:229–243.

Millar, R. H., and J. M. Scott. 1968. An effect of water quality on the growth of cultured larvae of the oyster *Ostrea edulis* L. J. Cons. Cons. Int. Explor. Mer 32:123–130.

Miller, J. W., and J. E. Mazur. 1974. Gallon-jar marine aquaria. Am. Biol. Teacher 36(1):40–41.

Miller, M. R. 1978. Blue crab larval culture: Methods and management. U.S. Natl. Mar. Fish. Mar. Fish. Rev. 40(11):10–17.

Miller, R. L. 1966. Chemotaxis during fertilization in the hydroid *Campanularia.* J. Exp. Zool. 162:23–44.

Miller, R. L. 1976. Some observations on sexual reproduction in *Tubularia.* Pp. 299–308 in G. O. Mackie, ed. Coelenterate Ecology and Behavior. Plenum Press, New York and London.

Minasian, L. L., Jr. 1976. Characteristics of asexual reproduction in the sea anemone, *Haliplanella luciae* (Verrill), reared in the laboratory. Pp. 289–298 in G. O. Mackie, ed. Coelenterate Ecology and Behavior. Plenum Press, New York and London.

Minchin, E. A. 1896. Note on the larva and the postlarval development of *Leucosolenia variabilis,* H. sp., with remarks on the development of other Asconidae. Proc. R. Soc. Lond. 60:42–52.

Miner, R. W. 1950. Field Book of Seashore Life. G. P. Putnam and Sons, New York. 888 pp.

Mirkes, D. Z., W. B. Vernberg, and P. J. DeCoursey. 1978. Effects of cadmium and mercury on the behavioral responses and development of *Eurypanopeus depressus* larvae. Mar. Biol. (Berlin) 47:143–147.

Mirkes, P. E. 1972. Polysomes and protein synthesis during development of *Ilyanassa obsoleta.* Exp. Cell Res. 74:503–508.

Missler, S. R. 1980. Maturation of penaeid shrimp: Dietary fatty acids. Diss. Abstr. Int. B Sci. Eng. 40:5652.

Modin, J. C., and K. W. Cox. 1967. Post-embryonic development of laboratory-reared ocean shrimp, *Pandalus jordani* Rathbun. Crustaceana (Leiden) 13:197–219.

Molenock, J., and E. D. Gomez. 1972. Larval stages and settlement of the barnacle *Balanus* (*Conopea*) *galeatus* (L.) (Cirripedia Thoracica). Crustaceana (Leiden) 23:100–108.

Moore, D. 1960. A modified filtration and aeration unit for experimental snail aquaria. J. Parasitol. 46:767.

Moore, H. B. 1958. Marine Ecology. John Wiley & Sons, New York, London, and Sydney. xi + 493 pp.

Moore, N. M., and A. R. D. Stebbing. 1976. The quantitative cytochemical effects of three metal ions on a lysosomal hydrolase of a hydroid. J. Mar. Biol. Assoc. U.K. 56:995–1005.

Morgan, T. H. 1933. The formation of the antipolar lobe in *Ilyanassa*. J. Exp. Zool. 64:433–467.

Morin, J. G., and I. M. Cooke. 1971. Behavioural physiology of the colonial hydroid *Obelia*. I. Spontaneous movements and correlated electrical activity. J. Exp. Biol. 54:689–706.

Morris, R. H., D. P. Abbott, and E. C. Haderlie. 1980. Intertidal Invertebrates of California. Stanford University Press. xiii + 690 pp.

Morse, D. E., H. Duncan, N. Hooker, and A. Morse. 1977. An inexpensive chemical method for the control and synchronous induction of spawning and reproduction in molluscan species important as protein-rich food resources. FAO Fish. Rep. 200:291–300.

Morse, D. E., N. Hooker, and A. Morse. 1978. Chemical control of reproduction in bivalve and gastropod molluscs. III. An inexpensive technique for mariculture of many species. Pp. 543–547 in J. W. Avault, Jr., ed. Proc. Ninth Annu. Meet. World Mariculture Soc. La. State Univ. Div. Continuing Educ., Baton Rouge.

Morse, D. E., N. Hooker, H. Duncan, and L. Jensen. 1979a. γ-aminobutyric acid, a neurotransmitter, induces planktonic abalone larvae to settle and begin metamorphosis. Science 204:407–410.

Morse, D. E., N. Hooker, L. Jensen, and H. Duncan. 1979b. Induction of larval abalone settling and metamorphosis by gamma-aminobutyric acid and its congeners from crustose red algae. II. Application to cultivation, seed-production and bioassays; principal causes of mortality and interference. Pp. 81–91 in J. W. Avault, Jr., ed. Proc. Tenth Annu. Meet. World Mariculture Soc. La. State Univ. Div. Continuing Educ., Baton Rouge.

Mortensen, T. H. 1938. Contributions to the study of the development and larval forms of echinoderms. IV. Kgl. Danske Vidensk. Selsk. Skr. Naturv. Math. Ser. 9 7(3):1–59.

Morton, B. 1978. Feeding and digestion in shipworms. Oceanogr. Mar. Biol. Annu. Rev. 16:107–144.

Mowbray, R. C., and G. G. Brown. 1974. Fertilization studies with egg sections of the horseshoe crab, *Limulus polyphemus* L.: The effects of bivalent and univalent anti-egg antibodies on sperm-egg attachment. Biol. Reprod. 10:62–68.

Moyse, J. 1960. Mass rearing of barnacle cyprids in the laboratory. Nature (Lond.) 185:120.

Mullin, M. M., and E. R. Brooks. 1967. Laboratory culture, growth rate, and feeding behavior of a planktonic marine copepod. Limnol. Oceanogr. 12:657–666.

Mullin, M. M., and E. R. Brooks. 1973. Growth and metabolism of two planktonic, marine copepods as influenced by temperature and type of food. Pp. 74–95 in J. H. Steele, ed. Marine Food Chains. Oliver & Boyd, Edinburgh.

Murano, M., S. Segawa, and M. Kato. 1979. Moult and growth of the Antarctic krill in laboratory. Trans. Tokyo Univ. Fish. No. 3:99–106.

Murchelano, R. A., and A. Rosenfield. 1978. Diseases of North American marine fishes, crustaceans and molluscs. Contract report prepared for Bureau of Land Management Interagency Agreement A 550-1A7-35.

Murphy, H. E. 1923. The life cycle of *Oithona nana,* reared experimentally. Univ. Calif. Publ. Zool. 22:449–454.

Muscatine, L. 1961. Symbiosis in marine and fresh water coelenterates. Pp. 255–268 in H. M. Lenhoff and W. F. Loomis, eds. The Biology of Hydra and of Some Other Coelenterates: 1961. University of Miami Press, Coral Gables.

Muscatine, L., and H. M. Lenhoff. 1963. Symbiosis: On the role of algae symbiotic with hydra. Science 142:956–958.

Myers, A. A. 1971. Breeding and growth in laboratory-reared *Microdeutopus gryllotalpa* Costa (Amphipoda: Gammaridea). J. Nat. Hist. 5:271–277.

Nachum, R., S. W. Watson, J. D. Sullivan, Jr., and S. E. Siegel. 1979. Antimicrobial defense mechanisms in the horseshoe crab, *Limulus polyphemus*: Preliminary observations with heat-derived extracts of *Limulus* amoebocyte lysate. J. Invertebr. Pathol. 33:290–299.

Nagao, Z. 1964. The life cycle of the hydromedusa, *Nemopsis dofleini* Maas, with a supplementary note on the life-history of *Bougainvillia superciliaris* (L. Agassiz). Annot. Zool. Jpn. 37:153–162.

Nair, K. K. C., and K. Anger. 1979a. Life cycle of *Corophium insidiosum* (Crustacea, Amphipoda) in laboratory culture. Helgol. Wiss. Meeresunters. 32:279–294.

Nair, K. K. C., and K. Anger. 1979b. Experimental studies on the life cycle of *Jassa falcata* (Crustacea, Amphipoda). Helgol. Wiss. Meeresunters. 32:444–452.

Nair, K. K. C., I. C. Gopalakrishnan, M. G. Peter, and T. S. S. Rao. 1978. A closed sea water circulating system for the cultivating of marine and estuarine organisms in the laboratory. Ind. J. Mar. Sci. 7:159–162.

Nakauchi, M., A. Osaki, and R. Okamoto. 1979. Inland culture of the colonial ascidian *Symplegma reptans.* Rep. Usa Mar. Biol. Inst. Kochi Univ. No. 1:59–64. (in Japanese with English summary)

Nassogne, A. 1969. La coltura dei copepodi in laboratorio. Pubbl. Stn. Zool. Napoli 37 (Suppl.):203–218.

Nassogne, A. 1970. Influence of food organisms on the development and culture of pelagic copepods. Helgol. Wiss. Meeresunters. 20:333–345.

Neal, R. A. 1969. Methods of marking shrimp. Food and Agricultural Organization of the U.N., Fish. Rep. 57, 3:1149–1165.

Needham, J. G., P. S. Galtsoff, F. E. Lutz, and P. S. Welch, eds. 1937. Culture Methods for Invertebrate Animals. Comstock Publishing Co., Ithaca, N.Y. (Reprinted in 1959, Dover Publications, New York.) xxxii + 590 pp.

Neff, J. M., and C. S. Giam. 1977. Effects of Aroclor 1016 and Halowax 1099 on juvenile horseshoe crabs *Limulus polyphemus.* Pp. 21–35 in F. J. Vernberg, A. Calabrese, F. P. Thurberg, and W. B. Vernberg, eds. Physiological Responses of Marine Biota to Pollutants. Academic Press, New York, San Francisco, and London.

Neudecker, T. 1977. A method for tagging oysters. Arch. Fischwiss. 28:143–147.

Neunes, H. W., and G. F. Pongolini. 1965. Breeding a pelagic copepod, *Euterpina acutifrons* (Dana), in the laboratory. Nature (Lond.) 208:571–573.

New, M. B. 1976. A review of dietary studies with shrimp and prawns. Aquaculture 9:101–144.

New, M. B., J. P. Scholl, J. C. McCarty, and J. P. Bennett. 1974. A recirculation system for experimental aquaria. Aquaculture 3:95–103.

Newrock, K. M., and R. A. Raff. 1975. Polar lobe specific regulation of translation in embryos of *Ilyanassa obsoleta.* Dev. Biol. 42:242–261.

Nicholas, W. L. 1975. The Biology of Free-living Nematodes. Oxford University Press, London. viii + 219 pp.

Nichols, D. 1969. Echinoderms. Hutchinson and Co. Ltd., London. 200 pp.

Nicol, J. A. C. 1967. The Biology of Marine Animals, 2d ed. Isaac Pitman and Sons, London. xi + 699 pp.

Nielsen, C. 1971. Entoproct life cycles and the entoproct/ectoproct relationship. Ophelia 9:209–341.

Nimmo, D. R., L. H. Bahner, R. A. Rigby, J. M. Sheppard, and A. J. Wilson, Jr. 1977. *Mysidopsis bahia*: An estuarine species suitable for life-cycle toxicity tests to determine the effects of a pollutant. Pp. 109–116 in F. L. Mayer and J. L. Hamelink, eds. Aquatic Toxicology and Hazard Evaluation. American Society for Testing and Materials, Philadelphia. (STP 634)

Nimmo, D. R., T. L. Hamaker, and C. A. Sommers. 1978a. Culturing the mysid (*Mysidopsis bahia*) in flowing sea water or a static system. Pp. 59–60 in Bioassay Procedures for the Ocean Disposal Permit Program. Environmental Research Laboratory, U.S. Environmental Protection Agency, Gulf Breeze, Fla.

Nimmo, D. R., T. L. Hamaker, and C. A. Sommers. 1978b. Entire life cycle toxicity test using mysids (*Mysidopsis bahia*) in flowing water. Pp. 64–68 in Bioassay Procedures for the Ocean Disposal Permit Program. Environmental Research Laboratory, U.S. Environmental Protection Agency, Gulf Breeze, Fla.

Nishimura, K., M. Miki, S. Ito, and T. Shioya. 1969. Studies on the aquaculture of *Sulculus diversicolor diversicolor*. I. Development and growth. Bull. Jpn. Soc. Sci. Fish. 35:336–341.

Nixon, M., and J. B. Messenger, eds. 1977. The Biology of Cephalopods. Symp. Zool. Soc. Lond. No. 38. xviii + 615 pp.

Numakunai, T. 1965. A simple method for collecting the tadpole larvae of the ascidian, *Halocynthia roretzi* (V. Drasche). Bull. Mar. Biol. Stn. Asamushi 12:173–174.

O'Dor, R. K. 1978. Laboratory experiments with *Illex illecebrosus*. Pp. 18.1–18.10 in N. Balch, T. Amaratunga, and R. K. O'Dor, eds. Proceedings of the Workshop on the Squid *Illex illecebrosus*. Can. Fish. Mar. Serv. Tech. Rep. No. 833.

O'Dor, R. K., R. D. Durward, and N. Balch. 1977. Maintenance and maturation of squid (*Illex illecebrosus*) in a 15 meter circular pool. Biol. Bull. (Woods Hole, Mass.) 153:322–335.

Odum, E. P. 1971. Fundamentals of Ecology. W. B. Saunders Co., Philadelphia. 574 pp.

Ohshima, Y., and S. Choe. 1961. On the rearing of young cuttlefish and squid. Bull. Jpn. Soc. Sci. Fish 27:979–986. (in Japanese with English summary)

Okazaki, K. 1975. Normal development to metamorphosis. Pp. 177–232 in G. Czihak, ed. The Sea Urchin Embryo. Springer-Verlag, New York.

Okubo, K., and T. Okubo. 1962. Study on the bio-assay method for the evaluation of water pollution. II. Use of the fertilized eggs of sea urchins and bivalves. Bull. Tokai Reg. Fish. Res. Lab. No. 32:131–140. (in Japanese with English summary)

Olson, C. S. 1979. Timing of developmental events in *Artemia salina* (L.) (Anostraca). Crustaceana (Leiden) 36:302–308.

Olson, M., and G. Newton. 1979. A simple, rapid method for marking individual sea urchins. Calif. Fish Game 65:58–62.

Omori, M. 1973. Cultivation of marine copepods. Bull. Plankton Soc. Jpn. 20:3–11.

Onbé, T. 1974. Studies on the ecology of marine cladocerans. J. Fac. Fish. Anim. Husb. Hiroshima Univ. 13:83–179. (in Japanese with English summary)

Onbé, T. 1977. The biology of marine cladocerans in a warm temperate water. Pp. 383–398 in Proceedings of the Symposium on Warm Water Zooplankton. Spec. Publ. Natl. Inst. Oceanogr. (Goa, India). xi + 722 pp.

Opresko, L., and R. F. Thomas. 1975. Observations on *Octopus joubini*: Some aspects of reproductive biology and growth. Mar. Biol. (Berlin) 31:51–61.

Osanai, K. 1975. Handling Japanese sea urchins and their embryos. Pp. 26–40 in G. Czihak and R. Peter, eds. The Sea Urchin Embryo, Biochemistry and Morphogenesis. Springer-Verlag, Berlin, Heidelberg, and New York.

Oshida, P. S. 1977. A safe level of hexavalent chromium for a marine polychaete. Pp. 55–60 in So. Calif. Coast. Water. Res. Proj. Annu. Rep. El Segundo, Calif.

Oshida, P. S., and D. J. Reish. 1975. Effects of chromium on reproduction in polychaetes. Pp. 55–60 in So. Calif. Coast. Water Res. Proj. Annu. Rep. El Segundo, Calif.

Oshida, P. S., A. J. Mearns, D. J. Reish, and C. S. Word. 1976. The effects of hexavalent and trivalent chromium on *Neanthes arenaceodentata* (Polychaeta: Annelida). So. Calif. Coast. Water Res. Proj. TM No. 225. 58 pp.

Oswald, R. F. 1977. Immobilization of decapod crustacea for experimental procedures. J. Mar. Biol. Assoc. U.K. 57:715–721.

Owen, G. 1955. Use of propylene phenoxetol as a relaxing agent. Nature (Lond.) 175:434.

Packard, A. S. 1885. On the embryology of *Limulus polyphemus* III. Am. Nat. 19:722–727.

Paffenhöfer, G. A. 1970. Cultivation of *Calanus helgolandicus* under controlled conditions. Helgol. Wiss. Meeresunters. 20:346–359.

Paffenhöfer, G. A. 1973. The cultivation of an appendicularian through numerous generations. Mar. Biol. (Berlin) 22:183–185.

Paffenhöfer, G. A., and R. P. Harris. 1976. Feeding, growth and reproduction of the marine planktonic copepod *Pseudocalanus elongatus*. J. Mar. Biol. Assoc. U.K. 56:327–344.

Paffenhöfer, G. A., and R. P. Harris. 1979. Laboratory culture of marine holozooplankton and its contribution to studies of marine planktonic food webs. Adv. Mar. Biol. 16:211–308.

Pantin, C. F. A. 1942. The excitation of nematocysts. J. Exp. Biol. 19:294–310.

Pardy, R. L. 1971. The feeding biology of the gymnoblastic hydroid *Pennaria tiarella*. Pp. 84–91 in H. M. Lenhoff, L. Muscatine, and L. V. Davis, eds. Experimental Coelenterate Biology. University of Hawaii Press, Honolulu.

Parrish, F. K., and J. W. Parrish. 1962. A method of determining spatial differences in cell surface permeability. Exp. Cell Res. 27:317–320.

Parrish, K. K., and D. F. Wilson. 1978. Fecundity studies on *Acartia tonsa* (Copepoda: Calanoida) in standardized culture. Mar. Biol. (Berlin) 46:65–81.

Parrish, K. M., and J. D. Bultman. 1978. Navy research on marine borers and the laboratory culturing of limnorians. Pp. 92–98 in Oceans '78, "The Ocean Challenge." Marine Technology Society, Washington, D.C. xxiii + 743 pp.

Pascual, E. 1978. Crecimiento y alimentacion de tres generaciónes de *Sepia officinalis* en cultivo. Invest. Pesq. 42:421–442.

Patten, W. 1896. Variations in the development of *Limulus polyphemus*. J. Morphol. 12:17–148.

Paul, A. J., J. M. Paul, P. A. Shoemaker, and H. M. Feder. 1979. Prey concentrations and feeding response in laboratory-reared stage-one zoeae of king crab, snow crab, and pink shrimp. Trans. Am. Fish. Soc. 108:440–443.

Pauley, G. B., and M. R. Tripp, eds. 1975. Diseases of crustaceans. NOAA (Natl. Ocean. Atmos. Adm.) Mar. Fish. Rev. 37(5/6):1–64.

Payen, G. G., and J. D. Costlow. 1977. Effects of a juvenile hormone mimic on male and female gametogenesis of the mud-crab, *Rhithropanopeus harrisii* (Gould) (Brachyura: Xanthidae). Biol. Bull. (Woods Hole, Mass.) 152:199–208.

Pearse, J. S. 1979. Polyplacophora. Pp. 27–85 in A. C. Giese and J. S. Pearse, eds.

Reproduction of Marine Invertebrates, vol. 5. Academic Press, New York, London, Toronto, Sydney, and San Francisco.

Pearson, F. C., III, and M. Weary. 1980. The *Limulus* amebocyte lysate test for endotoxin. BioScience 30:461–464.

Pearson, F. C., and E. Woodland. 1979. The *in vitro* cultivation of *Limulus* amoebocytes. Pp. 93–102 in E. Cohen *et al.*, eds. Biomedical Applications of the Horseshoe Crab (Limulidae). Alan R. Liss, New York.

Pechenik, J. A. 1975. The escape of veligers form the egg capsules of *Nassarius obsoletus* and *Nassarius trivittatus* (Gastropoda, Prosobranchia). Biol. Bull. (Woods Hole, Mass.) 149:580–589.

Pechenik, J. A., F. E. Perron, and R. D. Turner. 1979. The role of phytoplankton in the diets of adult and larval shipworms, *Lyrodus pedicellatus* (Bivalvia: Teredinidae). Estuaries 2:58–60.

Peltier, W. 1978. Methods for Measuring the Acute Toxicity of Effluents to Aquatic Organisms. Environmental Monitoring & Support Laboratory, U.S. Environmental Protection Agency, Cincinnati. EPA-600/4-78-012. ix + 52 pp.

Peponnet, F., and J. M. Quiot. 1971. Cell cultures of Crustacea, Arachnida, and Merostomacea. Pp. 341–360 in C. Vago, ed. Invertebrate Tissue Culture, vol. I. Academic Press, New York and London. xiv + 441 pp.

Perkins, E. J. 1972. Some methods of assessment of toxic effects upon marine invertebrates. Proc. Soc. Anal. Chem. 9:105–114.

Perkins, F. O., ed. 1979. Haplosporidian and haplosporidian-like diseases of shellfish. NOAA (Natl. Ocean. Atmos. Adm.) Mar. Fish. Rev. 41 (1/2):1–72.

Perron, F. E. 1980. Laboratory culture of the larvae of *Conus textile* Linne (Gastropoda: Toxoglossa). J. Exp. Mar. Biol. Ecol. 42:27–38.

Perron, F. E., and R. D. Turner. 1977. Development, metamorphosis, and natural history of the nudibranch *Doridella obscura* Verrill (Corambidae: Opisthobranchia). J. Exp. Mar. Biol. Ecol. 27:171–185.

Person Le-Ruyet, J. 1975. Élevage de copépodes calanoïdes biologie et dynamique des populations: Premiers résultats. Ann. Inst. Oceanogr. 51:203–221.

Person Le-Ruyet, J. 1976. Élevage larvaire d'*Artemia salina* (Branchiopoda) sur nourriture inerte: *Spirulina maxima* (Cyanophycée). Aquaculture 8:157–167.

Persoone, G., and P. Sorgeloos. 1972. An improved separator box for *Artemia* nauplii and other phototactic invertebrates. Helgol. Wiss. Meeresunters. 23:243–247.

Pesch, G. G., and C. E. Pesch. 1980. *Neanthes arenaceodentata* (Polychaeta: Annelida), a proposed cytogenetic model for marine genetic toxicology. Can. J. Fish. Aquatic Sci. 37:1225–1228.

Petrich, F. M., and D. J. Reish. 1979. Effects of aluminum and nickel on survival and reproduction in polychaetous annelids. Bull. Environ. Contam. Toxicol. 23:698–702.

Pettibone, M. H. 1963. Marine Polychaete Worms of the New England Region. 1. Aphroditidae to Trochochaetidae. U.S. Natl. Mus. Bull. No. 227. U.S. National Museum, Washington, D.C. 356 pp.

Pezalla, P. D., R. M. Dores, and W. S. Herman. 1978. Separation and partial purification of central nervous system peptides from *Limulus polyphemus* with hyperglycemic and chromatophorotropic activity in crustaceans. Biol. Bull. (Woods Hole, Mass.) 154:148–156.

Phillips, G. C. 1971. Marking shrimps and prawns by latex injection. Crustaceana (Leiden) 22:84–86.

Picken, L. E. R., and R. J. Skaer. 1966. A review of researches on nematocysts. Pp. 19–50 in W. J. Rees, ed. The Cnidaria and Their Evolution. Symp. Zool. Soc. Lond. No. 16. Academic Press, London.

Pilger, J. 1978. Settlement and metamorphosis in the Echiura: A review. Pp. 103–112 in F. Chia and M. E. Rice, eds. Settlement and Metamorphosis of Marine Invertebrate Larvae. Elsevier, New York and Oxford.

Pilkington, M. C., and V. Fretter. 1970. Some factors affecting the growth of prosobranch veligers. Helgol. Wiss. Meeresunters. 20:576–593.

Poizat, C. 1972. Méthodes d'élevage des gastéropodes opisthobranches de petites et moyennes dimensions. Mise au point d'un circuit fermé en eau de mer. Premiers résultats. Tethys 4:251–267.

Portman, J. E. 1972a. Results of acute toxicity tests with marine organisms, using a standard method. Pp. 212–217 in M. Ruivo, ed. Marine Pollution and Sea Life. Fishing News (Books) Ltd., West Byfleet, Surrey, and London.

Portman, J. E. 1972b. Toxicity-testing with particular reference to oil-removing materials and heavy metals. Pp. 217–222 in M. Ruivo, ed. Marine Pollution and Sea Life. Fishing News (Books) Ltd., West Byfleet, Surrey, and London.

Prentice, E. F., and J. E. Rensel. 1977. Tag retention of the spot prawn, *Pandalus platyceros,* injected with coded wire tags. J. Fish. Res. Bd. Can. 34:2199–2203.

Prior, M. G. 1979. Evaluation of brine shrimp (*Artemia salina*) larvae as a bioassay for mycotoxins in animal feedstuffs. Can. J. Comp. Med. 43:352–355.

Prosser, C. L., ed. 1973. Comparative Animal Physiology, 3d ed. Holt, Rinehart and Winston, New York. xxii + 966 + xiv pp.

Prosser, C. L., J. Koester, E. Mayeri, G. Liebeswar, E. R. Kandel, W. H. Hildemann, A. L. Reddy, M. Fingerman, R. C. Fay, V. C. Goldizen, E. T. LaRoe, A. O. D. Willows, J. C. Harshbarger, C. J. Dawe, and S. Wolf. 1973. Animal models for biomedical research V—invertebrates. Fed. Proc. 32:2177–2230.

Provasoli, L. 1968. Media and prospects for the cultivation of marine algae. Pp. 66–75 in A. Watanabe and A. Hattori, eds. Culture and Collections of Algae. Proc. U.S.–Jpn. Conf., Japan. Soc. Plant Physiol., Hakone.

Provasoli, L. 1976. Nutritional aspects of crustacean aquaculture. Pp. 13–21 in K. S. Price, Jr., W. N. Shaw, and K. S. Danberg, eds. Proc. First Int. Conf. Aquaculture Nutr. Coll. Mar. Stud. Univ. Del., Newark.

Provasoli, L. 1977. Cultivation of animals. Axenic cultivation. Pp. 1295–1319 in O. Kinne, ed. Marine Ecology, a Comprehensive, Integrated Treatise on Life in Oceans and Coastal Waters. Vol. III: Cultivation. John Wiley & Sons, Chichester, New York, Brisbane, and Toronto.

Provasoli, L., and A. D'Agostino. 1969. Development of artificial media for *Artemia salina.* Biol. Bull. (Woods Hole, Mass.) 136:435–453.

Provasoli, L., J. J. A. McLaughlin, and M. R. Droop. 1957. The development of artificial media for marine algae. Arch. Mikrobiol. 25:392–428.

Provasoli, L., K. Shiraishi, and J. R. Lance. 1959. Nutritional idiosyncrasies of *Artemia* and *Tigriopus* in monoxenic culture. Ann. N.Y. Acad. Sci. 77:250–261.

Provasoli, L., T. Yamasu, and I. Manton. 1968. Experiments on the resynthesis of symbiosis in *Convoluta roscoffensis* with different flagellate cultures. J. Mar. Biol. Assoc. U.K. 48:465–478.

Provasoli, L., D. E. Conklin, and A. S. D'Agostino. 1970. Factors inducing fertility in aseptic Crustacea. Helgol. Wiss. Meeresunters. 20:443–454.

Provenzano, A. J., Jr. 1967. Recent advances in the laboratory culture of decapod larvae. Symp. Ser. Mar. Biol. Assoc. India. Proc. Symp. Crustacea Ernakulam 1965, pt. 2:940–945.

Pucci-Minafra, L., S. Minafra, and J. R. Collier. 1969. Distribution of ribosomes in the egg of *Ilyanassa obsoleta*. Exp. Cell Res. 57:167–178.

Quetin, L. B., and J. J. Childress. 1980. Observations on the swimming activity of two bathypelagic mysid species maintained at high hydrostatic pressure. Deep-Sea Res. 27A:383–391.

Quinn, R. H., and D. R. Fielder. 1978. A laboratory beach system for prolonged maintenance of sand crabs, *Mictyris* Latreille, 1806 and *Scopimera* de Haan, 1883 (Decapoda, Brachyura). Crustaceana (Leiden) 34:310–313.

Raff, R. A. 1972. Polar lobe formation by embryos of *Ilyanassa obsoleta*. Exp. Cell Res. 71:455–459.

Raff, R. A., K. M. Newrock, R. D. Secrist, and F. D. Turner. 1976. Regulation of protein synthesis in embryos of *Ilyanassa obsoleta*. Am. Zool. 16:529–546.

Rand, M. C., A. E. Greenberg, M. J. Taras, and M. A. Franson, eds. 1976. Standard Methods for the Examination of Water and Wastewater, 14th ed. American Public Health Association, Washington, D.C. xxxix + 1193 pp.

Rannou, M. 1971. Cell culture of invertebrates other than mollusks and arthropods. Pp. 385–410 in C. Vago, ed. Invertebrate Tissue Culture, vol. I. Academic Press, New York and London. xiv + 441 pp.

Raps, M. E., and D. J. Reish. 1971. The effects of varying dissolved oxygen concentrations on the hemoglobin levels of the polychaetous annelid *Neanthes arenaceodentata*. Mar. Biol. (Berlin) 11:363–368.

Ray, D. L. 1958. Some marine invertebrates useful for genetic research. Pp. 497–512 in A. A. Buzzati-Traverso, ed. Perspectives in Marine Biology. University of California Press, Berkeley and Los Angeles.

Rebach, S. 1977. Simultaneous activity recording of multiple isolated marine organisms in an artificial saltwater recirculating system. J. Fish. Res. Bd. Can. 34:1426–1430.

Rechcigl, M., Jr., ed. 1977. CRC Handbook Series in Nutrition and Food. Section G: Diets, Culture Media, Food Supplements. Vol. II. Food Habits of, and Diets for Invertebrates and Vertebrates—Zoo Diets. CRC Press, Cleveland, Ohio. 462 pp.

Reddy, A. L., B. Bryan, and W. H. Hildemann. 1975. Integumentary allograft versus autograft reactions in *Ciona intestinalis*: A protochordate species of solitary tunicate. Immunogenetics 1:584–590.

Reed, P. H. 1969. Culture methods and effects of temperature and salinity on survival and growth of Dungeness crab (*Cancer magister*) larvae in the laboratory. J. Fish. Res. Bd. Can. 26:389–397.

Reed, S. A. 1971. Techniques for raising the planula larvae and newly settled polyps of *Pocillopora damicornis*. Pp. 66–72 in H. M. Lenhoff, L. Muscatine, and L. V. Davis, eds. Experimental Coelenterate Biology. University of Hawaii Press, Honolulu.

Rees, J. 1971. Paths and rates of food distribution in the colonial hydroid *Pennaria*. Pp. 119–128 in H. M. Lenhoff, L. Muscatine, and L. V. Davis, eds. Experimental Coelenterate Biology. University of Hawaii Press, Honolulu.

Rees, J. T. 1978. Laboratory and field studies on *Eutonina indicans* (Coelenterata: Hydrozoa), a common leptomedusa of Bodega Bay, California. Wasmann J. Biol. 36:201–209.

Rees, W. J., and F. S. Russell. 1937. On rearing the hydroids of certain medusae, with an account of the methods used. J. Mar. Biol. Assoc. U.K. 22:61–82.

Reeve, M. R. 1969. The Laboratory Culture of the Prawn *Palaemon serratus*. Fish. Invest. Ser. II Mar. Fish. G.B. Minist. Agric. Fish. Food 26 (1). vi + 38 pp.

Reeve, M. R. 1970. Complete cycle of development of a pelagic chaetognath in culture. Nature (Lond.) 227:381.

Reeve, M. R. 1977. The effect of laboratory conditions on the extrapolation of experimental measurements to the ecology of marine zooplankton. V. A review. Pp. 528–537 in Proceedings of the Symposium on Warm Water Zooplankton. Spec. Publ. Natl. Inst. Oceanogr. (Goa, India). xi + 722 pp.

Reeve, M. R., and M. A. Walter. 1972. Conditions of culture, food-size selection, and the effects of temperature and salinity on growth rate and generation time in *Sagitta hispida* Conant. J. Exp. Mar. Biol. Ecol. 9:191–200.

Reish, D. J. 1953. Description of a new technique for rearing polychaetous annelids to sexual maturity. Science 118:363–364.

Reish, D. J. 1954. The life history and ecology of the polychaetous annelid *Nereis grubei* (Kinberg). Allan Hancock Foundation Publications, University of California Press, 14:1–75.

Reish, D. J. 1957. The life history of the polychaetous annelid *Neanthes caudata* (delle Chiaje), including a summary of development in the family Nereidae. Pac. Sci. 11:216–228.

Reish, D. J. 1961. The use of the sediment bottle collector for monitoring polluted marine waters. Calif. Fish Game 47:261–272.

Reish, D. J. 1970. The effects of varying concentrations of nutrients, chlorinity, and dissolved oxygen on polychaetous annelids. Water Res. 4:721–735.

Reish, D. J. 1974. The establishment of laboratory colonies of polychaetous annelids. Thalassia Jugosl. 10:181–195.

Reish, D. J. 1977a. Effects of chromium on the life history of *Capitella capitata* (Annelida: Polychaeta). Pp. 199–207 in F. J. Vernberg, A. Calabrese, F. P. Thurberg, and W. B. Vernberg, eds. Physiological Responses of Marine Biota to Pollutants. Academic Press, New York, San Francisco, and London.

Reish, D. J. 1977b. The role of life history in polychaete systematics. Pp. 461–476 in D. J. Reish and K. Fauchald, eds. Essays in Memory of Dr. Olga Hartman. Allan Hancock Foundation, University of Southern California, Los Angeles.

Reish, D. J., and J. L. Barnard. 1960. Field toxicity tests in marine waters utilizing the polychaetous annelid *Capitella capitata* (Fabricius). Pac. Nat. 1:1–8.

Reish, D. J., and R. S. Carr. 1978. The effect of heavy metals on the survival, reproduction, development, and life cycles for two species of polychaetes. Mar. Pollut. Bull. 9:24–27.

Reish, D. J., and T. L. Richards. 1966. A culture method for maintaining large populations of polychaetous annelids in the laboratory. Turtox News 44:16–17.

Reish, D. J., and G. S. Stephens. 1969. Uptake of organic material by aquatic invertebrates. V. The influence of age on the uptake of glycine-C^{14} by the polychaete *Neanthes arenaceodentata*. Mar. Biol. 3:352–355.

Reish, D. J., F. M. Piltz, J. M. Martin, and J. Q. Word. 1974a. Induction of abnormal polychaete larvae by heavy metals. Mar. Pollut. Bull. 5:125–126.

Reish, D. J., J. M. Martin, F. M. Piltz, and J. Q. Word. 1974b. The effect of heavy metals on laboratory populations of two polychaetes with comparison to the water quality standards in southern California waters. Water Res. 10:299–302.

Reish, D. J., J. M. Martin, F. M. Piltz, and J. Q. Word. 1976. The effect of heavy metals on laboratory populations of two polychaetes with comparisons to the water quality conditions and standards in southern California marine waters. Water Res. 10:299–302.

Reish, D. J., C. E. Pesch, J. H. Gentile, G. Bellan, and D. Bellan-Santini. 1978. Interlaboratory calibration experiments using the polychaetous annelid *Capitella capitata*. Mar. Environ. Res. 1:109–118.

Renzoni, A. 1974. Influence of toxicants on marine invertebrate larvae. Thalassia Jugosl. 10:197–212.

Reverberi, G. 1971. Experimental embryology of marine and fresh water invertebrates. American Elsevier Publishing Co., New York. 587 pp.

Rhodes, E. W., A. Calabrese, W. D. Cable, and W. S. Landers. 1975. The development of methods for rearing the coot clam, *Mulinia lateralis,* and three species of coastal bivalves in the laboratory. Pp. 273–282 in W. L. Smith and M. H. Chanley, eds. Culture of Marine Invertebrate Animals. Plenum Press, New York and London.

Rice, A. L., and D. I. Williamson. 1970. Methods for rearing larval decapod Crustacea. Helgol. Wiss. Meeresunters. 20:417–434.

Rice, M. E. 1967. A comparative study of the development of *Phascolosoma agassizii, Golfingia pugettensis,* and *Themiste pyroides* with a discussion of developmental patterns in the Sipuncula. Ophelia 4:143–171.

Rice, M. E. 1976. Larval development and metamorphosis in Sipuncula. Am. Zool. 16:563–571.

Rice, M. E. 1978. Morphological and behavioral changes at metamorphosis in the Sipuncula. Pp. 83–102 in F. Chia and M. E. Rice, eds. Settlement and Metamorphosis in Marine Invertebrate Larvae. Elsevier, New York and Oxford.

Rice, M. E., and M. Todorovic. 1975. Proceedings of the International Symposium on the Biology of the Sipuncula and Echiura, vol. 1. Naucno Delo Press, Belgrade. 355 pp.

Rice, M. E., and M. Todorovic. 1976. Proceedings of the International Symposium on the Biology of the Sipuncula and Echiura, vol. 2. Naucno Delo Press, Belgrade. 204 pp.

Richard, A. 1976. L'élevage de la seiche (*Sepia officinalis* L., mollusque céphalopode). Pp. 359–380 in G. Persoone and E. Jaspers, eds. Proc. 10th Europ. Symp. Mar. Biol. Universa Press, Welteren, Belgium.

Richards, T. L. 1967. Reproduction and development of the polychaete *Stauronereis rudolphi,* including a summary of development in the superfamily Eunicea. Mar. Biol. (Berlin) 1:124–133.

Richter, K. O. 1973. Freeze-branding for individually marking the banana slug: *Ariolimax columbianus* G. Northw. Sci. 47:109–113.

Richet, C. 1902. Du poison pruritogene et urticant dans les tantacules d'actinies. C.R. Soc. Biol. Paris 54:1438.

Ricketts, E. F., and J. Calvin. 1968. Between Pacific Tides (*revised by J. W. Hedgpeth*), 4th ed. Stanford University Press, Stanford, Calif. 614 pp.

Rieper, M. 1978. Bacteria as food for marine harpacticoid copepods. Mar. Biol. (Berlin) 45:337–345.

Riley, J. D. 1973. Induced spawning of the mussel *Mytilus edulis* L. and its uses in larval fish feeding. Proc. Challenger Soc. 4:116.

Rio, G. J., M. F. Stempien, Jr., R. F. Nigrelli, and G. D. Ruggieri. 1965. Echinoderm toxins. I. Some biochemical and physiological properties of toxins from several species of Asteroidea. Toxicon 3:147–155.

Roberts, D. 1975. The effect of pesticides on byssus formation in the common mussel, *Mytilus edulis.* Environ. Pollut. 8:241–254.

Roberts, M. H., Jr. 1975. Culture techniques for decapod crustacean larvae. Pp. 209–220 in W. L. Smith and M. H. Chanley, eds. Culture of Marine Invertebrate Animals. Plenum Press, New York and London.

Robinson, A. B., K. F. Manly, M. P. Anthony, J. F. Catchpool, and L. Pauling. 1965. Anesthesia of Artemia larvae: Method for quantitative study. Science 149:1255–1258.

Rollefsen, G. 1939. Artificial rearing of fry of seawater fish. Preliminary communication. Rapp. P.-V. Reun. Cons. Perm. Int. Explor. Mer 109:133.

Roonwal, M. L. 1944. Some observations on the breeding biology, and on the swelling, weight, water-content and embryonic movements in the developing eggs of the moluscan king crab, *Tachypleus gigas* (Müller) [Arthropoda, Xiphosurida]. Proc. Indian Acad. Sci. Sect. B 20:115–129.

Roosen-Runge, E. C. 1970. Life cycle of the hydromedusa *Phialidium gregarium* (A. Agassiz, 1862) in the laboratory. Biol. Bull. (Woods Hole, Mass.) 139:203–221.

Ropes, J. W., and A. S. Merrill. 1970. Marking surf clams. Proc. Natl. Shellfish. Assoc. 60:99–106.

Rosenberg, P. 1973. The giant axon of the squid: A useful preparation for neurochemical and pharmacological studies. Pp. 97–160 in R. Fried, ed. Methods of Neurochemistry, vol. 4. Marcel Dekker, New York. 332 pp.

Rossi, S. S., and J. W. Anderson. 1976. Toxicity of water-soluble fractions of No. 2 fuel oil and South Louisiana crude oil to selected stages in the life history of the polychaete, *Neanthes arenaeceodentata.* Bull. Environ. Contam. Toxicol. 16:18–21.

Rossi, S. S., J. W. Anderson, and G. S. Ward. 1976. Toxicity of water-soluble fractions of four test oils for the polychaetous annelids, *Neanthes arenaeceodentata* and *Capitella capitata.* Environ. Pollut. 10:9–18.

Rothstein, M., and W. L. Nicholas. 1969. Culture methods and nutrition of nematodes and Acanthocephala. Pp. 289–328 in M. Florkin and B. T. Scheer, eds. Chemical Zoology, vol. III. Academic Press, New York and London.

Ruggieri, G. D. 1975a. Aquatic animals in biomedical research. Ann. N.Y. Acad. Sci. 245:39–56.

Ruggieri, G. D. 1975b. Echinodermata. Pp. 229–243 in W. L. Smith and M. H. Chanley, eds. Culture of Marine Invertebrate Animals. Plenum Press, New York and London.

Runham, N. W., K. Isarankura, and B. J. Smith. 1965. Methods for narcotizing and anesthetizing gastropods. Malacologia 2:231–238.

Russell, F. S., and W. J. Rees. 1936. On rearing the hydroid *Zanclea implexa* (Alder) and its medusa *Zanclea gemmosa* McCrady, with a review of the genus *Zanclea.* J. Mar. Biol. Assoc. U.K. 21:107–130.

Ryland, R. S. 1974. Behaviour, settlement and metamorphosis of bryozoan larvae: A review. Thalassia Jugosl. 10:239–262.

Saeki, A. 1964. Studies on fish culture in filtered closed-circulation aquaria. Fundamental theory and system design standards. Directorate of Scientific Information Services, Defence Research Board, Canada T 77 J. 15 pp.

Sagara, J. 1958. Artificial discharge of reproductive elements of certain bivalves caused by treatment of sea water and by injection with NH_4OH. Bull. Jpn. Soc. Sci. Fish. 23:505–510.

Saito, Y., and N. Nakamura. 1961. Biology of the sea-hare, *Aplysia juliana*, as a predator of the brown seaweed, *Undaria pinnatifida.* I. The feeding habit. Bull. Jpn. Soc. Sci. Fish. 27:395–400.

Salmon, M. L., K. Horch, and G. W. Hyatt. 1977. Barth's myochordotonal organ as a receptor, or auditory and vibrational stimuli in fiddler crabs (*Uca pugilator* and *U. minax*). Mar. Behav. Physiol. 4:187-194.

Salser, B. R., and C. R. Mock. 1973. An air-lift circulator for algal culture tanks. Pp. 295–298 in J. W. Avault, Jr., ed. Proc. Fourth Annu. Meet. World Mariculture Soc. La. State Univ. Div. Continuing Educ., Baton Rouge.

Samoiloff, M. R., S. Schulz, Y. Jordan, K. Denich, and E. Arnott. 1980. A rapid simple long-term toxicity assay for aquatic contaminants using the nematode *Panagrellus redivivus.* Can. J. Fish. Aquatic Sci. 37:1167–1174.

Sandeen, M. I., and F. A. Brown, Jr. 1952. Responses of the distal retinal pigment of *Palaemonetes* to illumination. Physiol. Zool. 25:222–230.

Sander, E., and H. Rosenthal. 1975. Application of ozone in water treatment for home aquaria, public aquaria and for aquaculture purposes. Pp. 103–114 in W. J. Blogoslawski and R. G. Rice, eds. Aquatic Applications of Ozone. International Ozone Institute, Syracuse, N.Y.

Sandifer, P. A., and T. I. J. Smith. 1979. A method for artificial insemination of *Macrobrachium* prawns and its potential use in inheritance and hybridization studies. Pp. 403–418 in J. W. Avault, Jr., ed. Proc. Tenth Annu. Meet. World Mariculture Soc. La. State Univ. Div. Continuing Educ., Baton Rouge.

Sandifer, P. A., and W. A. Van Engle. 1971. Larval development of the spider crab, *Libinia dubia* H. Milne-Edwards (Brachyura, Majidae, Pisinae) reared in laboratory culture. Chesapeake Sci. 12:18–25.

Sandifer, P. A., T. I. J. Smith, and D. R. Calder. 1974a. Hydrozoans as pests in closed-system culture of larval decapod crustaceans. Aquaculture 4:55–59.

Sandifer, P. A., P. B. Zielinski, and W. E. Castro. 1974b. A simple airlift-operated tank for closed-system culture of decapod crustacean larvae and other small aquatic animals. Helgol. Wiss. Meeresunters. 26:82–87.

Sandifer, P. A., P. B. Zielinski, and W. E. Castro. 1975. Enhanced survival of larval grass shrimp in dilute solutions of the synthetic polymer, polyethylene oxide. U.S. Natl. Mar. Fish. Serv. Fish. Bull. 73:678–680.

San Feliu, J. M., F. Muñoz, and M. Alcaraz. 1973. Techniques of artificial rearing of crustaceans. Stud. Rev. Gen. Fish. Counc. Medit. No. 52:105–121.

San Feliu, J. M., F. Muñoz, F. Amat, J. Ramos, J. Peffa, and A. Sanz. 1976. Techniques de stimulation de la ponte et d'élevage de larves de crustaces et de poissons. Stud. Rev. Gen. Fish. Counc. Medit. 55:1–34.

Sarver, D. 1979. Larval culture of the shrimp *Thor amboinensis* (De Man, 1888) with reference to its symbiosis with the anemone *Antheopsis papillosa* (Kwietniewski, 1898). Crustaceana Suppl. (Leiden) 5:176–184.

Sastry, A. N. 1970. Culture of brachyuran crab larvae using a re-circulating sea water system in the laboratory. Helgol. Wiss. Meeresunters. 20:406–416.

Sastry, A. N. 1971. Effect of temperature on egg capsule deposition in the mud snail, *Nassarius obsoletus* (Say). Veliger 13:339–341.

Sastry, A. N. 1976. An experimental culture-research facility for the American lobster, *Homarus americanus*. Pp. 419–435 in G. Persoone and E. Jaspers, eds. Proc. 10th Europ. Symp. Mar. Biol., vol. 1. Universa Press, Welteren, Belgium.

Scarpelli, D. G., and A. Rosenfield, eds. 1976. Molluscan pathology. NOAA (Natl. Ocean. Atmos. Adm.) Mar. Fish. Rev. 38(10):1–50.

Scheltema, R. S. 1961. Metamorphosis of the veliger larvae of *Nassarius obsoletus* (Gastropoda) in response to bottom sediment. Biol. Bull. (Woods Hole, Mass.) 120:92–109.

Scheltema, R. S. 1962. Pelagic larvae of New England intertidal gastropods. I. *Nassarius obsoletus* Say and *Nassarius vibex* Say. Trans. Am. Microsc. Soc. 81:1–11.

Scheltema, R. S. 1964. Feeding habits and growth in the mud-snail *Nassarius obsoletus*. Chesapeake Sci. 5:161–166.

Scheltema, R. S. 1967. The relationship of temperature to the larval development of *Nassarius obsoletus* (Gastropoda). Biol. Bull. (Woods Hole, Mass.) 132:253–265.

Scheltema, R. S., and A. H. Scheltema. 1963. Pelagic larvae of New England intertidal gastropods. II. Anachis avara. Hydrobiologia 22:85–91.

Scherer, E. 1977. Behavioural assays—principles, results and problems. Pp.33–40 in Proceedings of the 3rd Aquatic Toxicity Workshop, Halifax, N.S., 1976. Environ. Prot. Serv. Tech. Rep. (EPS-5AR-77-1)

Schiedges, K. L. 1979. Reproductive biology and ontogenesis in the polychaete genus *Autolytus* (Annelida: Syllidae): Observations on laboratory-cultured individuals. Mar. Biol. (Berlin) 54:239–250.

Schilansky, M. M., N. L. Levin, and G. H. Fried. 1977. Metabolic implications of glucose-6-phosphate dehydrogenase and lactic dehydrogenase in two marine gastropods. Comp. Biochem. Physiol. 56B:1–4.

Schlichter, D. 1978. The extraction of specific proteins for the simultaneous ectodermal absorption of charged and neutral amino acids by *Anemonia sulcata* (Coelenterata, Anthozoa). Pp. 155–163 in D. S. McLusky and A. J. Berry, eds. Physiology and Behaviour of Marine Organisms. Pergamon Press, Oxford, New York

Schmidt-Nielsen, K. 1979. Animal Physiology: Adaptation and Environment, 2d ed. Cambridge University Press, London, New York, and Melbourne. xi + 560 pp.

Scholl, G. 1977. Beiträge zur Embryonalentwicklung on *Limulus polyphemus* L. (Chilcerata, Xiophosura). Zoomorphologie 86:99–154.

Schrank, W. W., R. L. Shoger, L. M. Schechtman, and D. W. Bishop. 1967. Electrically induced spawning in the male and female horseshoe crab, *Limulus polyphemus*. Biol. Bull. (Woods Hole, Mass.) 133:453. (Abstr.)

Schroeder, P. S., and C. O. Hermans. 1975. Annelida: Polychaeta. Pp. 1–213 in A. G. Giese and J. S. Pearse, eds. Reproduction of Marine Invertebrates. Vol. 3: Annelida and Echiurans. Academic Press, New York.

Schulte, E. H. 1976. The laboratory culture of the palaemonid prawn *Leander squilla*. Pp. 437–454 in G. Persoone and E. Jaspers, eds. Proc. 10th Europ. Symp. Mar. Biol., vol. 1. Universa Press, Welteren, Belgium.

Schwab, W. E. 1977. The ontogeny of swimming behavior in the scyphozoan, *Aurelia aurita*. I. Electrophysiological analysis. Biol. Bull. (Woods Hole, Mass.) 152:233–250.

Schwartz, F. J. 1966. Use of M.S. 222 in anesthetizing and transporting the sand shrimp. Prog. Fish Cult. 28:232–234.

Schwartz, F. J. 1977. Evaluation of colored Floy anchor tags on white shrimp, *Penaeus setiferus*, tagged in Cape Fear River, North Carolina 1973–1975. Fla. Sci. 40:22–27.

Scofield, V., and I. Weissman. 1981. Allorecognition in biological systems. Developmental and Comparative Immunology, in press.

Sebens, K. P. 1976. Individual marking of soft-bodied intertidal invertebrates *in situ*: A vital stain technique applied to the sea anemone, *Anthopleura xanthogrammica*. J. Fish. Res. Bd. Can. 33:1407–1410.

Seelemann, U. 1967. Rearing experiments on the amphibian slug *Alderia modesta*. Helgol. Wiss. Meeresunters. 15:128–134.

Segrove, F. 1941. The development of the serpulid *Pomatoceros triqueter*. Q. J. Microsc. Sci. N.S. 82:467–540.

Seki, H. 1966. Studies on microbial participation to food cycle in the sea. III. Trial cultivation of brine shrimp to adult in a chemostat (1). J. Oceanogr. Soc. Jpn. 22:105–110.

Sekiguchi, K. 1960. Embryonic development of the horse-shoe crab studied by vital staining. Bull. Mar. Biol. Stn. Asamuchi Tohuku Univ. 10:161–164.

Sekiguchi, K. 1973. A normal plate of the development of the Japanese horse-shoe crab, *Tachypleus tridentatus*. Sci. Rep. Tokyo Kyoiku Daigaku B. 15:153–162.

Sekiguchi, K., and K. Nakamura. 1979. Ecology of the extant horseshoe crabs. Pp. 37–45 in E. Cohen, ed. Biomedical Applications of the Horseshoe Crab (Limulidae). Alan R. Liss, New York.

Sellner, B. W. 1976. Survival and metabolism of the harpacticoid copepod *Thompsonula hyaenae* (Thompson) fed on different diatoms. Hydrobiologia 50:233–238.

Sepers, A. B. J. 1977. The utilization of dissolved organic compounds in aquatic environments. Hydrobiologia 52:39–54.

Serfling, S. A., and R. F. Ford. 1975. Laboratory culture of juvenile stages of the California spiny lobster *Panulirus interruptus* (Randall) at elevated temperatures. Aquaculture 6:377–387.

Serfling, S. A., J. C. Van Olst, and R. F. Ford. 1974. A recirculating culture system for larvae of the American lobster, *Homarus americanus*. Aquaculture 3:303–309.

Sheader, M. 1977. Breeding and marsupial development in laboratory-maintained *Parathemisto gaudichaudi* (Amphipoda). J. Mar. Biol. Assoc. U.K. 57:943–954.

Shields, R. 1977. Laboratory maintenance of a marine parasitic copepod. Wiad. Parazytol. 23:189–193.

Shiraishi, K., and L. Provasoli. 1959. Growth factors as supplements to inadequate algal foods for *Tigriopus japonicus*. Tohoku J. Agric. Res. 10:89–96.

Shirgur, G. A., and A. A. Naik. 1977. Observations on morphology, taxonomy, ephippial hatching and laboratory culture of a new species of *Alona* (*Alona taraporevalae* Shirgur & Naik), a chydorid cladoceran from Back Bay, Bombay. Pp. 48–59 in Proceedings of the Symposium on Warm Water Zooplankton. Spec. Publ. Natl. Inst. Oceanogr. (Goa, India). xi + 722 pp.

Shoger, R. L., and G. G. Brown. 1970. Ultrastructural study of sperm-egg interactions of the horseshoe crab, *Limulus polyphemus* L. (Merostomata: Xiphosura). J. Submicrosc. Cytol. 2:167–179.

Shuster, C. N., Jr. 1950. Observations on the natural history of the American horseshoe crab, *Limulus polyphemus*. Third Rep. Invest. Methods Improv. Shellfish Resourc. Mass. Woods Hole Oceanogr. Inst.: 18–23.

Shuster, C. N., Jr. 1979. Distribution of the American horseshoe "crab," *Limulus polyphemus* (L). Pp. 3–26 in E. Cohen, ed. Biomedical Applications of the Horseshoe Crab (Limulidae). Alan R. Liss, New York.

Silberzahn, N. 1977. Élevage en laboratoire de larves de crépidule: Premiers résultats concernant l'apparition de la gonade. Haliotis 6:261–266.

Silverstone, M., T. R. Toteson, and C. E. Cutress. 1977. The effect of iodide and various iodocompounds on initiation of strobilation in *Aurelia*. Gen. Comp. Endocrinol. 32:108–113.

Simpson, K. L. 1979. Focusing on the modest and minute brine shrimp. Maritimes 23(4):9–11.

Sindermann, C. J. 1960. Ecological studies of marine dermatitis-producing schistosome larvae in northern New England. Ecology 41:785–790.

Sindermann, C. J. 1970. Principal Diseases of Marine Fish and Shellfish. Academic Press, New York and London. x + 369 pp.

Sindermann, C. J., ed. 1977. Disease Diagnosis and Control in North American Marine Aquaculture. Dev. Aquaculture Fish. Sci. 6. Elsevier Sci. Publ. Co., Amsterdam, Oxford, and New York. xi + 329 pp.

Sindermann, C., and R. Gibbs. 1953. A dermatitis producing schistosome which causes "Clam Diggers Itch" along the central Maine coast. Maine Dep. Sea Shore Fish. Res. Bull. No. 12:1–20.

Skinner, D. M., and D. E. Graham. 1972. Loss of limbs as a stimulus to ecdysis in Brachyura (true crabs). Biol. Bull. (Woods Hole, Mass.) 143:222–233.

Skinner, D. M., D. J. Marsh, and J. S. Cook. 1965. Physiological salt solution for the land crab, *Gecarcinus lateralis*. Biol. Bull. (Woods Hole, Mass.) 129:355–365.

Slautterback, D. B., 1967. The cnidoblast–musculoepithelial cell complex in the tentacles of Hydra. Z. Zellforsch. Mikrosk. Anat. 79:296–318.

Smith, N. L., and H. M. Lenhoff. 1976. Regulation of frequency of pedal laceration in sea anemones. Pp. 117–129 in G. Mackie, ed. Ecology and behavior of coelenterates. Plenum Press, New York.

Smith, R. I., ed. 1964. Keys to Marine Invertebrates of the Woods Hole Region. Marine Biological Laboratory, Woods Hole, Mass. 208 pp.

Smith, R. I., and J. T. Carlton, eds. 1975. Light's Manual: Intertidal Invertebrates on the Central California Coast, 3d ed. University of California Press, Berkeley., 716 pp.

Smith, S. T., and T. H. Carefoot. 1967. Induced maturation of gonads in *Aplysia punctata* Cuvier. Nature 215:652–653.

Smith, T. I. J., J. S. Hopkins, and P. A. Sandifer. 1978. Development of a large-scale *Artemia* hatching system utilizing recirculated water. Pp. 701–714 in J. W. Avault, Jr., ed. Proc. Ninth Annu. Meet. World Mariculture Soc. La. State Univ. Div. Continuing Educ., Baton Rouge.

Smith, W. L., and M. H. Chanley, eds. 1975. Culture of Marine Invertebrate Animals. Plenum Press, New York and London. viii + 338 pp.

Solangi, M. A., and J. T. Ogle. 1977. A selected bibliography on the mass propagation of rotifers with emphasis on the biology and culture of *Brachionus plicatilis*. Gulf Res. Rep. 6:59–68.

Sorgeloos, P., and G. Persoone. 1972. Three simple culture devices for aquatic invertebrates and fish larvae with continous recirculation of the medium. Mar. Biol. (Berlin) 15:251–254.

Sorgeloos, P., and G. Persoone. 1973. A culture system for *Artemia, Daphnia*, and other invertebrates, with continuous separation of the larvae. Arch. Hydrobiol. 72:133–138.

Sorgeloos, P., and G. Persoone. 1975. Technological improvements for the cultivation of invertebrates as food for fishes and crustaceans. II. Hatching and culturing of the brine shrimp, *Artemia salina* L. Aquaculture 6:303–317.

Sorgeloos, P., M. Baeza-Mesa, F. Benijts, and G. Persoone. 1976. Research on the culturing of the brine shrimp *Artemia salina* L. at the State University of Ghent (Belgium). Pp. 473–495 in G. Persoone and E. Jaspers, eds. Proc. 10th Europ. Symp. Mar. Biol., vol. 1. Universa Press, Welteren, Belgium.

Sorgeloos, P., G. Persoone, F. De Winter, E. Bossuyt, and N. De Pauw. 1977a. Airwater pumps as cheap and convenient tools for high density culturing of microscopic algae. Pp. 173–183 in J. W. Avault, Jr., ed. Proc. Eighth Annu. Meet. World Mariculture Soc. La. State Univ. Div. Continuing Educ., Baton Rouge.

Sorgeloos, P., E. Bossuyt, E. Laviña, M. Baeza-Mesa, and G. Persoone. 1977b. Decapsulation of *Artemia* cysts: A simple technique for the improvement of the use of brine shrimp in aquaculture. Aquaculture 12:311–315.

Sorgeloos, P., C. Remiche-Van der Wielen, and G. Persoone. 1978. The use of *Artemia* nauplii for toxicity tests—a critical analysis. Ecotoxicol. Environ. Safety 2:249–255.

Sosnowski, S. L., and J. H. Gentile. 1978. Toxicological comparison of natural and cultured populations of *Acartia tonsa* to cadmium, copper, and mercury. J. Fish. Res. Bd. Can. 35:1366–1369.

Southward, E. C., and A. J. Southward. 1977. Natural and synthetic diets for Pogonophora. Pg. 93 in M. Rechcigl, Jr., ed. CRC Handbook Series in Nutrition and Food. Section G: Diets, Culture Media, Food Supplements. Volume II: Food Habits of, and Diets for Invertebrates and Vertebrates—Zoo Diets. CRC Press, Cleveland.

Spangenberg, D. B. 1965. Cultivation of the life stages of *Aurelia aurita* under controlled conditions. J. Exp. Zool. 159:303–318.

Spangenberg, D. B. 1967. Iodine induction of metamorphosis in *Aurelia*. J. Exp. Zool. 165:441–450.

Spangenberg, D. B. 1969. Recent studies of strobilation in jellyfish. Oceanogr. Mar. Biol. Annu. Rev. 6:231–247.

Sparks, A. K. 1972. Invertebrate Pathology. Noncommunicable Diseases. Academic Press. New York and London. xvi + 387 pp.

Spiegel, M., and E. S. Spiegel. 1975. The reaggregation of dissociated embryonic sea urchin cells. Am. Zool. 15:583–606.

Spight, T. M. 1974. Sizes of populations of a marine snail. Ecology 55:712–729.

Spoon, D. M., and R. S. Blanquet. 1978. Life cycle and ecology of the minute hydrozoon *Microhydrula*. Trans. Am. Microsc. Soc. 97:208–216.

Spotte, S. 1973. Marine Aquarium Keeping the Science, Animals, and Art. John Wiley & Sons, New York, Chichester, Brisbane, and Toronto. xv + 171 pp.

Spotte, S. 1979a. Fish and Invertebrate Culture, Water Management in Closed Systems, 2d ed. John Wiley & Sons, New York, Chichester, Brisbane, and Toronto. xvi + 179 pp.

Spotte, S. 1979b. Seawater Aquariums, the Captive Environment. John Wiley & Sons, New York, Chichester, Brisbane, and Toronto. xxii + 413 pp.

Sprague, J. B. 1976. Current status of sublethal tests of pollutants in aquatic organisms. J. Fish. Res. Bd. Can. 33:1988–1992.

Stearns, L. W. 1974. Sea Urchin Development. Dowden, Hutchinson and Ross, Stroudsburg, Pa. 352 pp.

Stebbing, A. R. D. 1976. The effects of low metal levels on a clonal hydroid. J. Mar. Biol. Assoc. U.K. 56:977–994.

Stebbing, A. R. D. 1979. An experimental approach to the determinants of biological water quality. Philos. Trans. R. Soc. Lond. B Biol. Sci. 286:465–481.

Stein, J. R., ed. 1973. Handbook of Phycological Methods: Culture Methods and Growth Measurements. Cambridge University Press, New York. 448 pp.

Steinbeck, J., and E. F. Ricketts. 1941. Sea of Cortez a Leisurely Journal of Travel and Research with a Scientific Appendix Comprising Materials for a Source Book on the Marine Animals of the Panamic Faunal Province. Viking Press, New York. x + 598 pp.

Stephen, A. C., and S. J. Edmonds. 1972. The Phyla Sipuncula and Echiura. British Museum (Natural History), London. 528 pp.

Stephens, L. L., and J. E. Blankenship. 1974. A technique for rearing opisthobranch larvae. Echo Abstr. Proc. Sixth Annu. Meet. West. Soc. Malacologists 6:28–29.

Stewart, M. G. 1978. The uptake and utilization of dissolved amino acids by the bivalve *Mya arenaria* (L.). Pp. 165–171 in D. S. McLusky and A. J. Berry, eds. Physiology and Behaviour of Marine Organisms. Pergamon Press, Oxford, New York.

Stewart, M. G. 1979. Absorption of dissolved organic nutrients by marine invertebrates. Oceanogr. Mar. Biol. Annu. Rev. 17:163–192.

Stiles, S. S. 1978. Conventional and experimental approaches to hybridization and inbreeding research in the oyster. Pp. 577–586 in J. W. Avault, Jr., ed. Proc. Ninth Annu. Meet. World Mariculture Soc. La. State Univ. Div. Continuing Educ., Baton Rouge.

Stober, Q. J., P. A. Dinnel, S. C. Crumley, and C. Olds. 1979. Echinoderm fertilization bioassay. 1978 research in fisheries. Annu. Rep. Coll. Fish. Univ. Wash. Univ. Wash. Coll. Fish. Contr. No. 500:40.

Stokes, D. R., 1974. Morphological substrates of conduction in the colonial hydroid *Hydractinia echinata*. I. An ectodermal nerve net. J. Exp. Zool. 190:19–46.

Stonehouse, B., ed. 1978. Animal Marking. Recognition of Animals in Research. University Park Press, Baltimore. viii + 257 pp.

Stora, G. 1972. Contribution à l'étude de la notion de concentration léthale limite moyenne (CL 50) appliquée à des invertébrés marins. 1. Étude méthodologique. Tethys 4:597–644.

Strathmann, R. R. 1975. Larval feeding in echinoderms. Am. Zool. 15:717–730.

Strenth, N. E., and J. E. Blankenship. 1978. Laboratory culture, metamorphosis and development of *Aplysia brasiliana* Rang, 1828 (Gastropoda: Opisthobranchia). Veliger 21:99–103.

Strickland, J. D. H., and T. R. Parsons. 1972. A Practical Handbook of Seawater Analysis, 2d ed. Bull. Fish. Res. Bd. Can. No. 167. 310 pp.

Struhsaker, J. W., and J. D. Costlow, Jr. 1969. Some environmental effects on the larval development of *Littorina picta* (Mesogastropoda), reared in the laboratory. Malacologia 9:403–419.

Strumwasser, F., J. W. Jacket, and R. B. Alvarez. 1969. A seasonal rhythm in the neural extract induction of behavioral egg-laying in *Aplysia*. Comp. Biochem. Physiol. 29:197–206.

Stunkard, H., and M. Hinchliffe. 1952. The morphology and life history of *Microbitharzia variglandis* (Miller and Northrup 1926) Stunkard and Hinchliffe 1951, avian blood flukes whose larvae cause "Swimmer's Itch" of ocean beaches. J. Parasitol. 38:248–265.

Styron, C. E. 1967. Effects on development of inhibiting polar lobe formation by compression in *Ilyanassa obsoleta* Stimpson. Acta Embryol. Morphol. Exp. 9:246–254.

Sulkin, S. D. 1975. The significance of diet in the growth and development of larvae of the blue crab, *Callinectes sapidus* Rathbun, under laboratory conditions. J. Exp. Mar. Biol. Ecol. 20:119–135.

Sulkin, S. D. 1978. Nutritional requirements during larval development of the portunid crab, *Callinectes sapidus* Rathbun. J. Exp. Mar. Biol. Ecol. 34:29–41.

Sulkin, S. D., and C. E. Epifanio. 1975. Comparison of rotifers and other diets for rearing early larvae of the blue crab, *Callinectes sapidus* Rathbun. Estuarine Coastal Mar. Sci. 3:109–113.

Sulkin, S. D., and L. L. Minasian. 1973. Synthetic sea water as a medium for raising crab larvae. Helgol. Wiss. Meeresunters. 25:126–134.

Sulkin, S. D., and K. Norman. 1976. A comparison of two diets in the laboratory culture of the zoeal stages of the brachyuran crabs *Rhithropanopeus harrisii* and *Neopanope* sp. Helgol. Wiss. Meeresunters. 28:183–190.

Sulkin, S. D., and D. L. Pickett. 1973. Effect of agitation in the culture of the zoeal stages of the mud crab, *Rhithropanopeus harrisii*. Chesapeake Sci. 14:292–294.

Sulkin, S. D., E. S. Branscomb, and R. E. Miller. 1976. Induced winter spawning and culture of larvae of the blue crab, *Callinectes sapidus* Rathbun. Aquaculture 8:103–113.

Summers, W. C., and J. J. McMahon. 1974. Studies on the maintenance of adult squid (*Loligo pealei*). I. Factorial survey. Biol. Bull. (Woods Hole, Mass.). 146:279–290.

Summers, W. C., J. J. McMahon, and G. N. P. A. Ruppert. 1974. Studies on the maintenance of adult squid (*Loligo pealei*). II. Empirical extensions. Biol. Bull. (Woods Hole, Mass.). 146:291–301.

Swedmark, M., Å. Granmo, and S. Kollberg. 1976. Toxicity testing at Kristineberg Marine Biology Station. Pp. 65–74 in Lectures Presented at the Second FAO/SIDA Training Course on Marine Pollution in Relation to Protection of Living Resources. (FAO/SIDA/TF 108, Suppl. 1)

Switzer-Dunlap, M. 1978. Larval biology and metamorphosis of aplysiid gastropods. Pp. 197–206 in F. S. Chia and M. E. Rice, eds. Settlement and Metamorphosis of Marine Invertebrate Larvae. Elsevier North-Holland, New York.

Switzer-Dunlap, M., and M. G. Hadfield. 1977. Observations on development, larval growth and metamorphosis of four species of Aplysiidae (Gastropoda: Opisthobranchia) in laboratory culture. J. Exp. Mar. Biol. Ecol. 29:245–261.

Switzer-Dunlap, M., and M. G. Hadfield. 1979. Reproductive patterns of Hawaiian aplysiid gastropods. In S. E. Stancyk, ed. Reproductive Ecology of Marine Invertebrates. University of South Carolina Press, Columbia.

Tabb, D. D., W. T. Yang, Y. Hirono, and J. Helmen. 1972. A manual for culture of pink shrimp, *Penaeus duorarum*, from eggs to post larvae suitable for stocking. Sea Grant Spec. Bull. No. 7:1–59.

Takami, A., and H. Iwasaki. 1978. Cultivation of marine Cladocera, *Penilia avirostris* Dana. Bull. Jpn. Soc. Sci. Fish. 44:393.

Takami, A., H. Iwasaki, and M. Nagoshi. 1978. Studies on the cultivation of marine Cladocera. II. Cultivation of *Penilia avirostris* Dana. Bull. Fac. Fish. Mie Univ. No. 5:47–63. (in Japanese with English summary)

Takano, H., 1971a. Breeding experiments of a marine littoral copepod, *Tigriopus japonicus* Mori. Bull. Tokai Reg. Fish. Res. Lab. No. 64:71–80.

Takano, H. 1971b. Notes on the raising of an estuarine copepod *Gladioferens imparipes* Thomson. Bull. Tokai Reg. Fish. Res. Lab. No. 64:81–87.

Taki, I. 1941. On keeping octopods in an aquarium for physiological experiments, with remarks on some operative techniques. Venus Jpn. J. Malacol. 10:140–156.

Tarpley, W. A. 1958. Studies on the use of the brine shrimp *Artemia salina* (Leach) as a test organism for bioassay. J. Econ. Entomol. 51:780–783.

Tatem, H. E., J. W. Anderson, and J. M. Neff. 1976. Seasonal and laboratory variations in the health of grass shrimp *Palaemonetes pugio*: Dodecyl sodium sulfate bioassay. Bull. Environ. Contam. Toxicol. 16:368–375.

Taylor, G. T., and E. Anderson. 1969. Cytochemical and fine structural analysis of oogenesis in the gastropod, *Ilyanassa obsoleta*. J. Morphol. 129:211–248.

Taylor, W. R. 1962. Marine Algae of the Northeastern Coast of North America, 2d rev. ed. University of Michigan Press, Ann Arbor. 509 pp.

Theilacker, G. H., and M. F. MacMaster. 1971. Mass culture of the rotifer *Brachionus plicatilis* and its evaluation as a food for larval anchovies. Mar. Biol. (Berlin) 10:183–188.

Thomas, L. 1976. Marine models in modern medicine. Oceanus 19(2):2–5.

Thomas, R. F., and L. Opresko. 1973. Observations on *Octopus joubini*: Four laboratory reared generations. Nautilus 87:61–65.

Thompson, C. H., R. A. A. Blackman, S. Genovese, P. G. Jeffery, A. B. Jernelöv, E. M. Levy, O. G. Mironov, K. H. Palmork, G. Tomczak, and S. L. D. Young. 1977. Impact of Oil on the Marine Environment. IMCO–FAO–UNESCO–WMO–WHO–IAEA–UN Jt. Group Experts Sci. Aspects Mar. Pollut.-Gesamp.-Rep. Stud. No. 6. ix + 250 pp.

Thompson, T. E., and A. Bebbington. 1969. Structure and function of the reproductive organs of three species of *Aplysia* (Gastropoda, Opisthobranchia). Malacologia 7:347–380.

Thorson, G. 1946. Reproduction and larval development of Danish marine bottom invertebrates with special reference to the planktonic larvae in the Sound (Oresund). Medd. Dan. Fisk. Havunders. Ser. Plankton. 4:1–523.

Thorson, K. N. 1967. A new high-speed tagging device. Calif. Fish Game 53:289–292.

Thun, W. von. 1966. Eine Methode zur Kultivierung der Mikrofauna. Veroeff. Inst. Meeresforsch. Bremerhaven Suppl. 2:277–280.

Tietjen, J. H. 1967. Observations on the ecology of the marine nematode *Monhystera filicaudata* Allgen, 1929. Trans. Am. Microsc. Soc. 86:304–306.

Tietjen, J. H., and J. J. Lee. 1972. Life cycles of marine nematodes. Influence of temperature and salinity on the development of *Monhystera denticulata* Timm. Oecologia (Berlin) 10:167–176.

Tietjen, J. H., and J. J. Lee. 1973. Life history and feeding habits of the marine nematode, *Chromadora macrolaimoides* Steiner. Oecologia (Berlin) 12:303–314.

Tietjen, J. H., and J. J. Lee. 1977a. Feeding behavior of marine nematodes. Pp. 21–35 in B. C. Coull, ed. Ecology of Marine Benthos. University of South Carolina Press, Columbia.

Tietjen, J. H., and J. J. Lee. 1977b. Life histories of marine nematodes. Influence of temperature and salinity on the reproductive potential of *Chromadorina germanica* Bütschli. Mikrofauna Meeresboden 61:263–270.

Tietjen, J. H., J. J. Lee, J. Rullman, A. Greengart, and J. Trompeter. 1970. Gnotobiotic culture and physiological ecology of the marine nematode *Rhabditis marina* Bastian. Limnol. Oceanogr. 15:535–543.

Tighe-Ford, D. J., M. J. D. Power, and D. C. Vaile. 1970. Laboratory rearing of barnacle larvae for antifouling research. Helgol. Wiss. Meeresunters. 20:393–405.

Tilney, L. G. 1975. Actin filaments in the acrosomal reaction of *Limulus* sperm. J. Cell Biol. 64:289–310.

Tilney, L. G., J. G. Clain, and M. S. Tilney. 1979. Membrane events in the acrosomal reaction of *Limulus* sperm. Membrane fusion, filament-membrane particle attachment, and the source and formation of new membrane surface. J. Cell Biol. 81:229–253.

Timourian, H. 1968. The effect of zinc on sea-urchin morphogenesis. J. Exp. Zool. 169:121–132.

Tomey, W. A. 1978. New techniques with brine shrimp hatchings. Pet Fish Mon. 12:411–412.

Toth, S. E. 1965. Cultivation of marine hydroids. Bios 36:63–65.

Towle, A., and A. C. Giese. 1966. Biochemical changes during reproduction and starvation in the sipunculid worm *Phascolosoma agassizii*. Comp. Biochem. Physiol. 19:667–680.

Townsley, P. M., R. A. Richy, and P. C. Trussell. 1966. The laboratory rearing of the shipworm, Bankia setacea (Tryon). Proc. Natl. Shellfish. Assoc. 56:49–52.

Tranter, D. J., and O. Augustine. 1973. Observations on the life history of the blue-ringed octopus *Hapalochlaena maculosa*. Mar. Biol. (Berlin) 18:115–128.

Traut, W. 1968. Genetische Analyse zweier Mutanten von *Dinophilus gyrociliatus* (Archiannelida) mit veränderter Eigrösse. Helgol. Wiss. Meeresunters. 18:296–316.

Trider, D. J., E. G. Mason, and J. D. Castell. 1979. Survival and growth of juvenile American lobsters (*Homarus americanus*) after eyestalk ablation. J. Fish. Res. Bd. Can. 36:93–97.

Trieff, N. M., M. McShan, D. Grajcer, and M. Alam. 1973. Biological assay of *Gymnodinium breve* toxin using brine shrimp (*Artemia salina*). Tex. Rep. Biol. Med. 31:409–422.

Trinkaus-Randall, V., and J. E. Mittenthal. 1978. Intra- and interspecific transplantation of limb buds in Crustacea: A new method for studying central and peripheral interactions in crustacean limbs. J. Exp. Zool. 204:275–281.

Tubiash, H. S. 1971. Soft-shell clam, *Mya arenaria*, a convenient laboratory animal for screening pathogens of bivalve mollusks. Appl. Microbiol. 22:321–324.

Tusov, J., and L. V. Davis. 1971. Influence of environmental factors on the growth of *Bougainvillia* sp. Pp. 52–65 in H. M. Lenhoff, L. Muscatine, and L. V. Davis, eds. Experimental Coelenterate Biology. University of Hawaii Press, Honolulu.

Tyler, A. 1949. A simple, non-injurious method for inducing spawning of sea urchins and sand dollars. Collectors Net 19:19–20.

Tyler, A., and B. S. Tyler. 1966. The gametes; some procedures and properties. Pp. 639–682 in R. A. Boolootian, ed. Physiology of Echinodermata. John Wiley Interscience, New York.

Tyler-Schroeder, D. B. 1978a. Culture of the grass shrimp (*Palaemonetes pugio*) in the laboratory. Pp. 69–72 in Bioassay Procedures for the Ocean Disposal Permit Program. Environmental Research Laboratory, U.S. Environmental Protection Agency, Gulf Breeze, Fla.

Tyler-Schroeder, D. B. 1978b. Static bioassay procedure using grass shrimp (*Palaemonetes* sp.) larvae. Pp. 73–82 in Bioassay Procedures for the Ocean Disposal Permit Program. Environmental Research Laboratory, U.S. Environmental Protection Agency, Gulf Breeze, Fla.

Tyler-Schroeder, D. B. 1978c. Entire life-cycle toxicity test using grass shrimp (*Palaemonetes pugio* Holthuis). Pp. 83–88 in Bioassay Procedures for the Ocean Disposal Permit Program. Environmental Research Laboratory, U.S. Environmental Protection Agency, Gulf Breeze, Fla.

Tyler-Schroeder, D. B. 1979. Use of the grass shrimp (*Palaemonetes pugio*) in a life-cycle toxicity test. Pp. 159–170 in L. L. Marking and R. A. Kimerle, eds. Aquatic Toxicology. Proc. 2d Annu. Symp. Aquatic Toxicol. American Society for Testing and Materials, Philadelphia. (STP 667)

Uchida, T., and Y. Dotsu. 1973. Collection of the T. S. Nagasaki Maru of Nagasaki University. III. The transportation, rearing and larva hatching of the spiny lobster, *Panulirus polyghagus*. Bull. Fac. Fish. Nagasaki Univ. No. 35:1–9. (in Japanese with English summary)

Ukeles, R. 1976a. Views on bivalve larvae nutrition. Pp. 127–162 in K. S. Price, Jr., W. N. Shaw, and K. S. Danberg, eds. Proc. First Int. Conf. Aquaculture Nutr. Coll. Mar. Stud. Univ. Del., Newark.

Ukeles, R. 1976b. Cultivation of plants: Unicellular plants. Pp. 368–466 in O. Kinne, ed. Marine Ecology. Vol III: Cultivation. John Wiley & Sons, London.

U.S. Environmental Protection Agency. 1976. Bioassay Procedures for the Ocean Disposal Permit Program. Environmental Research Laboratory, Gulf Breeze–Narragansett–Corvallis. 109 pp. (EPA-600/9-76-010)

U.S. Environmental Protection Agency. 1978. Bioassay Procedures for the Ocean Disposal Permit Program. Environmental Research Laboratory, U.S. Environmental Protection Agency, Gulf Breeze, Fla., x + 121 pp. (EPA-600/9-78-010)

U.S. Environmental Protection Agency and U.S. Army Corps of Engineers. 1977. Ecological Evaluation of Proposed Discharge of Dredged Material into Ocean Waters: Implementation Manual for Section 103 of Public Law 92–533 (Marine Protection Research and Sanctuary Act) of 1972. Report of the U.S. Environmental Protection Agency Technical Committee on Criteria for Dredged and Fill Materials and U.S. Army Corps of Engineers. Environmental Effects Laboratory, U.S. Army Engineer Waterways Experimental Station, Vicksburg, Miss. 24 pp. and Appendices A–H.

U.S. Waterways Experiment Station. 1976. Ecological Evaluation of Proposed Discharge of Dredged or Fill Material into Navigable Waters: Interim Guidance for Implementation of Section 404 (B) (1) of Public Law 92–500 (Federal Water Pollution Control Acts Amendment) of 1972. Environmental Effects Laboratory, U.S. Army Engineer Waterways Experimental Station, Vicksburg, Miss. 33 pp. and Appendices A–E. (Miscellaneous Paper D-76-17).

Usuki, I. 1970. Studies on the life history of Aplysiae and their allies in the Sado district of the Japan Sea. Sci. Rep. Niigata Univ. Ser. D (Biol.) 7:91–105.

Vanderhorst, J. R., C. I. Gibson, and L. J. Moore. 1976. The role of dispersion in fuel oil bioassay. Bull. Environ. Contam. Toxicol. 15:93–100.

Van Heukelem, W. F. 1977. Laboratory maintenance, breeding, rearing, and biomedical research potential of the Yucatan octopus (*Octopus maya*). Lab. Anim. Sci. 27:852–859.

Vannucci, M. 1959. Catalogue of Marine Larvae. Universidade de Saõ Paulo, Instituto Oceanografico, Saõ Paulo, Brazil. 44 pp.

Veitch, F. P., and H. Hidu. 1971. Gregarious setting in the American oyster *Crassostrea virginica* Gmelin. I. Properties of a partially purified "setting factor." Chesapeake Sci. 12:173–178.

Venkataramiah, A., G. J. Lakshmi, and G. Gunter. 1976. A review of the effects of some environmental and nutritional factors on brown shrimp, *Penaeus aztecus* Ives in laboratory cultures. Pp. 523–547 in G. Persoone and E. Jaspers, eds. Proc. 10th Europ. Symp. Mar. Biol., vol. 1. Universa Press, Welteren, Belgium.

Vernberg, W. B., and F. J. Vernberg. 1975. The physiological ecology of larval *Nassarius obsoletus* (Say). Pp. 179–190 in Harold Barnes, ed. Ninth Mar. Biol. Symp. Aberdeen University Press, Aberdeen.

Vernberg, W. B., F. J. Vernberg, and F. W. Beckerdite, Jr. 1969. Larval trematodes: Double infections in common mud-flat snail. Science 164:1287–1288.

Viglierchio, D. R., and R. N. Johnson. 1971. On the maintenance of *Deontostoma californicum*. J. Nematol. 3:86–88.

Voss, G. L. 1976. Seashore Life of Florida and the Caribbean a Guide to the Common Marine Invertebrates of the Atlantic from Bermuda to the West Indies and of the Gulf of Mexico. E. A. Seemann Publ. Co., Miami. 168 pp.

Vuillemin, S. 1968. Élevage de serpulinés (Annélides Polychètes). Vie Milieu A 19:195–199.

Wagner, A. 1960. Maintenance of a marine snail in the laboratory. J. Parasitol. 46:186.

Waldichuk, M. 1974. Some biological concerns in heavy metals pollution. Pp. 1–57 in F. John and Winona B. Vernberg, eds. Pollution and Physiology of Marine Organisms. Academic Press, New York.

Walker, J. J., N. Longo, and M. E. Bitterman. 1970. The octopus in the laboratory. Handling, maintenance, training. Behav. Res. Methods Instrum. 2:15–18.

Walne, P. R. 1964: The culture of marine bivalve larvae. Pp. 197–210 in K. M. Wilbur and C. M. Yonge, eds. Physiology of Mollusca, vol. I. Academic Press, New York.

Walne, P. R. 1966. Experiments in the large-scale culture of the larvae of *Ostrea edulis* L. Fish. Invest. Minist. Agric. Fish. Food (G. Br.) Ser. II Salmon Freshwater Fish. 25 (4):1–53.

Walne, P. R. 1970. Present problems in the culture of the larvae of *Ostrea edulis*. Helgol. Wiss. Meeresunters. 20:514–525.

Walne, P. R. 1974a. Shellfish culture. Pp. 379–398 in F. R. Harden Jones, ed. Sea Fisheries Research. John Wiley & Sons, New York.

Walne, P. R. 1974b. Culture of Bivalve Mollusks: 50 Years' Experience at Conwy. Fishing News (Books) Ltd., Surrey, England. 173 pp.

Walter, M. A. 1977. Feeding and nutrition of Ctenophora. Pp. 13–17 in M. Rechcigl, Jr., ed. CRC Handbook Series in Nutrition and Food. Section G: Diets, Culture Media, Food Supplements. Volume II: Food Habits of, and Diets for Invertebrates and Vertebrates—Zoo Diets. CRC Press, Cleveland.

Ward, T. J., E. D. Rider, and D. A. Drozdowski. 1979. A chronic toxicity test with the marine copepod *Acartia tonsa*. Pp. 148–158 in L. L. Marking and R. A. Kimerle, eds. Aquatic Toxicology. Proc. 2d Annu. Symp. Aquatic Toxicol. American Society for Testing and Materials, Philadelphia. (STP 667)

Ward, W. W. 1974. Aquarium systems for the maintenance of ctenophores and jellyfish and for the hatching and harvesting of brine shrimp (*Artemia salina*) larvae. Cheasapeake Sci. 15:116–118.

Warren, L. M. 1976a. Acute toxicity of manganese and mercury to *Capitella*. Mar. Pollut. Bull. 7:69.

Warren, L. M. 1976b. A review of the genus *Capitella* (Polychaeta Capitelladae). J. Zool. Lond. 180:195–209.

Watanabe, J. M., and L. R. Cox. 1975. Spawning behavior and larval development in *Mopalia lignosa* and *Mopalia mucosa* (Mollusca: Polyplacophora) in central California. Veliger 18 (Suppl.):18–27.

Waterman, T. H. 1958. On the doubtful validity of *Tachypleus hoeveni* Pocock, an Indonesian horseshoe crab (Xiphosura). Postilla Yale Peabody Mus. Nat. Hist. No. 36:1–17.

Waterman, T. H., ed. 1960. The Physiology of Crustacea. Vol. I: Metabolism and Growth. Academic Press, New York. 670 pp.

Waterman, T. H., ed. 1961. The Physiology of Crustacea. Vol. II: Sense Organs, Integration, and Behavior. Academic Press, New York. 681 pp.

Watling, L., and D. Mauer. 1973. Guide to the Macroscopic Estuarine and Marine Invertebrates of the Delaware Bay Region. Del. Bay Rep. Ser. Univ. Del. 5. 178 pp.

Wear, R. G., and J. A. Santiago. 1976. Induction of maturity and spawning in *Penaeus monodon* Fabricius, 1798, by unilateral eyestalk ablation (Decapoda, Natantia). Crustaceana (Leiden) 31:218–220.

Weiler-Stolt, B. 1960. Über die Bedeutung der interstitiellen Zellen für die Entwicklung und Fortpflanzung mariner Hydroiden. Wilhelm Roux' Arch. Entwicklungsmech. Org. 152:398–454.

Weis, J. S. 1977. Limb regeneration in fiddler crabs: Species differences and effects of methyl-mercury. Biol. Bull. (Woods Hole, Mass.) 152:263–274.

Welch, J., and S. D. Sulkin. 1974. Effect of diet concentration on mortality and rate of development of zoeae of the Xanthid crab, *Rhithropanopeus harrisii* (Gould). J. Elisha Mitchell Sci. Soc. 90:69–72.

Wells, M. J. 1978. Octopus: Physiology and Behaviour of an Advanced Invertebrate. Chapman & Hall, London. John Wiley & Sons, New York. xvi + 417 pp.

Wells, M. J., and J. Wells. 1972. Sexual displays and mating of *Octopus vulgaris* Cuvier and *O. cyanea* Gray and attempts to alter performance by manipulating the glandular condition of the animals. Anim. Behav. 20:293–308.

Wells, R. M. G., and L. M. Warren. 1975. The function of the cellular haemoglobins in *Capitella capitata* (Fabricius) and *Notomastus latericeus* Sars (Capitellidae: Polychaeta). Comp. Biochem. Physiol. 51A:737–740.

Werner, B. 1955. Über die Anatomie, die Entwicklung und Biologie des Veligers und der Veliconcha von *Crepidula fornicata* L. (Gastropoda Prosobranchia). Helgol. Wiss. Meeresunters. 5:169–217.

Werner, B. 1968. Polypengeneration und Entwicklungsgeschichte von *Eucheilota maculata* (Thecata—Leptomedusae). Mit einem Beitrag zur Methoden der Kultur mariner Hydroiden. Helgol. Wiss. Meeresunters. 18:136–168.

Werner, B. 1975. Bau und Lebensgeschichte des Polypen von *Tripedalia cystophora* (Cubozoa, class nov., Carybdeidae) und seine Bedeutung für die Evolution de Cnidaria. Helgol. Wiss. Meeresunters. 27:461–504.

West, D. L., and R. W. Renshaw. 1970. The life cycle of *Clytia attenuata* (Calyptoblastea: Campanulariidae). Mar. Biol. (Berlin) 7:332–339.

West, W. Q. B., and K. K. Chew. 1968. Application of the Bergman-Jefferts tag on the spot shrimp, *Pandalus platyceros* Brandt. Proc. Natl. Shellfish. Assoc. 58:93–100.

Westernhagen, H. von. 1976. Some aspects of the biology of the hyperiid amphipod, *Hyperoche medusarum*. Helgol. Wiss. Meeresunters. 28:43–50.

Westfall, J. A. 1970. The nematocyte complex in a hydromedusan, *Gonionemus vertens*. Z. Zellforsch. Mikrosk. Anat. 110:457–470.

Wickins, J. F. 1972. Developments in the laboratory culture of the common prawn, *Palaemon serratus* Pennant. Fish. Invest. Ser. II. Mar. Fish. G. Br. Minist. Agric. Fish. Food 27(4):1–23.

Wickins, J. F. 1976a. The tolerance of warm-water prawns to recirculated water. Aquaculture 9:19–37.

Wickins, J. F. 1976b. Prawn biology and culture. Oceanogr. Mar. Biol. Annu. Rev. 14:435–507.

Widdows, J. 1978. Physiological indices of stress in *Mytilus edulis*. J. Mar. Biol. Assoc. U.K. 58:125–142.

Wilber, C. G. 1947. The effect of prolonged starvation on the lipids in *Phascolosoma gouldii*. J. Cell. Comp. Physiol. 29:179–183.

Williams, A. B. 1965. Marine decapod crustaceans of the Carolinas. Fish. Bull. Fish Wildl. Serv. (U.S.) 65:1–298.

Williams, B. G. 1968. Laboratory rearing of the larval stages of *Carcinus maenas* (L.) [Crustacea: Decapoda]. J. Nat. Hist. 2:121–126.

Willows, A. O. D. 1973. Gastropod nervous system as a model experimental system in neurobiological research. Fed. Proc. 32:2215–2223.

Wilson, D. P. 1968. Long-term effects of low concentrations of an oil-spill remover ("detergent"): Studies with the larvae of *Sabellaria spinulosa*. J. Mar. Biol. Assoc. U.K. 48:177–186.

Wilson, D. P. 1970. Additional observations on larval growth and settlement of *Sabellaria alveolata*. J. Mar. Biol. Assoc. U.K. 50:1–31.

Wilson, E. B. 1904. Experimental studies on germinal localization. I. The germ regions in the egg of *Dentalium*. J. Exp. Zool. 1:1–72.

Wilson, H. V. 1937. Notes on the cultivation and growth of sponges from reduction bodies, dissociated cells, and larvae. Pp. 137–139 in J. G. Needham *et al.*, eds. Culture Methods for Invertebrate Animals. Comstock Publishing Co., Ithaca, N.Y.

Wilt, F. H., and N. K. Wessels, eds. 1967. Methods in Developmental Biology. Thomas Y. Crowell Co., New York. 813 pp.

Winget, R. R., D. Maurer, and L. Anderson. 1973. The feasibility of closed system mariculture: Preliminary experiments with crab molting. Proc. Natl. Shellfish. Assoc. 63:88–92.

Winkler, L. R. 1959. A mechanism of color variation operating in the West Coast sea hare, *Aplysia californica* Cooper. Pac. Sci. 13:63–66.

Winston, J. E. 1976. Experimental culture of the estuarine ectoproct *Conopeum tenuissimum* from Chesapeake Bay. Biol. Bull. (Woods Hole, Mass.). 150:318–335.

Winston, J. E. 1977. Feeding in marine bryozoans. Pp. 233–271 in R. M. Woollacott and R. L. Zimmer, eds. Biology of Bryozoans. Academic Press, New York, San Francisco, and London.

Wisely, B. 1960. Experiments on rearing the barnacle *Elminius modestus* Darwin to the settling stage in the laboratory. Aust. J. Mar. Freshwater Res. 2:42–54.

Wodinsky, J. 1971. Movement as a necessary stimulus of *Octopus* predation. Nature (Lond.) 229:493–494.

Woelke, C. E. 1965. Bioassays of pulp mill wastes with oysters. Pp. 67–77 in C. M. Tarzwell, ed. Biological Problems in Water Pollution Third Seminar 1962. Division of Water Supply and Pollution Control, U.S. Public Health Service, Cincinnati. (PHS Publ. 999-WP-25)

Woelke, C. E. 1968. Application of shellfish bioassay results to the Puget Sound pulp mill pollution problem. Northwest Sci. 42:125–133.

Woelke, C. E. 1972. Development of a Receiving Water Quality Bioassay Criterion Based on the 48-Hour Pacific Oyster (*Crassostrea gigas*) Embryo. Wash. Dep. Fish. Tech. Rep. No. 9. vii + 93 pp.

Worthman, S. G. 1974. A method of raising clones of the hydroid *Phialidium gregarium* (A. Agassiz, 1862) in the laboratory. Am. Zool. 14:821–824.

Wyttenbach, C. R. 1973. The role of hydroplasmic pressure in stolonic growth movements in the hydroid, *Bougainvillia*. J. Exp. Zool. 186:79–90.

Yamamichi, Y., and K. Sekiguchi. 1974. Embryo and organ cultures of the horseshoe crab, *Tachypleus tridentatus*. Dev. Growth Differ. 16:295–304.

Yanagita, T. M. 1959. Physiological mechanism of nematocyst responses in sea anemone. VII. Extrusion of resting cnidae—its nature and its possible bearing on the normal nettling response. J. Exp. Biol. 36:478–494.

Yayanos, A. A. 1978. Recovery and maintenance of live amphipods at a pressure of 580 bars from an ocean depth of 5700 meters. Science 200:1056–1059.

Yoshida, S. 1976. Acceleration of maturation and spawning of the nereid worm *Perinereis nuntia* var. *vallata*. Bull. Jpn. Soc. Sci. Fish. 42:1199–1203. (in Japanese with English summary)

Young, J. Z. 1977. What Squids and Octopuses Tell Us about Brains and Memories. Forty-sixth James Arthur Lecture on the Evolution of the Human Brain. American Museum of Natural History, New York. 27 pp.

Zaroogian, G. E., G. Pesch, and G. Morrison. 1969. Formulation of an artificial sea water media suitable for oyster larvae development. Am. Zool. 9:1144.

Zerbe, W. B., and C. B. Taylor. 1953. Sea Water Temperature and Density Reduction Tables. Special Publication 298, U.S. Department of Commerce. U.S. Government Printing Office, Washington, D.C.

Zillioux, E. J. 1969. A continuous recirculating culture system for planktonic copepods. Mar. Biol. (Berlin) 4:215–218.

Zillioux, E. J., and N. F. Lackie. 1970. Advances in the continuous culture of planktonic copepods. Helgol. Wiss. Meeresunters. 20:325–332.

Zillioux, E. J., and D. F. Wilson. 1966. Culture of a planktonic calanoid copepod through multiple generations. Science 151:996–997.

Zillioux, E. J., H. R. Foulk, J. C. Prager, and J. A. Cardin. 1973. Using Artemia to assay oil dispersant toxicities. J. Water Pollut. Control Fed. 45:2389–2396.

Zobell, C. E., and D. Q. Anderson. 1936. Observation on the multiplication of bacteria in different volumes of stored sea water and the influence of oxygen tension and solid surfaces. Biol. Bull. (Woods Hole, Mass.) 71:324–342.

Zurlini, G., I. Ferrari, and A. Nassogne. 1978. Reproduction and growth of *Euterpina acutifrons* (Copepoda: Harpacticoida) under experimental conditions. Mar. Biol. (Berlin) 46:59–64.

APPENDIX **A**

Sources of Permits and Licenses— State Fish and Game Departments

Department of Conservation
Game and Fish Division
Montgomery, AL 36104

Commissioner
Department of Fish and Game
Subport Building
Juneau, AK 99801

Arizona Game and Fish Commission
2222 West Grenway Road
Phoenix, AZ 95023

State of California
The Resources Agency
Department of Fish and Game
1416 Ninth Street
Sacramento, CA 95814

Director
Division of Wildlife
6060 Broadway
Denver, CO 80216

State of Connecticut
Department of Environmental Protection
State Office Building
Hartford, CT 06115

Department of Natural Resources and Environmental Control
Tatnall Building
Dover, DE 19901

Florida Game and Fresh Water Fish Commission
Farris Bryant Building
620 South Meridian Street
Tallahassee, FL 32304

Department of Natural Resources
270 Washington Street, S.W.
Atlanta, GA 30334

Department of Land & Natural Resources
Division of Fish and Game
1179 Punchbowl Street
Honolulu, HI 96813

Director
Idaho Fish & Game Department
P.O. Box 25
600 South Walnut Street
Boise, ID 83707

Department of Conservation
100½ East Washington
Springfield, IL 62706

License Section
Indiana Department of Natural Resources
Division of Fish and Game
607 State Office Building
Indianapolis, IN 46204

License Section
Iowa Conservation Commission
300 Fourth Street
Des Moines, IA 50319

Kansas Forestry, Fish & Game Commission
P.O. Box 1028
Pratt, KS 67124

Director
Division of Fish & Wildlife Resources
Capital Plaza Tower
Frankfort, KY 40601

Louisiana Wildlife & Fisheries Commission
P.O. Box 44095
Capitol Station
Baton Rouge, LA 70804

Department of Inland Fisheries & Game
Game Division
State Office Building
Augusta, ME 04330

Department of Wildlife Resources
Wildlife Administration
Tawes State Office Building
Annapolis, MD 21401

Director
Division of Fisheries & Game
Leverett Saltonstall Building, Government Center
100 Cambridge Street
Boston, MA 02202

Department of Natural Resources
Lansing, MI 48926

Section of Wildlife
Department of Natural Resources
Centennial Office Building
St. Paul, MN 55155

Chief, Game and Fisheries
Game and Fish Commission
P.O. Box 451
Jackson, MS 39205

Department of Conservation
Division of Wildlife
P.O. Box 180
Jefferson City, MO 65101

Department of Fish and Game
Helena, MT 59601

Director
Game and Parks Commission
P.O. Box 30370
Lincoln, NB 68503

Director
Division of Enforcement
Department of Fish & Game
P.O. Box 10678
Reno, NV 89510

Director
Fish & Game Department
34 Bridge Street
Concord, NH 03301

Department of Environmental Protection
Division of Fish, Game and Shell Fisheries
P.O. Box 1809
Trenton, NJ 08625

Department of Game and Fish
State Capitol
Santa Fe, NM 87503

Department of Environmental Conservation
50 Wolf Road
Albany, NY 12233

Wildlife Resources Commission
Raleigh, NC 27601

Commissioner
Game and Fish Department
Bismarck, ND 58501

Department of Natural Resources
Division of Wildlife
Foutain Square
Columbus, OH 43224

Department of Wildlife Conservation
Game Division
P.O. Box 53465
Oklahoma City, OK 73105

Oregon Wildlife Commission
P.O. Box 3503
Portland, OR 97208

Pennsylvania Fish Commission
R.D. 3, Box 70
Bellefonte, PA 16823

Department of Natural Resources
Division of Fish and Wildlife
83 Park Street
Providence, RI 02903

Director
South Carolina Wildlife & Marine Resources Department
P.O. Box 167
Columbia, SC 29202

Department of Game, Fish & Parks
Division of Game and Fish
Pierre, SD 57501

Tennessee Wildlife Resources Agency
Ellington Agricultural Center
P.O. Box 40747
Nashville, TN 37204

Texas Parks & Wildlife Department
John H. Reagon Building
Austin, TX 78701

Department of Natural Resources
Division of Fish & Wildlife
1596 West North Temple
Salt Lake City, UT 84116

Agency of Environmental Conservation
Fish & Game Department
Montpelier, VT 05603

Commission of Game & Inland Fisheries
P.O. Box 11104
Richmond, VA 23230

Department of Fisheries
Room 115, General Administration Building
Olympia, WA 98504

Director
Department of Natural Resources
State Capitol Building
Charleston, WV 25305

Department of Natural Resources
P.O. Box 450
Madison, WI 53701

Wyoming Game and Fish Department
P.O. Box 1589
Cheyenne, WY 82001

Secretary of Natural Resources
P.O. Box 5887
Puerta de Tierra, PR 00906

APPENDIX **B**

Sources of Equipment and Supplies

The following is not intended to be an exhaustive list of supplies and equipment that is available, but rather covers equipment that we have found to be reliable over many years of use. See also "Sources of commercial supply in index."

AQUARIA, FIBERGLASS
Pacific General
2487 Spring Street
Redwood City, CA 94063

Aquarium Systems
8141 Tyler Blvd.
Mentor, OH 44060

Fiberglass construction with plate glass front

Aquaria with built-in cooling and filtering

AQUARIUM FILTERS
Eugene G. Danner Mfg., Inc.
160 Oval Drive
Central Islip, NY 11722

"Supreme Aquaking," mounts on side of tank. Often available at aquarium supply stores.

CENTRIFUGAL WATER PUMPS
March Manufacturing Co.
1819 Pickwick Ave.
Glenview, IL 60025

Model MDX, also available from:

Cole-Parmer Instrument Co.
7425 North Oak Park Ave.
Chicago, IL 60648

as catalog number: 7004–10 and 7004–20

AIR PUMPS
General Rand Co.
100 Menlo Park
Edison, NJ 08817

Imports the Reciprotor Pump Model 506 R. Very quiet, enough air for about 10 aquaria.

Aquarium Pump Supply Co.
314 Wipple St.
Prescott, AZ 86301

"Silent Giant" air pump. Available in most aquarium stores.

REFRACTOMETERS

B and L hand refractometers (sugar refractometer, inexpensive)

AO salinity refractometer (fairly expensive)

Either can be purchased from scientific supply companies

COMPOUNDED SALT MIXTURES

Aquarium Systems
8141 Tyler Blvd.
Mentor, OH 44060

AQUARIA, ALL-GLASS, 6 QUART

Ward's Natural Science Establishment, Inc.
P.O. Box 1712
Rochester, NY 14603
Excellent for use in experiments where small numbers of specimens need to be kept.

APPENDIX **C**

Suppliers of Marine Organisms

The following list is current at the time of publication.

Aplysia Aquarium Collecting and Research Center
Billy D. and Laura G. Causey
Route 1, Box 429E
Big Pine Key, FL 33043
(305) 872-9508

Arrive Alive Bio Supply, Inc.
Bruce Folz
P.O. Box 458
Greenport, NY 11944
(516) 765-3174

ASI Lobster Hatchery
William C. Green
311 Chaffinch Island Road
P.O. Box 263
Guilford, CT 06437
(203) 453-3705

(postlarval and juvenile lobsters only)

Biofish Associates
Salvatore A. Testaverde
10 Riggs Street
Gloucester, MA 01930
(617) 283-3154

Carolina Biological Supply Co.
2700 York Road
Burlington, NC 27215
(919) 584-0381

Gulf Specimen Company
Jack J. Rudloe
P.O. Box 237
Panacea, FL 32346
(904) 984-5297

Gulf-South Biologicals, Inc.
Jim Bankston
P.O. Box 417
Ponchatoula, LA 70454
(504) 386-8250

(Preserved speciments only)

Maine Bait Company
Thomas H. Dinsmore
Academy Street
Newcastle, ME 04553
(207) 563-3000

(*Nereis virens* and *Glycera dibranchiata* only)

Marine Biomedical Institute
Roger T. Hanlon
200 University Boulevard
Galveston, TX 77550
(713) 765-2101

Marine Bilogical Laboratory
Supply Department
John Valois or Homer Smith
Woods Hole, MA 02543
(617) 548-3705

Northeast Marine Environmental Institution, Inc.
Paul A. Shave
P.O. Box 666
Monument Beach, MA 02553
(617) 759-4055

Pacific Bio-Marine Laboratories, Inc.
George B. Kelly
P.O. Box 536
Venice, CA 90291
(213) 822-5757

Sea Life Supply
Michael O. Morris
740 Tioga Avenue
Sand City, CA 93955
(408) 394-0828

Turtle Cove Laboratory, Inc.
Russell Miget
P.O. Box 219
Port Aransas, TX 78373
(512) 749-5643

Sources of Marine Unicellular Algae

Ukeles (1976b) lists the five major collections of the world as:

- Culture Centre of Algae and Protozoa, 36 Storey's Way, Cambridge CB3 ODT, England
- Sammulung von Algenkulturen, Pflanzenphysiologischen Instituts der Universität Göttingen, Nikolausbergerweg 18, Göttingen, West Germany
- Sbirka Kultur Autorofnich Organismu CSVA, Viničcá, 5, Praha, 2, Czechoslovakia
- Culture Collection of Algae, Institute of Applied Microbiology, University of Tokyo, Japan
- Culture Collection of Algae, Department of Botany, University of Texas, Austin

In addition, there are smaller collections in the following laboratories:

- W. H. Thomas, Scripps Institution of Oceanography, La Jolla, California
- Woods Hole Oceanographic Institution, Woods Hole, Massachusetts
- National Marine Fisheries Service Laboratory, Milford, Connecticut
- American Type Culture Collection, Washington, D.C.

APPENDIX D

List of Contributors

C. G. BOOKHOUT, Duke University Marine Laboratory, Beaufort, NC 28516

GEORGE GORDON BROWN, Hopkins Marine Station, Stanford University, Pacific Grove, CA 93950

MATOIRA H. CHANLEY, Indian River Mariculture Associates, P.O. Box 12, Grant, FL 32949

DAVID L. CLAPPER, Hopkins Marine Station, Stanford University, Pacific Grove, CA 93950

J. R. COLLIER, Department of Biology, Brooklyn College, Brooklyn, NY 11210

J. D. COSTLOW, JR., Duke University Marine Laboratory, Beaufort, NC 28516

MICHAEL G. HADFIELD, University of Hawaii, Pacific Biomedical Research Center, Kewalo Marine Laboratory, 41 Ahui Street, Honolulu, HA 96813

DAVID A. HESSINGER, Department of Physiology and Pharmacology, Loma Linda University School of Medicine, Loma Linda, CA 92350

JUDITH A. HESSINGER, Department of Physiology and Pharmacology, Loma Linda University School of Medicine, Loma Linda, CA 92350

DONALD J. REISH, California State University, Long Beach, Long Beach, CA 90840

MARY M. ROCHA TUZZI, Thimann Laboratories, University of Santa Cruz, Santa Cruz, CA 95064

MARILYN SWITZER-DUNLAP, University of Hawaii, Pacific Biomedical Research Center, Kewalo Marine Laboratory, 41 Ahui Street, Honolulu, HA 96813

Subject Index

B

C

D

E

F

Q

R

S